DEUTSCHE FORSCHUNGSANSTALT FÜR
LUFT- UND RAUMFAHRT E.V. (DLR)

SOLAR THERMAL ENERGY UTILIZATION

German Studies on Technology and Application

Volume 6:
Final Reports 1990

Editors: M. Becker, K.-H. Funken, G. Schneider

Springer-Verlag Berlin Heidelberg GmbH 1992

Dr.-Ing. Manfred Becker
Dr. rer. nat. Karl-Heinz Funken
Dipl.-Ing. Gernot Schneider
Deutsche Forschungsanstalt für Luft- und Raumfahrt e.V. (DLR),
Hauptabteilung Energietechnik, Köln

ISBN 978-3-540-54836-2   ISBN 978-3-662-09931-5 (eBook)
DOI 10.1007/978-3-662-09931-5

© Springer-Verlag Berlin Heidelberg 1992
Originally published by Springer-Verlag Berlin Heidelberg New York in 1992

2362/3020−543210

Preface

The German R+D program "Solares Testzentrum Almeria" (SOTA) provides the scientific basis for the realization of advanced solar technologies including facility modifica-tions, component tests and new lines of development. One of the working packages, WP 300, addresses the "Scientific Support" by the performance of pre-paratory studies, exploratory laboratory activities and quali-fied expertise. Universities, Research Institutes and Company R+D Entities in Germany are enabled to treat the following aspects:

* Meteorological, system and cost investigations,
* Development of important components as concentrator, receiver, storage,
* Utilization of solar energy for process heat and chemical reactions.

In 1990 the studies covered the developments of terminal concentrators, receivers, integrated receiver/reactors, and solar chemistry processes. The reports of the activities were finalized recently and collected in the present volume. The final reports were printed as received. The achieved results and the views ex-pressed within the reports were under the responsibility of the authors.

As organizers of these solar thermal activities and as editors of the corresponding final reports, we would like to thank BMFT and KFA-BEO for the farsighted funding (grant no. 0328823A) and the research groups for their engagement and consultations. We would also like to thank Ms S. Preusser and Ms U. Rachow, who among many others contributed to the realization of this volume.

Cologne, January 1992

M. Becker, K.-H. Funken, G. Schneider

Contents

Solar-Thermal Power Generation
with an Alkali Metal Thermo-Electric Converter
(AMTEC)

R. Knödler, H.-P. Boßmann, R. Krapf, F. Harbach,

Asea Brown Boveri AG,
Heidelberg

Contents

Summary

This study is intended to help to improve the performance and attractiveness of solar/dish systems for the direct conversion of solar energy into electrical energy.

Present solar/dish systems use Stirling engines which have several drawbacks (frequent maintenance, high volume, vibrations, noise), preventing a wider use. The present study investigates the use of the Alkali Metal Thermo-Electric Converter (AMTEC) instead of the Stirling engine. The AMTEC is a heat engine with sodium as working medium. Although AMTEC is available today only as small laboratory unit, it can be expected that it will be well suited for the replacement of Stirling engines in solar/dish systems, showing essentially the same efficiency as the Stirling but none of its drawbacks (as outlined above). The operation principle of AMTEC is described in chapter 3.1 and is shown in fig. 2.

The objective of this study was to design a 10 kW AMTEC module which fits into the focus of a parabolic dish and to estimate the performance data of this system for the insolation conditions of Almeria.

The study shows that a design is possible, allowing the daily movement of the module according to the suns position. The system design comprises a sodium pool receiver which supplies sodium in the form of saturated sodium vapor to the beta-alumina tubes of the AMTEC module. 3 tubes at a time will be electrically connected in parallel, and 37 of these groups will be connected in series, resulting in a voltage of 37 V at the nominal output of 10 kW. The efficiency of such a modul will be 42 %, leading to about 25 % for the entire system (including receiver, dish and power conditioning). There is only a small decrease of the efficiency at part-load conditions down to about 20 % of full load. This leads to a high annual system efficiency of about 23 % for Almeria. The power density of the designed module is high (module volume: about 78 l).

In summary, AMTEC will fit well into a solar dish system. The realization of the designed module and of the entire system would have a positive effect on the further development of solar thermal systems.

Zusammenfassung

Die vorliegende Studie soll einen Beitrag zur Erhöhung der Attraktivität von solarelektrischen Parabolspiegelsystemen liefern.

Heutige Parabolspielgelsysteme arbeiten mit Stirlingmotoren, die aber alle gewisse Nachteile aufweisen, wie häufige Wartung, großes Volumen, Vibrationen, Lärmbelästigung. Diese Nachteile verhindern eine weitere Verbreitung. Die vorliegende Studie untersucht den Einsatz des Alkalimetall Thermoelektrischen Converters (AMTEC) anstelle des Stirlingmotors. Der AMTEC ist eine Wärme-Kraftmaschine mit Natrium als Arbeitsmittel. Obwohl der AMTEC heute erst als kleine Laboreinheit verfügbar ist, kann man erwarten, daß er gut geeignet ist für einen Ersatz des Stirlingmotors in Parabolspiegelanlagen: Der AMTEC besitzt einen ähnlich hohen Wirkungsgrad wie der Stirlingmotor, ohne seine Nachteile aufzuweisen. Die Wirkungsweise des AMTEC ist in Kapitel 3.1. beschrieben, das Prinzip zeigt Bild 2.

Das Ziel der vorliegenden Studie war die Konstruktion eines 10 kW-AMTEC-Moduls, der in den Fokus eines Parabolspiegels paßt. Weiterhin sollen die Betriebsdaten eines solchen Moduls für die Einstrahlungsbedingungen von Almeria abgeschätzt werden.

Die Studie zeigt, daß eine Konstruktion möglich ist, die eine Nachführung entsprechend der täglichen Bewegung der Sonne erlaubt. Das System umfaßt einen Natrium-Pool-Receiver, der Natrium in Form von gesättigtem Natriumdampf in die Beta-Aluminiumoxidrohre des eigentlichen AMTEC-Moduls liefert. Je 3 Rohre werden parallel und 37 dieser Gruppen dann in Reihe geschaltet, was eine Spannung von 37 V bei der Nennleistung von 10 kW liefert. Der Wirkungsgrad eines solchen Moduls wird etwa 42 % betragen, der Wirkungsgrad des Gesamtsystems (einschließlich Receiver, Spiegel und Converter) etwa 25 %. Es wird nur einen kleinen Abfall des Wirkungsgrades im Teillastbetrieb geben (bis zu etwa 20 % der Vollast), was zu einem hohen jährlichen Systemwirkungsgrad von etwa 23 % für Almeria führt. Die Leistungsdichte des Moduls wird hoch sein (Modulvolumen etwa 78 l).

Zusammenfassend kann gesagt werden, daß sich der AMTEC vorteilhaft in ein Parbolspiegelsystem einfügen wird. Der Bau und Betrieb eine solchen Moduls und des gesamten Systems würde positive Auswirkungen auf die weitere Verbreitung der solarthermischen Erzeugung von Elektrizität haben.

1. Introduction

The generation of electric power directly from the sun is possible in two ways:
1) Via solar cells using the photoelectric effect ("Photovoltaics").
2) Via concentrators and heat engines ("Solar - Thermal").

Both systems have their inherent advantages and disadvantages:

	Photovoltaics	Solar-Thermal

Advantages	- can use diffuse
	  + direct radiation
	- long lifetime,
	  maintenance-free
		- high efficiency
		- no storage needed
		  (can switch to burner
		  at sun-free periods)[*]

Disadvantages	- very expensive
	- low efficiency
	- storage only in
	  expensive and bulky
	  batteries
		- can use only direct
		  radiation
		- tracking necessary

The most attractive advantage of "Solar-Thermal" is its higher efficiency (20 - 25 % as compared to about 10 % for photovoltaics). This enables the production of electricity at a reasonable cost.

Fig. 1 shows the components of the different approaches. In the case of "Solar-Thermal", heat engines are necessary in order to convert heat into mechanical energy. For large unit sizes (100 MW) and medium temperatures (500 °C) one approach is to use parabolic trough concentrators and steam turbines. A generator in turn transforms the resulting mechanical into electrical energy.

---

[*] This "hybrid" mode will be mentioned a couple of times in this study, but will not be discussed in detail!

As heat engines are subject to the Carnot process, it is desirable to use temperatures as high as possible in order to get a higher efficiency. This can be achieved by using parabolic dish concentrators which can produce temperatures of > 1000 °C /1/. As heat engine for these high temperatures Stirling engines are used at present. Dish/Stirling systems have reached overall efficiencies of 28 - 32 % /2/.

The Stirling engine, however, has several drawbacks, like:

- high frequency of maintenance
- high specific volume (especially due to the generator)
- noise, vibrations.

For a heat engine in the focus of a parabolic dish, it would, however, be desirable that it is maintenance-free, small , noiseless and vibration-free. If these requirements could be met, the dish itself could be constructed in a considerably lighter and cheaper manner.

The Alkali Metal Thermo-Electric Converter (AMTEC), which will be described later in detail, is a candidate for such a heat engine. Its main advantages are the following:

- high efficiency (of up to 45 %)
- high power density up to 200 W/l
- no moving parts (i.s. silent, maintenance-free)
- lower installation costs (about DM 350/kW without receiver and dish)
- modular design, therefore economical also in small units.

The AMTEC, originally intended for use in space power systems, has also been evaluated as heat engine for a solar dish system /3, 4, 5/. It turned out that the AMTEC would be an attractive candidate for this purpose. In these studies only performance and economics of a system with an AMTEC heat engine are evaluated, while assuming a black box as AMTEC. However, a concept of a module which can suit the special conditions of a parabolic dish (e.g. following the daily course of the sun, thermal cycling according to weather conditions) were never established.

Therefore, the task of the present study is to make a detailed design for an AMTEC module, able to be placed into the focus of a parabolic dish.

2. Objectives

A 10 kW AMTEC module will be designed which allows an operation in a parabolic dish. Based on this design, the complete system - consisting of dish, AMTEC, receiver, storage and power conditioning - will be evaluated. This evaluation will be done for the insolation and weather conditions of Almeria, Spain. These investigations will serve as first step towards construction and operation of a 10 kW AMTEC/dish system in Almeria.

3. The Alkali Metal Thermo-Electric Converter

3.1 Operation principle

The AMTEC (sometimes also called "sodium heat engine") is a heat engine /6/. Its working medium is sodium. The AMTEC essentially consists of two compartments: a high-pressure compartment, where the sodium is heated by a suitable heat source (e.g. combustion of coal or oil; nuclear or solar generated heat) to 800 to 1000 °C and a low-pressure compartment, where sodium vapour is condensed by an air- or water-cooled condenser at a temperature of about 300 °C, corresponding to a saturated vapour pressure of about 10^{-5} bar. A solid electrolyte, consisting of beta alumina, separates the high- and low-pressure compartments and allows only sodium ions to pass. An electromagnetic pump recirculates the sodium (see fig. 2).

The electric power output of the AMTEC is due to the isothermal expansion process of Na across the solid electrolyte. The driving force for this process is the pressure difference between the two compartments. Because the solid electrolyte allows only sodium ions to pass through, but no sodium atoms or electrons, a pressure equilibration can take place only if an electronically conductive external load is applied. In this case sodium ions flow from the high-pressure to the low-pressure compartment via the solid electrolyte, while electrons flow through the load, delivering usable electric power.

Separation and recombination of ions and electrons take place at electrodes on both sides of the solid electrolyte.

The open-circuit cell voltage can be calculated using the modified Nernst equation:

$$U_o = \frac{RT_2}{F} \cdot \ln \frac{P_2}{P_1 \sqrt{T_2 / T_1}}$$

where R is the gas constant, F the Faraday constant, T_2 the upper temperature, T_1 the lower temperature and P_1 and P_2 the corresponding partial pressures of the both compartments, respectively. With $T_2 = 1200$ K, $P_1 = 10^{-5}$ bar and $P_2 = 1$ bar an open circuit voltage of

$$U_o \approx 1.2 \text{ V}$$

is obtained. Under current flow there is a reduction in cell voltage caused by pressure drops and an IR-drop due to ohmic losses to about 0.8 - 1.0 V at a current density of about 1 A/cm^2.

For the realization of the process it is advisable to supply the hot sodium as saturated vapour to the solid electrolyte. By this, handling of hot sodium and series connection of single tubes to larger units is facilitated. This concept is different from cells of other research organizations which supply the sodium in liquid form.

The very heart of an AMTEC is the solid electrolyte separator. In today's laboratory test cells as well as in future AMTEC modules, beta alumina ceramic tubes will be used for this purpose, the same tubes that are being produced already in large quantities and sufficient quality by ABB for ABB's high-energy battery.

The theoretical maximum efficiency of an "ideal" AMTEC was calculated to be about 90 % of the Carnot efficiency. This means an upper efficiency limit of 53 % for an AMTEC running between 1000 °C and 300 °C. A "real"

AMTEC of an advanced, optimized design will have efficiencies between 35 and 45 % - strongly dependent upon the actual values for electrical and heat losses. As with other direct energy conversion systems, efficiency is practically independent of load over a wide range of current density. Even improved performance is to be expected below maximal load. The solid electrolyte allows high current to be drawn, which means that (especially at higher temperatures) power densities of more than 1 W/cm^2 can be achieved.

The maximum upper temperatures of 1000 °C is determined by material problems, the minimum lower limit of 300 °C by the pressure which the current itself produces.

As is shown in fig. 3, an AMTEC module - consisting of many single beta-alumina tubes - can be placed in the focus of a parabolic dish. Process heat can be obtained from cooling the condenser, which has a temperature of about 300 °C. During times of reduced insolation (clouds, night), the AMTEC can in principle generate electricity by burning fossil fuels ("hybrid" operation mode).

3.2 State of development

The AMTEC has been developed up to now mainly in the U.S., where small laboratory cells have been built and tested, confirming the potential for high efficiency and long lifetime.

ABB started AMTEC-development in 1989. By the end of 1990 test cells with a power of 25 W have been built and operated at 900 °C. The experimental arrangement of a single test cell shows fig. 4. The electrodes in the low pressure compartment (= outside the tube) were made by sputtering thin, porous Mo-layers onto the β"-Al$_2$O$_3$ tube. As current collector molybdenum wires and meshes were used. The electrical resistance of electrode und current collector is still about 0.6 Ωcm^2 at present. However, lower values (corresponding to higher power data) are possible, because the lower limit is determined by the β"-Al$_2$O$_3$ tube, its resistance being about 0.15 Ωcm^2 at 1000 °C.

The cell shown in fig. 4 is a so-called electrode test cell, in which liquid sodium is inside the tube. However, for larger modules this arrangement is not optimal, because in this case series connection of tubes will be difficult. Therefore, we tested also so-called vapor-fed cells, shown in fig. 5, where sodium is brought into the β"-Al$_2$O$_3$ tube as vapor. Although there are some difficulties with this cell type in reaching a homogeneous temperature distribution inside the tube, we could show that the cells worked in the same manner as the liquid-sodium type does. The vapor-fed design was selected for investigation also with respect to solar-thermal applications, because here the heat could be transported via a sodium pool receiver directly to the solid electrolyte tubes.

Fig. 6 and 7 show current-voltage and current-power relationships of single cells which were operated at temperatures up to 900 °C. The lifetime of the experimental cells is limited presently to < 200 h due to lack of sodium recirculation and due to short circuits through liquid sodium. The ceramic itself showed no signs of degradation whatever.

The efficiency of single cells is generally poor, because the heat insulation of such small cells (about 300 mm length, 50 mm diameter) cannot be made sufficiently good. In the US (FORD) values of 19 % for single cells have been reported. Single cells at ABB have not been optimized toward efficiency; the values are, therefore, presently below 10 %.

Larger modules will be built by the end of this year, a 1 kW-module is envisaged for 1992/1993.

4. Requirements for a 10 kW AMTEC system for solar applications

In order to design such a system for solar applications, a series of specific requirements must be met:

The AMTEC needs heat input at temperatures above approximately 750 °C for rational energy conversion, i.e. for efficiencies above 20 %. Present development goal is an operation temperature above 900 °C for engine efficiencies of around and above 35 % (electrical power out over heat accepted). For competitive application in solarthermal plants, efficiencies of 40 % or above would be helpful, which means an operation temperature of around 1000 °C.

4.1 Concentrator

In solarthermal operation high temperatures can only be attained, if high concentration ratios of the concentrating system can be reached. At the same time, high concentration factors are a must to reduce radiation losses back from the receiver to acceptable levels, if the receiver wall temperature reaches 900 °C or more.

Fig. 8 shows the dependence of receiver efficiency on receiver temperature for different concentration ratios (taken from /13/). For operation at 1000 °C, a concentration ratio above 2000 is strictly advisable.

For high annual efficiencies in areas with extended periods of reduced insolation, a high concentration factor is even more important because only with concentration ratios of 2000 or above, the AMTEC's high efficiency at partial load would not be counteracted too severely by the relative increase of the receiver radiation losses at constant receiver temperature but reduced heat input.

(Under system efficiency and cost considerations, it may be seen as advisable to operate the AMTEC with reduced temperature at times of low insolation. On the other hand, constant temperature operation through reduction of current and load is more appropriate to reduce thermal and thermomechanical stresses inside the AMTEC. A clear distinction between both modes of operation can be made only on the basis of specific component identification. In the context of this study, constant concentrator and receiver efficiencies will be used therefore.)

Unfortunately, fundamental trade-off exists between increasing the geometric concentration ratio and reducing the cost of the collector, because collectors with high concentration ratios must be manufactured precisely. Generally, a direct correlation exists between the accuracy of the concentrater and its cost.

Optimizing the concentrator under cost considerations (taking system, receiver and engine efficiency and meteorological data into account) is a necessary task before any decision is made for a specific concentrator design for a given localisation. This task has to be done in close cooperation between the manufactures of concentrator, receiver and power conversion unit (PCU). This task is not included within this study, which should concentrate on AMTEC design and performance data, i.e. on the PCU side of the system.

4.2. Receiver

The most important requirement is a high efficiency which decreases only little at partload conditions.

The receiver must stand a temperature of at least 1050 °C for several thousand hours. Therefore, materials must be used which are stable at this temperature in sodium- (inside) and air- (outside) atmosphere. For this purpose highly alloyed steels or ceramics may be suitable.

The heat losses of the receiver occur primarily as black body radiation of the aperture. This can be minimized by using highly concentrating dishes.

Low cost of the receiver can be achieved by using a common sodium loop AMTEC / receiver. By this way, additional heat exchangers (Na/Na or air/Na) are not necessary. This helps in keeping costs down and efficiency high.

4.3 AMTEC module

4.3.1 General design considerations

Conversion of dc in ac requires a minimum output voltage of about 25 V. Therefore being the voltage of a single tube 0.8 - 1.0 V, at least 30 tubes have to be put electrically in series.

Each tube will have a maximum power output of 50 to 100 W. Therefore, in order to reach a power output of 10 kW, several tubes have to be connected in parallel.

The resulting "mixed" connection mode (fig. 9a) will have the advantage of avoiding the failure of the complete module in the case of failure of only one tube. Such would be the case if there were only series connection (fig. 9b). Another benefit of a mixed connection mode is the need for fewer feedthroughs, because only series connections need feedthroughs. For these reasons, we will design the module in a "mixed" mode.

The geometrical arrangement of the tubes inside a module must be optimized towards a minimal inhibition of the sodium gas, streaming in the direction of the condenser. This is important, because such inhibitions cause an increase of the pressure on the "cold" side of the AMTEC, which in turn decreases the power output and the efficiency. The tubes, therefore, cannot be packed as close as possible in a module, but a certain distance must be kept in order to allow the sodium gas to reach the condenser with a minimal pressure drop.

The application in a parabolic dish requires that the AMTEC can be tilted to a certain extent in order to follow the daily and yearly movement of the sun. This, in turn, requires that the backflow of the condensed sodium must be possible also in tilted positions. These requirements can be fulfilled by arranging the receiver in an angle of 45 °C to the AMTEC and by turning the AMTEC around horizontally by 180 ° during passing of the meridian at noon. This principle /7/ is shown in fig. 10. Fig. 11 shows a schematic drawing of the different daily aspects of the solar-thermal system arrangement using a tilted receiver. It can be seen that with this arrangement, the condensed sodium can always flow back into the receiver.

4.3.2 Power output

The insolation data for Almeria show that the maximum direct radiation is 970 W/m^2 /8/. However, only about 5 % of the overall insolation reaches values of > 900 W/m^2 and an additional 17 % of between 800 and 900 W/m^2. Therefore, we will define a "nominal" electrical output of the AMTEC modul of 10 kW at 800 W/m^2 insolation. This insolation corresponds to AM

1.5. The maximum output will be higher than 10 kW. From this procedure the part-load behavior will benefit (i.e. higher efficiencies also during part-load.

The total power output will be reduced by the electronic controllers and converters (dc/dc and dc/ac), according to their efficiencies.

4.3.3 Efficiencies

The efficiency of the AMTEC-module itself will be determined according to the design described later. The efficiencies of dish, receiver and control equipment should be not lower than 85 %, 80 % and 90 %, respectively (also during part-load conditions).

4.3.4 Temperatures

The upper temperature of the AMTEC will be 1000 °C, which requires a receiver temperature of 1050 - 1100 °C. The temperature of 1000 °C must be controlled, because higher temperatures can cause material problems. Lower temperatures, however, will yield lower efficiencies and lower power densities at least as far as the AMTEC is concerned.

The control of the temperature during variations in the sun's radiation conditions can be done by regulating the electrical current of the AMTEC. The theoretical background for this is shown in fig. 12: there is an equilibrium between the sun's radiation (= heat input) and the sum of electrical output, waste heat and parasitic heat-losses. At a given radiation value and a given current value, one distinct temperature will develop in order to obtain the equilibrium in fig. 12. On the other hand, a desired temperature of, e.g., 1000 °C can be maintained by varying the current

according to the actual radiation values. This is shown in fig. 13: 1000 °C can be maintained, when the current is adjusted following the line in the middle of fig. 13. In the region above this line temperatures of > 1000 °C will develop, below this line of < 1000 °C.

It can be seen that at zero electrical output, about 0.5 W/cm^2 is needed in order to keep the 1000 °C (Note: cm^2 means the area of the beta-alumina tubes and not the area exposed to the sun). This value is the parasitic heat loss Q_{loss} and is calculated in chapter 7.2 in detail for the case of the newly designed module.

The lower temperature should be maintained at about 300 °C or lower. This can be reached by natural convection, if a proper design is made. As emergency cooling a fan has to be integrated.

A further use of the waste-heat will not be taken into account at present.

4.3.5 Sodium management

The sodium content of the module should be as small as possible because of safety reasons. In order to help to maintain a constant temperature, however, the amount of sodium should not be too small, either. At the nominal power of 10 kW a module will have a sodium transfer rate of 12 kg/h. This amount can probably be pumped about 2 times per hour. This means a sodium content of about 6 kg. This amount of sodium will be also sufficient to fill the pool receiver in fig. 23.

4.4. Transients

The average time length of cloudy conditions in Almeria is about 1.1 h. Therefore, an insulation or a storage device must be kept around the receiver, which allows the temperature to be maintained at > 800 °C during this time.

The transient phase from cloudiness to cloudlessness and vice versa in Almeria lasts 3 min. as an average. Because during cloudy periods the power of the AMTEC will be reduced in order to maintain 1000 °C, there may be an overheating when the sun starts to shine again. This overheating must be avoided by increasing the power output shortly beyond the maximum power and - if necessary - by cooling the condenser with a fan.

4.5 Safety

Hot sodium can burn if in contact with air, caused by leakages of the module or the recirculation system. This must be avoided by double-reinforcing all tubings which are outside the module. It must be noted here that the temperature of the sodium loop outside the module - including the pump - can be kept at relatively low temperatures (< 400 °C). This low temperature is releatively easy to manage.

If the module fails as a whole and power cannot be drawn any more, the sodium in the receiver would be overheated. This can be avoided by flooding the condenser with nitrogen (thus enhancing drastically the heat transfer to the condenser walls) and by cooling the condenser with a fan. This - relatively quick procedure - must then be followed by defocussing the dish.

5. Selection of the most suitable design for an 10 kW-AMTEC system for solar applications.

5.1 Concentrator

Of the following different concentrator systems currently under investigation or in development:
- parabolic troughs
- Fresnel lenses
- heliostat systems (central receiver systems)
- parabolic dishes

only the last three can yield the desired concentration ratio of 2000 or above and only with the last two such concentration ratios have already been approched or achieved up to now.

As far as the AMTEC is concerned, a central receiver heliostat system has a clear advantage over a parabolic dish system: receiver and power conversion unit (PCU) remain in a fixed position; thus it is much easier to support their weight and to control and ensure correct flow of the working medium or the working media (in case, different media are used for receiver and PCU) than with parabolic dishes.

A central receiver system need not to be a non-modular system. Small scale heliostat systems have already been considered and realized, e.g. in Italy /10/. It can be realized with a small tower of only 3 m height and correspondingly with clearly reduced maximum distance from the outermost mirrors to the receiver, thus reducing cosine and other optical losses to more acceptable levels than with large-dimension central receiver solar plants.

But any heliostat system has at least one clear system disadvantage against parabolic dishes. Cosine losses reduce the effective aperture of the concentrating system especially during morning and evening hours. These losses are responsible for a significant reduction of annual efficiencies for heliostat systems as compared to parabolic dishes with two-axes tracking. Other disadvantages of heliostat systems make a parabolic dish an even better choice: multiple need for tracking control systems and tracking mechanics, increased foot print, especially if shadowing effects are to be kept at a minium.

So-called Fix-Focus Concentrators /13/ are being discussed as having neither the disadvantages of central receiver systems nor the disadvantages of parabolic dishes, but showing their combined advantages. In fact, cosine losses of such reflectors are substantially reduced, while receiver and PCU remain in a fixed position during tracking. But up to now, no mean concentration ratios (CR) around or even above 2000 have been realized. (Presently highest mean CR values reached are 1500 /14/.)

Consequently, if only possible, a receiver-AMTEC-combination should be designed in such a way to fit into the focus of a parabolic dish and to be capable of a minimum motional range that is needed for two-axes tracking.

Due to a couple of reasons (ease of manufacturing, compliance with extreme wind forces etc.), a parabolic dish is modular by nature (just like the AMTEC itself). Largest sizes, that are acceptable from manufacturing and cost points of view have diameters below 15 m. On the other hand, cost for support structure and other size-independent balance-of-plant positions ask for a minimum size above about 5 m. At 1 kW/m^2 insolation, this leads to maximum solarthermal power collection in the range of 20 to 150 kW$_{th}$.

Together with PCU efficiencies of around 40 %, the maximum electrical power output would, thus, be between 8 to 60 kW_{el}.

From the point of market potential smaller sizes are to be preferred, especially for market entrance reasons.

The nominal electrical output should neither correspond to the point of highest power density (as would be the result from pure engine cost considerations) nor to some compromise between optimal efficiency and optimal power density (as does result from power generating cost consideration for combustion-based systems). Within a solarthermal plant, the parabolic dish is by far the most expensive component equivalent to about 40 % of total plant cost. Under these circumstances, the PCU should work at or near its highest point of efficiency - giving just some allowance for a flat decline or near-constant shape of the efficiency curve for periods of reduced insolation.

A size of 10 kW_{el} nominal output power was chosen for the AMTEC. This value corresponds to the need of an acceptable PCU size for market entrance and is based on the following assumptions:

- AMTEC efficiency 40 %
- concentrator diameter 8 m
- concentrator area 50 m^2
- concentrator efficiency 85 %
- receiver efficiency 80 %.

Other technical data characterizing the envisaged system are
- concentration ratio 2500
- receiver aperture area 200 cm^2,
  i.e. 16 cm aperture diameter.

5.2. Receiver

Three types of receiver fit the requirements:
• volumetric receiver with Na-filled absorber
• Na-pool receiver as cavity receiver
• Na heat pipe receiver as cavity receiver.

Volumetric receiver:

This receiver works with an absorber behind a quart glass plane at the front. Normally a compressor pumps cold air along housing and glass plane through the hot (ceramic) absorber into the heat exchanger of a heat engine. With an AMTEC as heat engine, this expensive heat exchanger is not necessary, if the absorber consists of Na heat pipes or Na condenser tubes. The air flow levels the absorber surface temperature especially at the start up period with frozen Na. During normal operation, the compressed air can be preheated with the AMTEC waste heat ($\approx$ 250 °C). The air flow is clearly smaller in comparison to a ceramic absorber, but in both designs, air of 1050 °C leaves the receiver without being used for electricity generation. This heat could be used, if a second AMTEC with working temperature of 600 °C would be added, thus increasing the overall efficiency.) The front glass plane reduces the radiation losses of the receiver.

Pool receiver:

In this type of receiver, liquid sodium covers the backside of the area which is heated up by concentrated solar radiation. The absorber for radiation is built as a cavity with a small aperture. The small heat transition resistance of Na and its high heat conductance smoothes the peak temperature of the concentrated radiation. If the peak heat flux is not extremely high, boiling retardation and hot spots do not occur. For starting this receiver with frozen Na, only slow heating is possible. This can be realized with a shutter in front of the receiver.
The heat losses of the receiver are primarily depending on the aperture and the temperature drop of the outside wall to Na.

Heat pipe receiver

The shape of a heat pipe receiver is very similar to that of a pool receiver, but the inside surface is covered with wick material and only a small quantitly of Na is inside. Liquid sodium flows through the wicks to the hottest places of the absorber. The heat transfer occurs by Na evaporation; the saturated Na vapor with its latent heat supplies the AMTEC with heat. The Na which condenses inside the high pressure compartment of the AMTEC is picked up by the wick, so that the inside Na loop is closed. This receiver is able to manage high peak heat fluxes if the wick is not blocked by frozen Na at the beginning of heating. The heat losses and the efficiency are in the same magnitude as those of the pool receiver.

Selection

For a single 10 kW-AMTEC module, the heat pipe and the pool receiver are equivalent; the volumetric receiver becomes more interesting if process heat is demanded.

Considering the costs, the pool receiver will be more attractive than the heat pipe type with complex wick structures inside.

The heating of a frozen Na-receiver is more difficult for a heat pipe receiver, because the Na transport via wicks occurs only with liquid Na. Also a pool receiver has to be heated up carefully, but the heat conductance of Na supports this procedure.

Considering the above facts, we will couple the AMTEC-module with a pool receiver.

5.3. AMTEC module

Fig. 14 shows the different possibilities for the design of a solar-operated AMTEC with 10 kW electric output. The AMTEC can be built as one single module (M) or it can consist of several submodules (S).

These submodules can be independent units (SA), which - if arranged in a funnel-shaped way - need no extra solar-receiver. The independent submodules can be put also onto a separate receiver (SAW). With these two concepts, a single submodule can be exchanged easily without disrupting a sodium loop. With the submodul design SW all submodules have a common sodium-loop, which is - however - separated from the receiver. Concept SI has two sodium-loops, both connected to a common sodium reservoir.

If the AMTEC is being designed as one single module, only two different arrangements are possible: with concept MW the module will be installed onto a separate receiver, with concept MI the AMTEC will be integrated into the receiver. In the following, the concepts described above will be explained in detail and compared with each other by stating their specific advantages and disadvantages.

The respective sketches in this and the following figures show the principal arrangement modes. A submodule is being represented as one or two single tubes for reasons of simplicity. The tubings, pumps and valves show the sodium loop and indicate the amount of plumbing needed. As heat input sodium-pool- or heat-pipe-receiver are assumed (if not stated otherwise).

5.3.1 Submodule-concepts

SA: independent submodules

Fig. 15 shows this arrangement. A submodule consists of 3 to 4 tubes, connected in parallel, an electromagnetic (EM) pump and its own condenser. The sodium-high pressure compartment will be designed as wick- or pool-receiver, so that the bottom of the submodule can serve directly as receiver for solar radiation. For an optimal utilization of the concentrated radiation, the submodules will be arranged in the shape of a funnel. The electrical series connection can be outside the submodules, so the current feedthroughs (one for each submodule) can be put into the cold part.

Disadvantages	Advantages
• Great amount of casing materials	- Small number of feedthroughs; which can be on the cold side.
• Heat losses through many casings	- Exchange of a failed submodule is easy
• Large number of pumps (costly)	- Submodules can be provided with a fuse (for failures)
• Temperature distribution not uniform	- Submodules are almost identical
• Large volume, leading to reduction of usable radiation	- Operation either with liquid sodium or with sodium vapor.

SAW: Submodules + separated heat input

Fig. 16 shows this arrangement, which is similar to concept SA. The inhomogeneous temperature distribution is avoided here by mounting the submodules on a common wick- or pool-receiver with homogeneous temperature conditions. Also in this configuration, feedthroughs are necessary only for the cold parts. However, the two-walled arrangement hampers the heat flow and leads to efficiency-losses.

Disadvantages	Advantages
• Temperature of AMTEC considerably lower than temperture of receiver	- High volumetric density
• Material consumption is high because of the double-walls	- Homogeneous temperature distribution
• Each submodule must have its own pump	- Only one "cold" feedthrough necessary per submodule
	- Fuses can bridge failed cells
	- Exchange of submodules is easy

SW: Submodules + separated heat input, but integrated sodium loop

Fig. 17 shows that in this concept liquid sodium will be pumped by only one EM-pump through long, thin tubes to the submodules. This causes a short circuit (bypass), which is, however, relatively small. Heat is transferred (like with SAW) from a separate receiver to the bottom of the submodules. For the exchange of failed submodules, a series of valves must be provided in the tubings.

Disadvantages	Advantages
• Temperature of AMTEC considerably lower than temperature of receiver	- Only one EM-Pump necessary
• High material consumption	- High volumetric density
• Reduction of power because of the bypass	- Homogeneous temperature distribution
• Valves cause additional problems	- Only one cold feedthrough per module
• Only liquid sodium can be used	- Fuses can bridge failed cells

SI: Submodules with integrated heat input

Fig. 18 shows that in this concept the sodium transport and at the same time the heat transport to the submodules is carried out by pumping sodium with a high-temperature pump. This transport mode allows a very low temperature gradient building up between receiver and submodule.

Disadvantages	Advantages
• High materials consumption	- Only one pump necessary
• High-temperature valves and a high-temperature sodium pump are necessary	- Homogeneous temperature distribution
• Exchange of failed submodules is difficult	- Fuses can bridge failed cells

5.3.2 Module-concepts

MW: Module with separated heat input

In this concept, shown in fig. 19, all tubes share the same condenser. Module and receiver are two separate units, so it will be possible to use solar receivers which are already available. The electrical series connection must be done inside the module (at 1000 °C).

Disadvantages	Advantages
• AMTEC-temperature will be considerably lower than the receiver-temperature	- Receiver and AMTEC can be exchanged separately
• Internal series connection difficult, because of 1000 °C	- Two separate sodium loops with different requirements concerning purity
• Heat flow between receiver and AMTEC requires large areas	- Less materials consumption than with submodules
• No exchange of failed cells possible	Compact design reduces heat losses

MI: Module with integrated heat input

Fig. 20 shows an integrated concept, where all tubes share the same condenser and no separate receiver is required. The temperatures of receiver and AMTEC are almost the same, because of the heat-pipe effect of this arrangement. This design is very compact and allows minimal heat losses.

Disadvantages	Advantages
	- Smallest material consumption of all cell concepts
• Internal series connection difficult, because of 1000 °C	- The compact design reduces heat losses
• No exchange of failed cells possible	- Only one pump necessary

5.3.3 Selection

The assessment of the concepts described above leads to the conclusion that the integrated module design (MI) is superior to all submodule-designs and also to the module-design with separated heat input.

This choice was made because the concept MI

- has the prospect of a high efficiency (compact design, efficient heat flow)
- offers low material consumption concerning pumps, tubings, plumbings, casings, valves, etc.

This conclusion is valid only for solar applications. For other applications different requirements must be met.

5.4 Storage device

Solar thermal systems generally require some form of energy storage /9/ to:

- provide continuous operation during periods of variable insolation
- extend operation into non-solar hours
- buffer potentially harmful system transients induced by abrupt insolation changes.

Whereas photvoltaic systems can use only batteries for this purpose, solar-thermal systems can use heat storage as well. This storage of heat is more efficient and considerably cheaper than battery storage and therefore, is to be considered also in the present study.

One possible approach is the storage as latent heat. This can be accomplished e.g. by using NaF (melting point 990 °C), which can be placed in a separate enclosure in the receiver close to the wall adjacent to the sodium pool /3/. The storage capability of NaF with respect to its latent heat is 0.3 kWh/kg. Thus, a one-hour "buffer" will need about 100 kg NaF for nominal power output. This is far more than the weight of the whole 10 kW-AMTEC.

Therefore, we decided to use only the sensible heat of the sodium in the pool to act as a small buffer, which will prevent thermoshocks from the beta-alumina tubes. The sodium content of about 6 kg (see 4.3.5) corresponds to a sensible heat of about 8 kJ/K. Allowing a temperature drop of 200 K, this corresponds to a storage capacity of about 500 Wh. At nominal power output, this will mean a "buffer-time" of only about 1 min. After this time (at 800 °C) the electric load must be removed in order to slow-down the cooling process. The remaining sensible heat of the sodium pool inside the receiver down to 100 °C (before sodium will freeze and heat input will become difficult) is about 1700 Wh.

This will allow a time of several hours till sodium freezes. However, if a shutter is placed before the receiver, it will last at least 12 hours until freezing. So, by using sodium itself as a storage medium, it will be possible to start-up in the morning with liquid sodium, which means no delay for melting sodium.

5.5. Power conditioning

Details of power conditioning systems are not the subject of this study, because they are described in a number of publications (e.g. /11/).

The AMTEC produces direct current. For consumer applications, a transformation into alternating current of 110 oder 220 V must be carried out. For a stand-alone operating mode (with no connection to an a.c. grid) dc/ac inverters are commercially available today with an efficiency of about 90 %. This is also valid for small units of 10 kW and input-voltages as low as 24 V. Because of this relatively low voltage, several AMTEC-tubes can be connected in parallel, as it will be shown in the following chapter.

The electric power output (and dependent on that also the temperature) will be controlled by a thyristor-equipped regulator. This regulator has generally an efficiency of > 98 %.

6. Design of a 10 kW AMTEC system

General

The general design considerations were made in 5.3. It was shown that the "integrated module design (MI)" is the best choice for solar applications. For this design a detailed plan was made which is described in the following. In addition to the specifications given in chapter 4 the module was optimized mainly towards:

- high efficiency: - short and thick connections to neighbor cells,
                        - compact arrangement
                        - free sodium gas flow to condenser
- small volume: - dense packed tubes
                        - small condenser area.

The basic data of ABB's β"-Al_2O_3 tubes are:
- length: $L = 22$ cm
- outer diameter: $d = 2.5$ cm
- useable surface area: $A = 140$ cm^2.

From recent experiments it was concluded that the ratio of the area of the condenser to the area of the β"-Al_2O_3 tubes should be at least $\varepsilon = 0.2$. Smaller values have the consequence of decreasing the power density because of pressure losses in the condenser area.

Number of tubes

Feedthroughs are critical parts in the AMTEC, because they can be the reasons for short circuits by sodium films. Feedthroughs are necessary inside the module only if tubes are to be series-connected. Thus, the number of feedthroughs can be reduced by parallel connection of some tubes. We decided to connect 3 tubes in parallel and to connect these "groups" in series. This is shown in fig. 21.

As one tube yields a maximum power output of about 90 W, 37 groups of 3 tubes each (i.e. overall: 111 tubes) are needed in order to arrive at the 10 kW-module. The optimal arrangement of the 37 groups is shown in fig. 22.

The most efficient arrangement of tubes with a ratio χ of condenser area to footprint of

$$\chi = \frac{3600}{1395} = 2.6$$

will be a sector of a sphere, or, more plainly speaking, the back of a hedgehog with the beta-alumina tubes as spines on it. This is illustrated in fig. 23. This is in principle the "integrated module design (MI)", described in chapter 5.3. The receiver is designed as sodium pool receiver, in which sodium is being evaporated and can move inside the tubes which have their opening at the bottom. The dimensions of such a module (without receiver) are shown in fig. 24. The volume of the 10 kW-module without receiver and thermal insulation can be calculated to be:

$$V = 60 \, l.$$

Fig. 25 shows details of internal connection and of metal/ceramic joinings. There are (as example) two groups with 3 tubes each (only 2 tubes are shown in this drawing). The tubes inside one group are being connected in parallel, the two groups are connected in series. As electric insulator between the groups an MgO-ring (which is sodium-resistant) will be used. In order to prevent sodium films, which can cause short circuits, the MgO-ring will be heated. As brazing for the ceramic metal joining, several active brazes can be applied.

Current collector on the inert part of the tubes will be a metallic cylinder which contains liquid sodium. This sodium is supplied by the evaporation of sodium from the pool-receiver. On the outside of the tubes a thin Mo-sputter layer and a nickel mesh will serve as current collector.

The condenser is placed adjacent to the closed end of the tubes. Because it is formed as part of a sphere, the condensed sodium can be captured in a trough and thus will flow to either end of the module and return to the receiver via an electromagnetic pump (see fig. 23).

A thermal insulation must be put around the module and receiver in order to minimize parasitic heat losses, which reduce the efficiency. The condenser area - however - must remain free in order to allow the heat of condensation to be removed.

7. Calculation of expected data for Almeria conditions

7.1. Solar radiation data for Almeria

The following data are extracted from the report on the GAST-technology program (gas-cooled solar-tower power plant) /8/:

1.	Direct radiation per year:	4380 h
2.	Direct radiation > 300 W/cm^2 per year:	2824 h
3.	Overall direct radiation energy per year:	1959 kWh/m^2
4.	Maximum possible overall direct radiation energy per year:	3266 kWh/m^2
5.	Global radiation on horizontal plane per year:	1776 kWh/m^2
6.	Diffuse radiation per year:	612 kWh/m^2
7.	Maximum direct radiation power:	970±30 W/m^2
8.	Most frequent direct radiation power:	ca. 800 W/m^2
9.	Average period of sunshine:	2.5 h
10.	Average period of cloudiness:	1.1 h
11.	Transient period between sun and cloud:	3 min

In fig. 26 the annual distribution of the direct radiation is shown. It can be seen, that about 1/3 of the direct radiation is in the region below 300 W/m^2, which is generally not very efficient for solar energy conversion. The data in fig. 26 yield an average value for the direct radiation of 510 W/m^2 for Almeria.

7.2 AMTEC module

Internal resistance:

Calculation for the present design yields a value of about 2.1 mΩ for the β''-Al_2O_3 tube including current collector at 1000 °C. This can be converted to an overall specific resistance of

$$R_o = 0.3 \ \Omega cm^2$$

The internal resistance of the 10 kW-module will then be about 26 mΩ.

Voltage/Current

The open circuit voltage U_0 of one tube is described in /6/:

$$U_o = \frac{RT_2}{F} \bullet \ln \left(\frac{P_2}{\sqrt{T_2 / T_1} \, P_1} \right)$$

This was already described in chapter 3.1 with a first rough estimation. For T_2 = 1000 °C and T_1 = 300 °C an open circuit voltage:

$$U_o = 1.27 \ V$$

results.

Thus, the whole module will have an open circuit voltage of

$$U_{OM} = 47 \ V$$

When current is being drawn (i.e. power is produced), the voltage drops due to ohmic losses and pressure drops. This is described also in /6/:

$$U = \frac{RT_2}{F} \bullet \ln\left(\frac{P_2}{\sqrt{T_2/T_1}\ P_1 + \sqrt{2\pi MRT_2} \bullet i/F}\right) - R_o i$$

where i is the current density, M the molecular weight of sodium and R_o the specific internal resistance. Using the value of $R_o = 0.3\ \Omega cm^2$ for the present module, the current-voltage and current-power relationship in fig. 27 was obtained.

The maximum power lies according to this graph beyond 15 kW. However - as shown further below - the efficiency reaches its maximum value at about 250 A (corresponding to a current density of 0.6 A/cm^2). At the nominal power of 10 kW there is:

$$I_N = 270\ A$$
$$U_N = 37\ V\ .$$

Heat losses
The parasitic heat losses Q across the low-pressure compartment are primarily radiation losses and are governed by the following equation:

$$Q_r = \sigma\left(T_2^4 - T_1^4\right) \bullet \varepsilon$$

where σ is the Boltzmann constant and ε the emissivity. At $T_2 = 1000$ °C and $T_1 = 300$ °C a value of $Q = 0.5$ W/cm^2 can be achieved by using highly reflecting materials with $\varepsilon = 4$ %. (Considerably lower values can be achieved if radiation shields are used in addition.) The above value means a heat loss of 1830 W across the condenser (3660 cm^2). The heat loss of a 5 cm thick conventional fiber insulation at the module walls outside the condenser adds about 700 W. If additional conductive heat losses are estimated to be about 200 W, then a total parasitic heat loss of

$$Q = 2700\ W$$

at 1000 °C must be expected.

For the following calculation the value of Q per cm^2 of electrode area is important. This value is here: q = 0.17 W/cm^2.

Efficiency

The efficiency has already been explained in fig. 12. The more detailed expression is:

$$\eta = \frac{U \cdot i}{U \cdot i + i\Delta h / F + iC_p(T_2 - T_1)/F + q}$$

where Δh is the molar heat of vaporization and C_p the molar specific heat for sodium.

Fig. 28 shows a calculation of the efficiency of the designed module. It can be seen that at the nominal power of 10 kW the efficiency is

$$\eta_N = 42 \%.$$

This is about the point of the maximum efficiency. However, even at the 5 kW-partload point the efficiency is still about 40 %. These figures are valid only for an optimized system. In the near term 30 - 35 % will be obtained, because of ohmic and parasitic heat losses.

The part-load behavior (of the module, without dish and receiver efficiencies being taken into account) is shown in fig. 29. The electric output becomes zero when the heat input is equal to the parasitic heat losses of 2.7 kW. This energy is needed to keep the 1000 °C. A heat input of < 2.7 kW leads to cooling down of the module.

7.3. AMTEC system

The complete system for the conversion of solar energy into ac electricity via dish and AMTEC is shown in fig. 30.

System efficiency

According to a study conducted by Battelle /12/, the following efficiencies can be reached today

−Dish: η_c = 85 %

- Receiver: η_R = 80 %

(These are rather conservative figures, which include also partload values.)

The efficiency of the converter (power conditioning) can be assumed to be (see chapter 5.5.):

– Converter: η_P = 90 %.

Therefore, assuming an efficiency of η_A = 42 % for an optimized AMTEC, a system efficiency of

$$\eta_{SM} = 25 \%$$

will be obtained for nominal power output.

Dish size

Taking a solar insolation of 800 W/m^2 for the nominal power output of 10 kW (see chapter 4.3.2), a dish area of

$$A_D = 50 \text{ m}^2$$

is needed. This means a diameter of about

$$\phi = 8 \text{ m} \quad \text{(without frames)}.$$

The maximum electrical output which can be achieved with this dish and the AMTEC module described above, will be at the maximum insolation of 970 W/m^2:

$$P_{max} = 12 \text{ kW}.$$

This corresponds to a thermal input of 29 kW (see fig. 29).

Performance data of the system for Almeria

The part-load behavior can be obtained by combining figs. 26 und 28. Table 1 shows the results.

Table 1: Calculation of annual efficiency for Almeria

Direct insolation W/m^2	Annual distribution (fig. 26) %	Efficiency AMTEC (fig. 28) %	Efficiency system %	"Relative" electric power output W/m^2
0 - 300	32	-	-	-
300 - 400	6	38	23.2	4.2
400 - 500	7	40	24.4	7.7
500 - 600	10	41	25.0	13.8
600 - 700	10	42	25.7	16.7
700 - 800	13	42	25.7	25.1
800 - 900	17	42	25.7	37.1
900 - 1000	5	42	25.7	12.2
Sum	100			117

Related to the average annual insolation of 510 W/m^2, the average annual power output of 117 W/m^2 means an annual efficiency of the system of

$$\eta = 23 \%.$$

The capacity factor is defined as:

$$c = \frac{\text{annual electric output } (kWh)}{\text{max. electric power} * \text{annual operation hours}}.$$

The annual electric output will be:

$$EA = 50 \ m^2 * 4380 \ h * 0.117 \ kW / m^2 = 25.6 \ MWh$$

This leads to a capacity factor of:

$$c \approx 25 \%.$$

This is in good agreement with other solar/dish systems /12/.

Power density:
The volume without insulation and receiver was calculated to be 60 l. This means a volumetric power density of

$$P_{max} = 200 \text{ W}/l$$

in the case of maximum power output (170 W/l at nominal conditions).
If - as in chapter 7.2 was described - an insulation of 5 cm thickness is used, the volume of the module will be about 78 l, leading to a maximum power density of 150 W/l (130 W/l at nominal conditions).

A very rough estimation of the weight of the module can be made as follows:

- Tubes with joining:	9 kg
- MgO rings	1 kg
- Condenser (steel)	4 kg
- Bottom	2 kg
- Current collectors	6 kg
- Sodium	4 kg
- OTHER (wicks, support for tubes etc.)	4 kg
Overall	30 kg

Thus, the gravimetric power density of the modul will be about 400 W/kg (without receiver) for maximum power output (300 W/kg at nominal conditions).

7.3. Cost calculations

Detailed cost calculations for complete solar-thermal system are contained in /12/. Because of the large number of unknow parameters, in the present study the calculations will be confined to the AMTEC module alone.

The following rough estimation of the costs is based on the pure material costs, multiplied by an empirical factor. This factor depends on the production size. A typical value known from series production is a factor of (3...4) for mass production, being several GW per year in the case of AMTEC production.

The material costs for the present AMTEC components can be estimated to be about DM 125.--/kW. Thus, production costs will amount to about DM 500.-/kW. Assuming an overhead of about DM 300 /kW, the end costs will be about DM 800/kW.

This value is rather low compared to Stirling - or even to Diesel engines. Thus, also the complete system will be competitive.

8. Comparison with competing systems and conclusions

The Dish-systems in general are more economic than Central Receiver - or Trough-Systems, according to /12/. Within the Dish system, there are a number of advantages of the AMTEC compared to the Stirling engine:

• Continuous high efficiency also during part load (until about 20 % of maximum load). This promises high annual efficiency, also in areas with medium insolation conditions.

• Low specific volume and weight. The reason for this is primarily the lack of rotating parts and the absence of an electric generator.

• No vibrations. This property - together with AMTEC's low weight - allows the construction of a considerably lighter and, therefore, cheaper dish.

• Silent. The dish/AMTEC system can be operated also in highly populated areas.

• No maintenance required. This is an important point, because especially the Stirling engine requires high O+M (= operating and maintencance) costs /12/, which affects the economy of the whole system.

- Low installation costs. The cost of the heat engine was recognized as a critical point with dish systems /12/. It was found that the (probably) high cost of Stirling engines would be a major drawback.

The results of the work carried out in the present study allows the conclusions that:

- An AMTEC-module can be built which meets the special requirements for solar applications.

- An AMTEC/dish system would have some important advantages over existing systems.

- It would be worthwile to develop the designed AMTEC-module and subsequently or in parallel a dish/AMTEC system.

9. Development plan for a realization of the designed 10 kW AMTEC system

According to the existing AMTEC program at ABB, a 10 kW module of a general design will be available by the end of 1994. However, this module will be stationary and not suited for solar applications. For adaption of the general 10 kW-module to the module designed in the present study, the following steps are necessary:

1. Detailed construction work
2. Design and construction of a receiver
3. Assembling of module and receiver
4. Integration into a dish
5. Power conditioning
6. Testing of the system in Köln and Almeria

Point 2, 3, 4 and 6 must be done in cooperation with DLR.

The proposed work can start in 1993 when ABB has built its 1 kW module. The 10 kW-"solar"- Module will then be built parallel to the "stationary" module developed bei ABB alone. 10 man-month will be required to carry out the work for the "solar" modul. A detailed proposal can be worked out by the end of 1992.

10. References

/1/ C.-J. Winter, J. Nitsch, Wasserstoff als Energieträger, Springer-Verlag Berlin, Heidelberg, New York, Tokio 1986

/2// Performance of the Vanguard Solar Dish-Stirling Engine Module, EPRI-Report AP-4608, July 1986

/3/ K. Subramanian, T.K. Hunt, Solar residential total energy system using the sodium heat engine - a concept study -, Proc. 17th IECEC (1982), 1474

/4/ L.L. Lukens, Dish electric systems heat engine assessment, Sandia Report SAND85-0522, June 1985

/5/ T.K. Hunt, Solar Thermal/Electric Conversion with the sodium heat engine, Sandia Report SAND84-2354, April 1985, p. 139

/6/ T. Cole, Thermoelectric Energy Conversion with Solid Electrolytes, Science 221 (1983) 915

/7/ W.B. Stine, Progress in Parabolic Dish Technology SERI-Publication SERI/SP-220-3237, June 1989, p. 29

/8/ Interatom, Asinel, Technologieprogramm GAST, Abschlußbericht 30.04.1988

/9/ G. Warfield, Solar Electric systems, Hemisphere Publishing Coporation, Washington, New York, 1984, p. 94

/10/ G. Bado, Ansaldo Research Institute, Genova, Italy: private communication, 1990

/11/ A. Räuber, F. Jäger, Photovoltaik, Verlag C.F. Müller, Karlsruhe, 1990

/12/ T.A. Williams, J.A. Dirks, D.R. Brown, M.K. Frost, Z.A. Antoniac, B.A. Ross (Battelle Northwest), Characterization of solar thermal concepts for electricity generation, PNL-6128, March 1987.

/13/ J. Kleinwächter: Paraboloid systeme; publication of Bomin Solar GmbH & Co. KG, 1987

/14/ J. Kleinwächter: private communication, May 1991

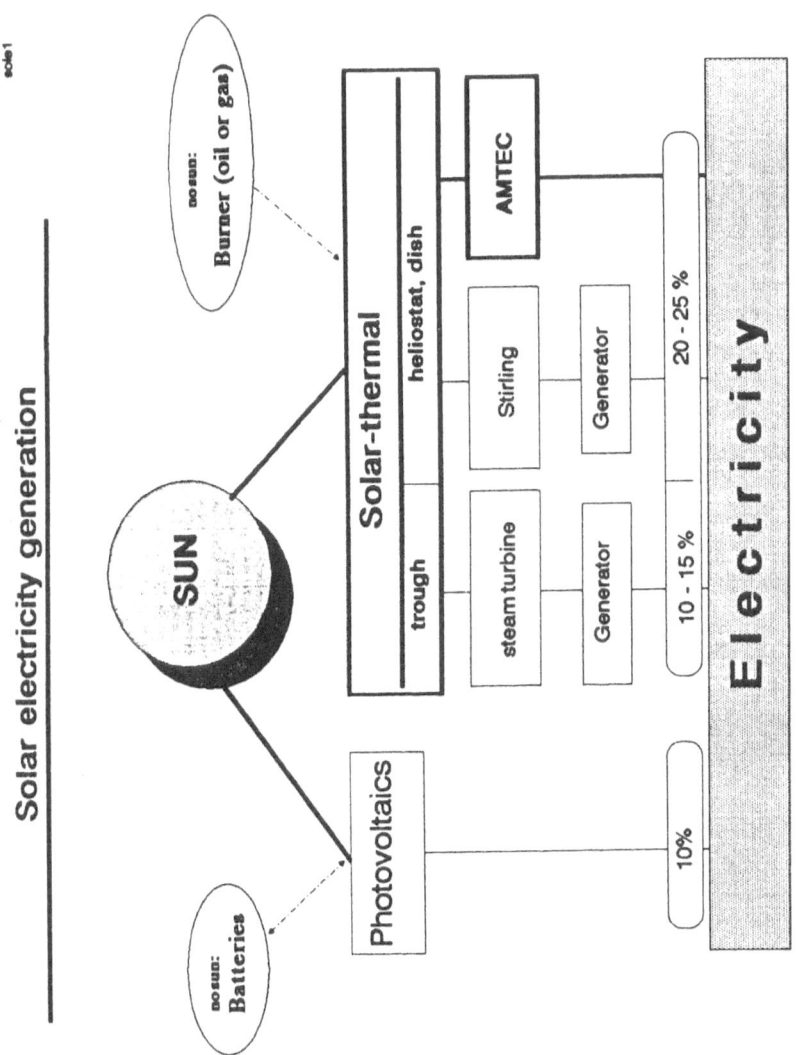

Fig. 1: Solar electricity generation

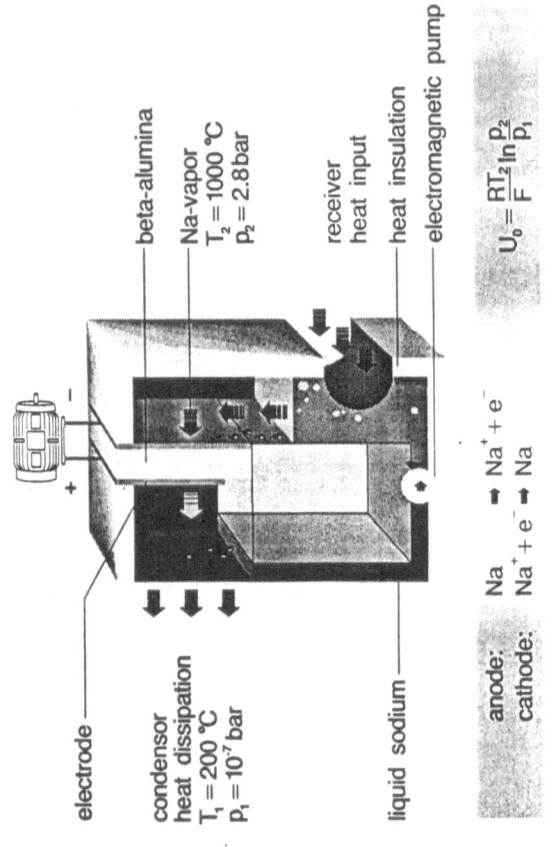

Fig. 2: AMTEC: Principle of operation

The following labels appear in the figure:

electrode

condensor
heat dissipation
$T_1 = 200\,°C$
$p_1 = 10^{-7}\,bar$

liquid sodium

beta-alumina

Na-vapor
$T_2 = 1000\,°C$
$p_2 = 2.8\,bar$

receiver
heat input

heat insulation

electromagnetic pump

anode: Na $\rightarrow$ $Na^+ + e^-$
cathode: $Na^+ + e^-$ $\rightarrow$ Na

$$U_0 = \frac{RT_2}{F}\ln\frac{p_2}{p_1}$$

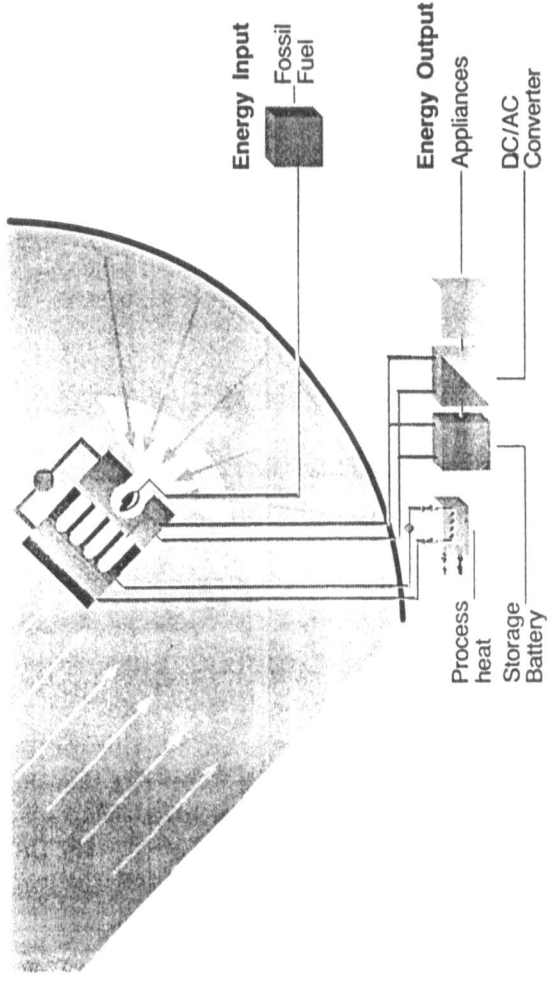

Fig. 3: Decentralized Hybrid (Fuel/Solar) Thermoelectric
Conveter Station

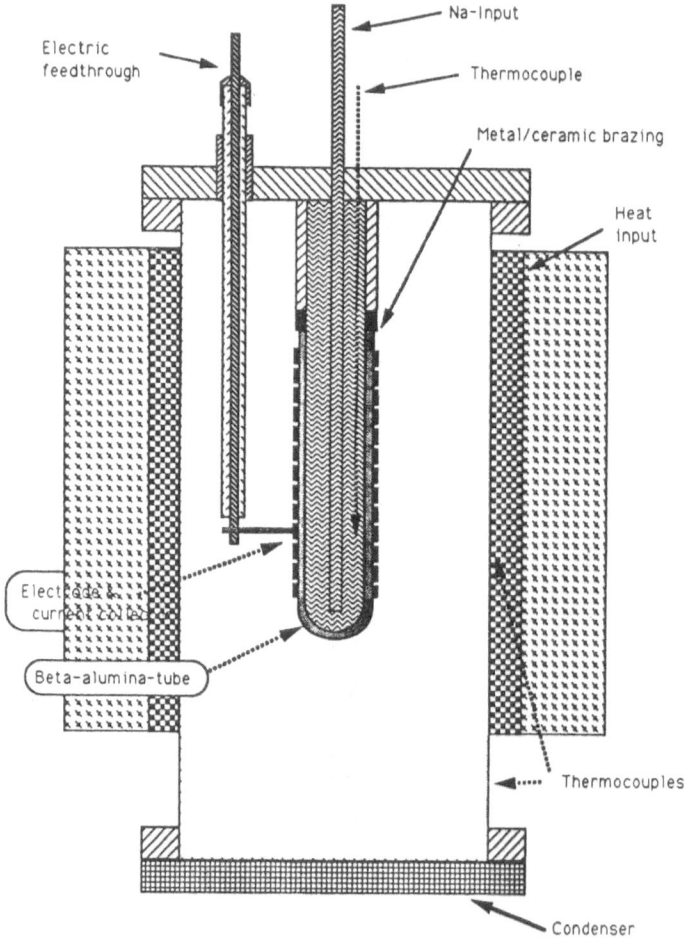

Electric
feedthrough

Na-Input

Thermocouple

Metal/ceramic brazing

Heat
Input

Electrode &
current collec.

Beta-alumina-tube

Thermocouples

Condenser

Fig. 4: Electrode test cell with liquid sodium (schematically)

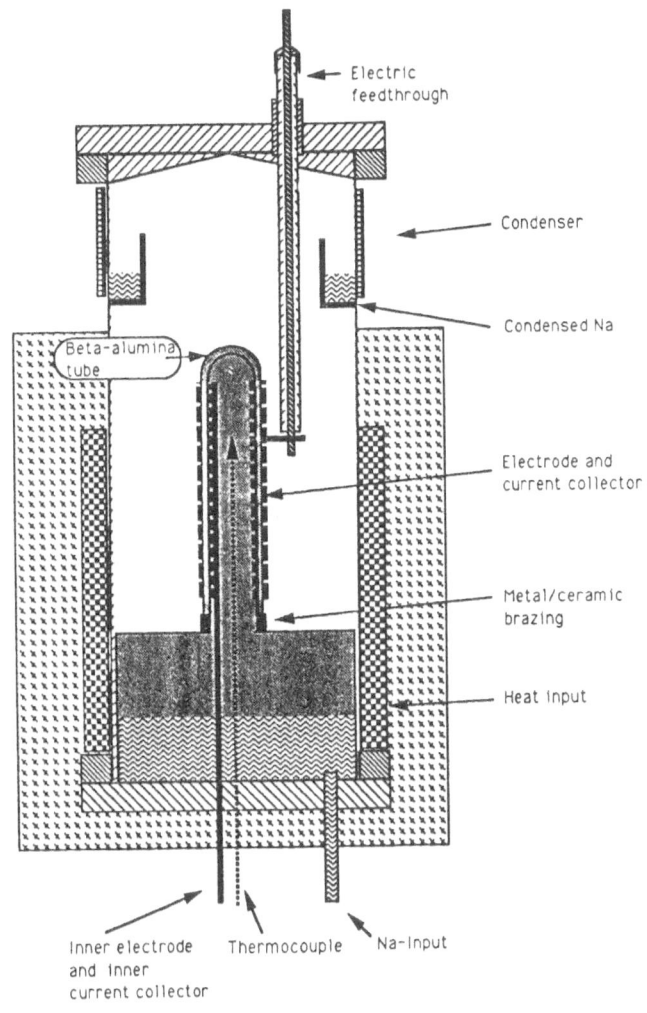

Fig. 5: Single cell with saturated sodium vapor as sodium supply

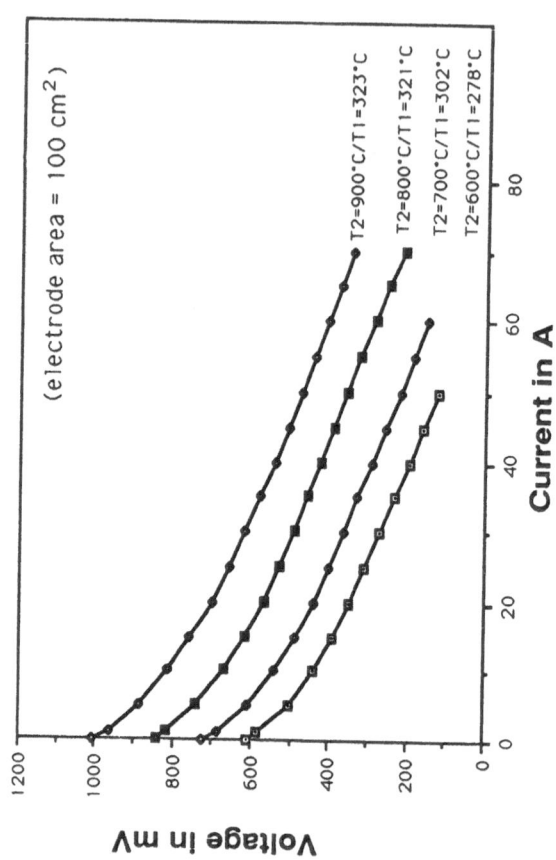

Fig. 6: Current/voltage relationships of an electrode test cell

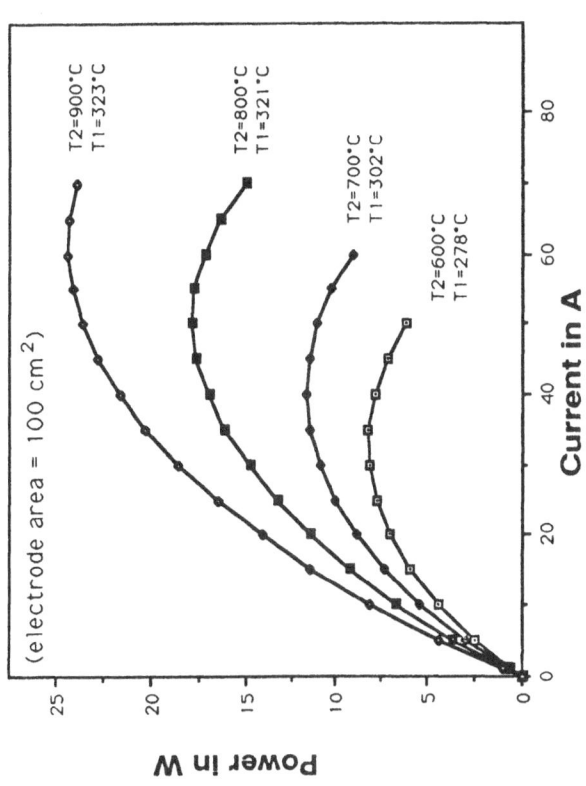

Fig. 7: Power/current relationship of an electrode test cell

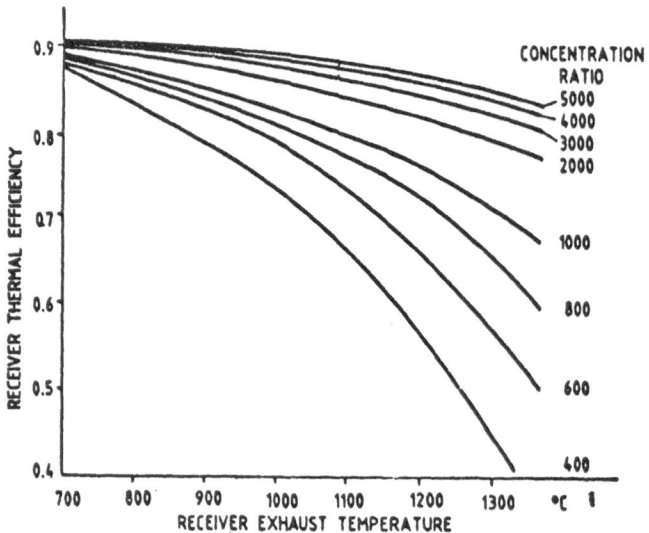

Fig. 8: Influence of the concentration ratio on the efficiency.

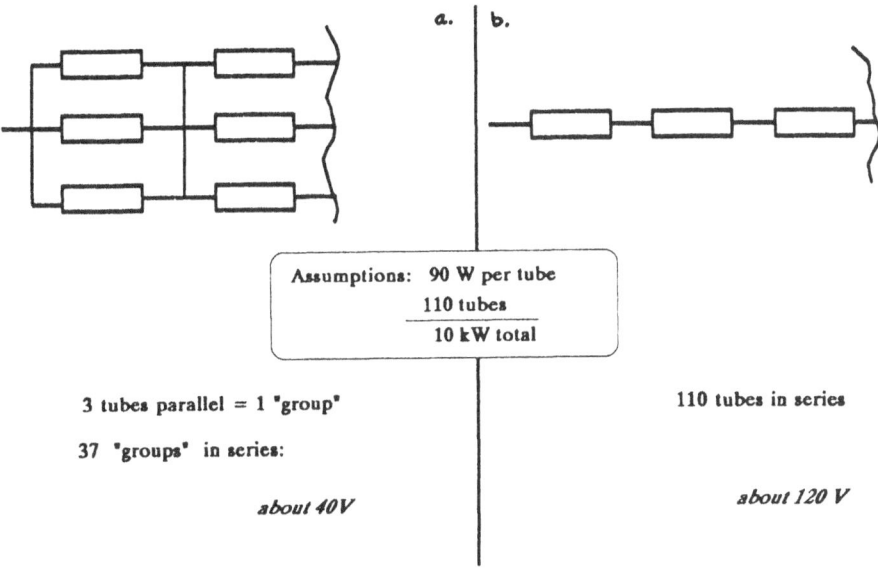

a. | b.

Assumptions: 90 W per tube
110 tubes
─────────
10 kW total

3 tubes parallel = 1 "group"   110 tubes in series

37 "groups" in series:

about 40V   about 120 V

Fig. 9a): Mixed parallel and series connection mode of tubes
Fig. 9b): Series connetion mode of tubes

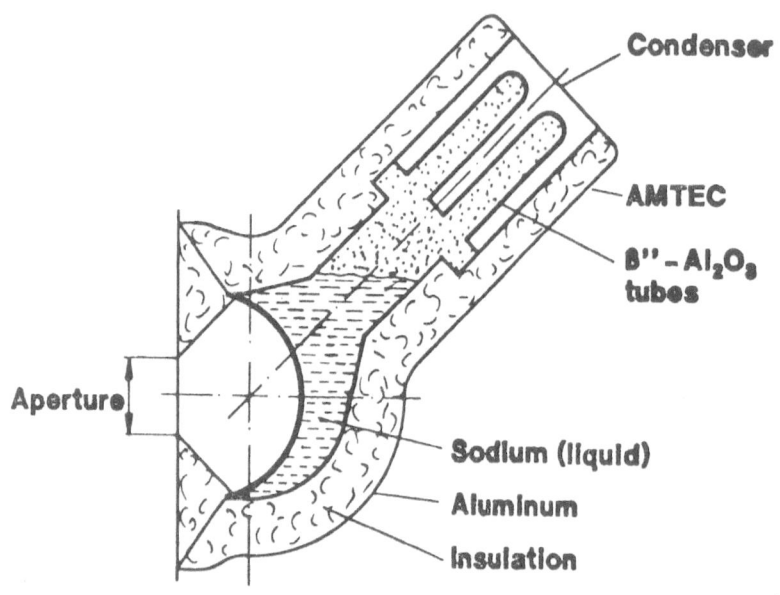

Fig. 10: Receiver of an AMTEC module, using liquid sodium as heat transfer medium. Adapted from a stirling engine concept /7/. Aperture: 16 cm, Na-content: 6 kg, average heat flux: 0.15 MW/m^2

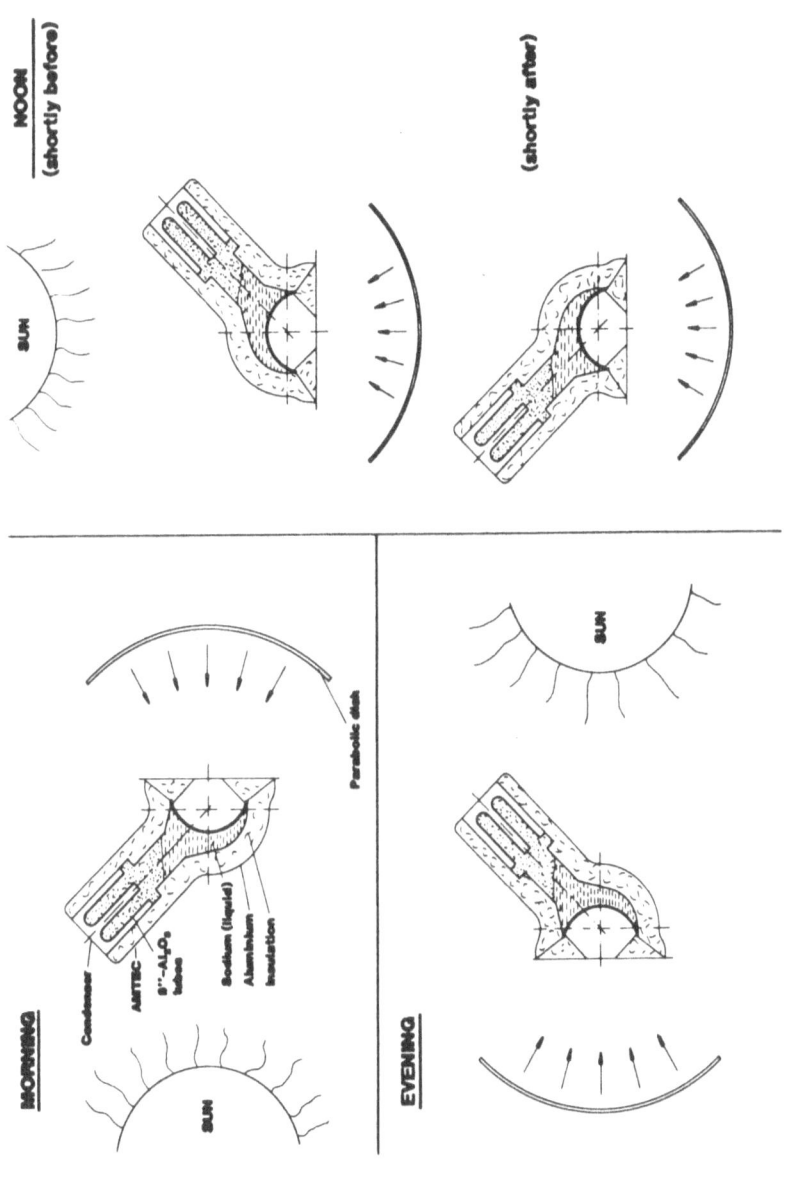

Fig11 .: Daily movement of a dish/AMTEC system

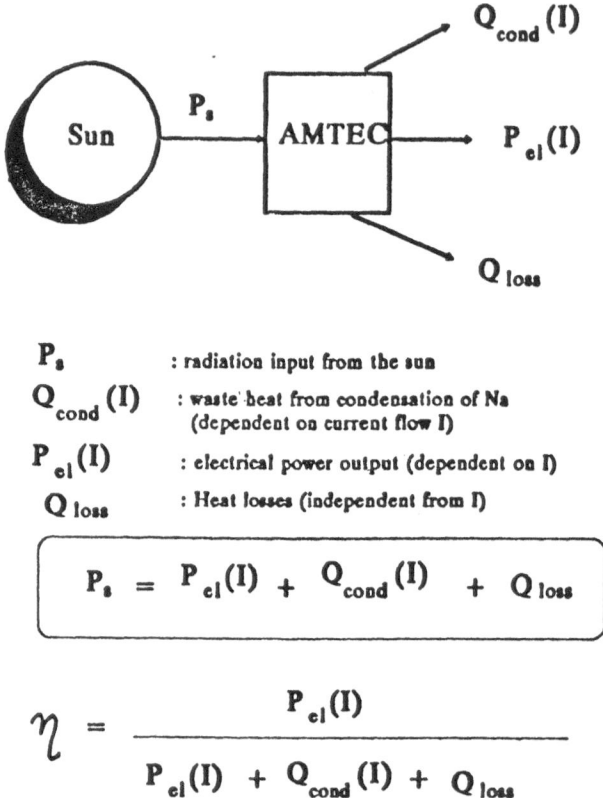

P_s : radiation input from the sun

$Q_{cond}(I)$: waste heat from condensation of Na
               (dependent on current flow I)

$P_{el}(I)$: electrical power output (dependent on I)

Q_{loss} : Heat losses (independent from I)

$$P_s = P_{el}(I) + Q_{cond}(I) + Q_{loss}$$

$$\eta = \frac{P_{el}(I)}{P_{el}(I) + Q_{cond}(I) + Q_{loss}}$$

Fig. 12: Energy flow in a solar-thermal AMTEC

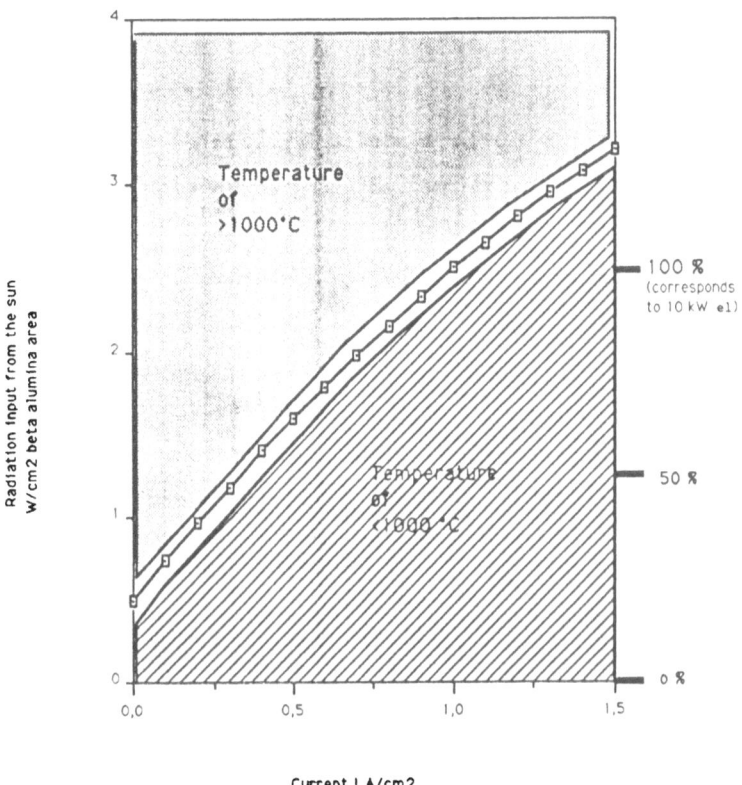

Fig. 13: Correlation between current density of a module and radiation input from the sun (calculated for a fictitious module, which is not the module in fig. 23).

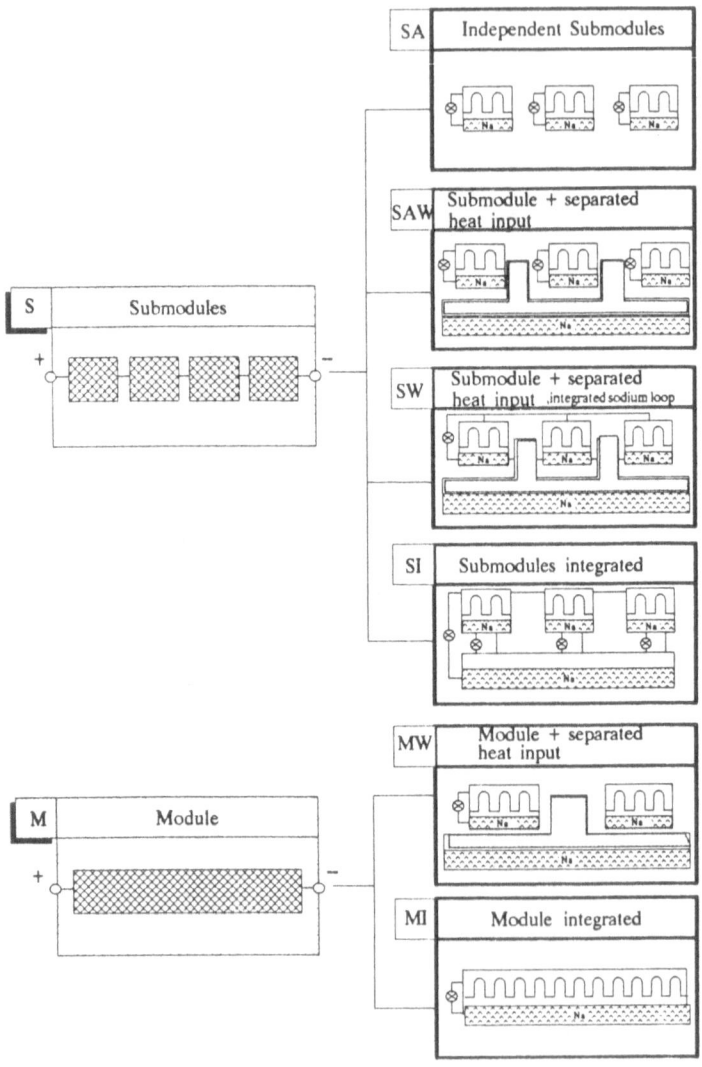

Fig. 14: Decision tree for a 10-kW-module concept

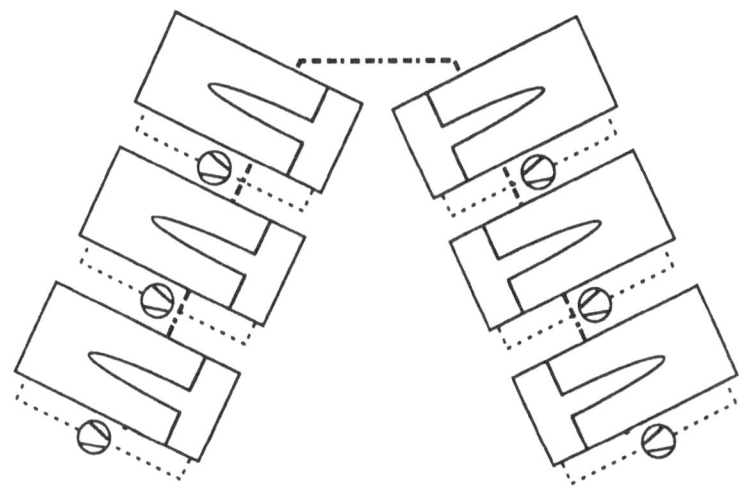

Fig. 15: SA: "Independent submodule"-concept

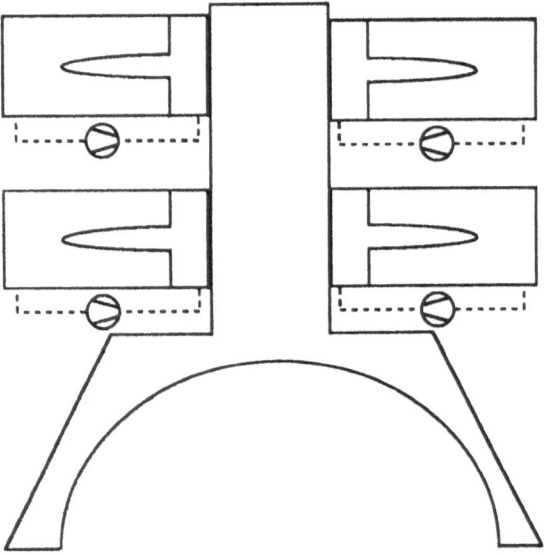

Fig. 16: SAW "Submodule + separated heat input"-concept

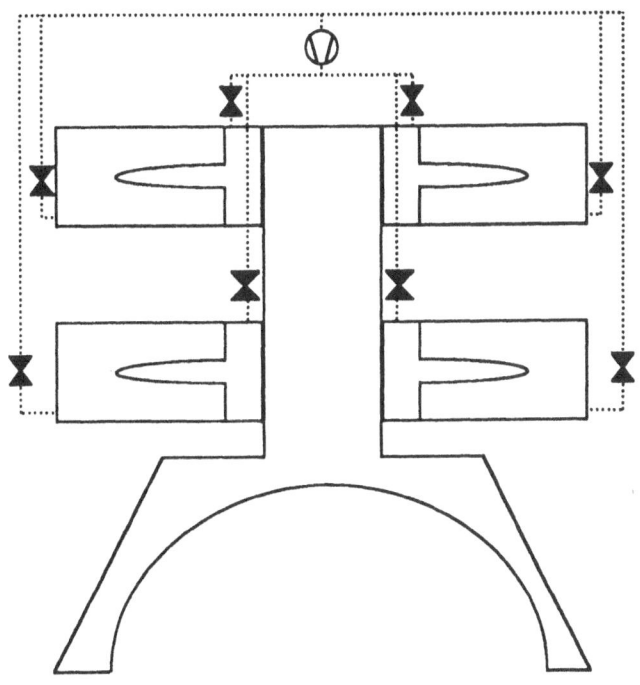

Fig. 17: SW: "Submodule + separated heat input"-concept

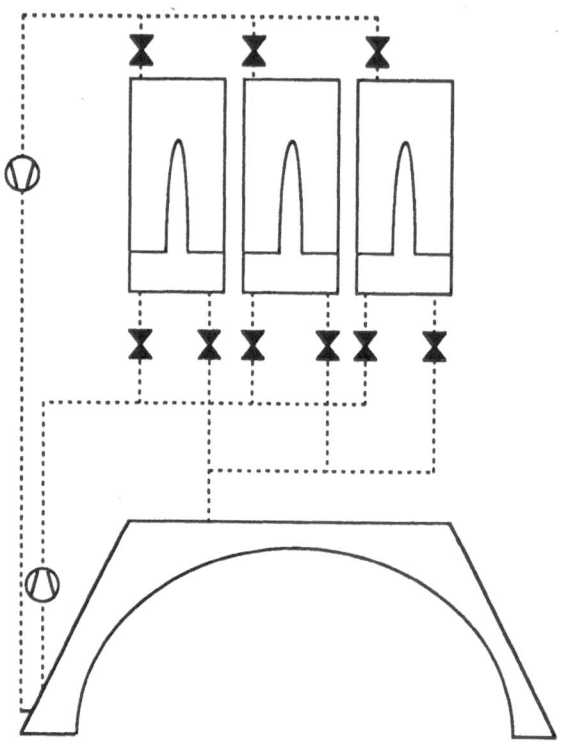

Fig. 18: SI: "Submodules integrated"-concept

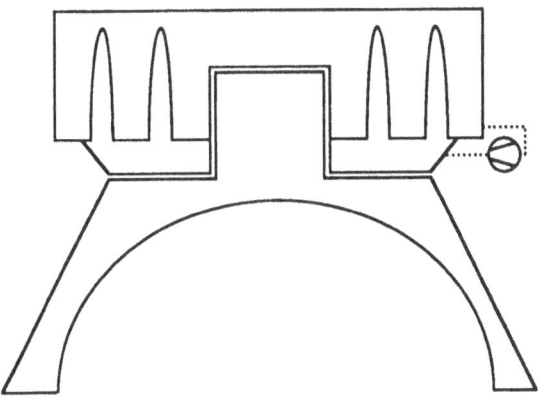

Fig. 19: MW: "Module + separated heat input"-concept

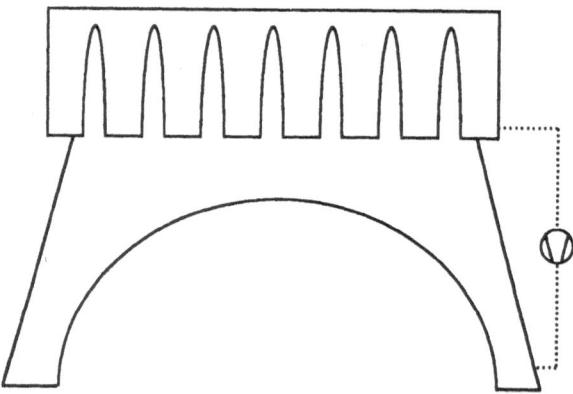

Fig. 20: MI: "Module integrated"-concept

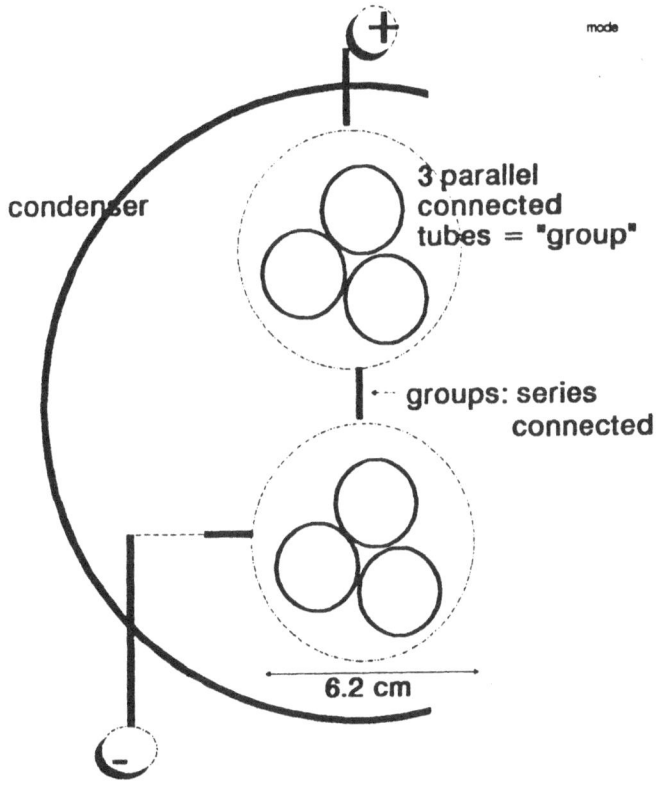

mode

condenser

3 parallel
connected
tubes = "group"

groups: series
connected

6.2 cm

+

−

Fig. 21: Connection mode inside the module (schematically)

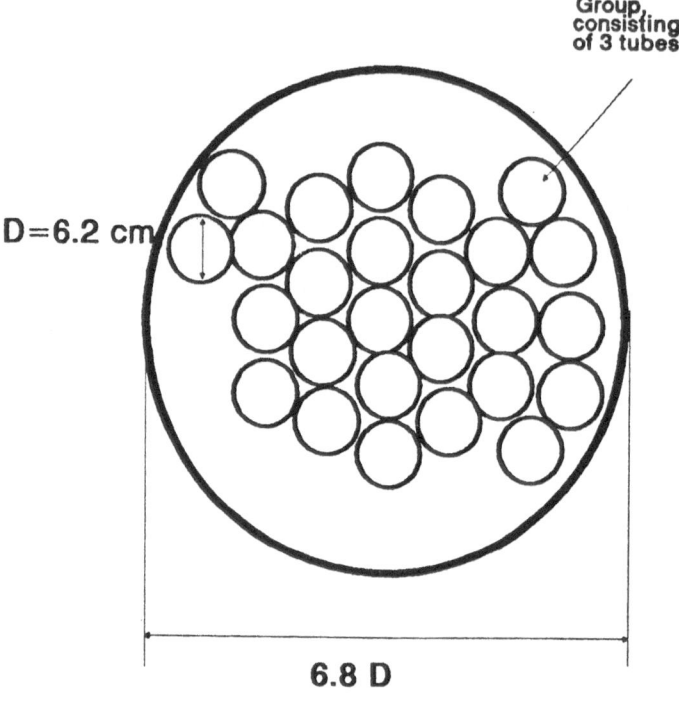

Group, consisting of 3 tubes

D=6.2 cm

6.8 D

Fig. 22: Principal arrangement of the 37 groups in the module

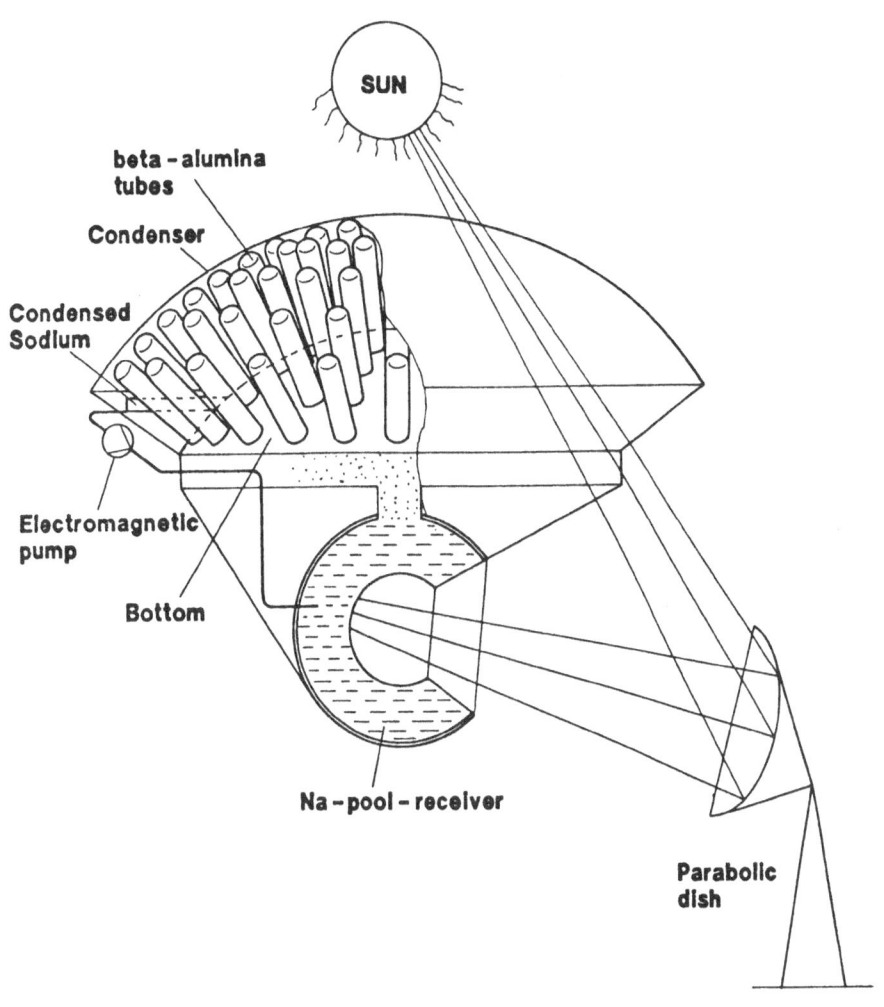

SUN

beta - alumina
tubes

Condenser

Condensed
Sodium

Electromagnetic
pump

Bottom

Na - pool - receiver

Parabolic
dish

Fig. 23: 10 kW-module design

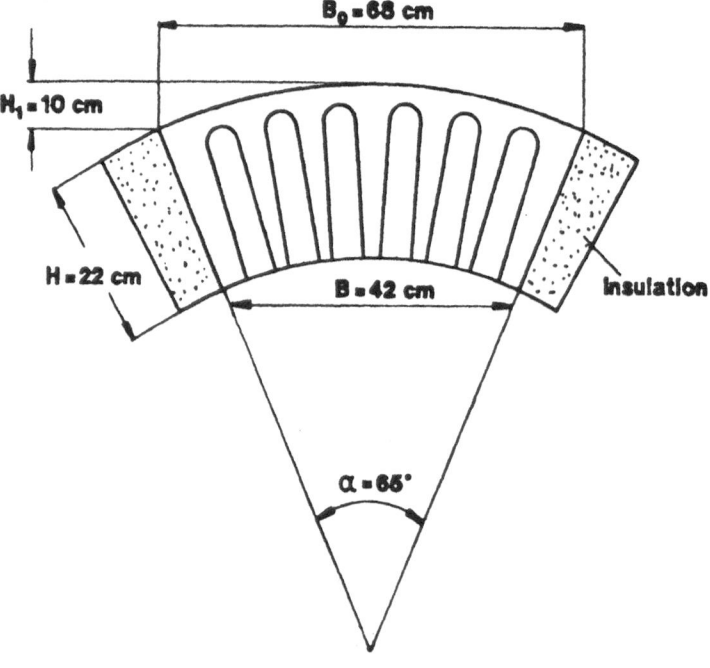

Fig. 24: Dimensions of the 10 kW-module

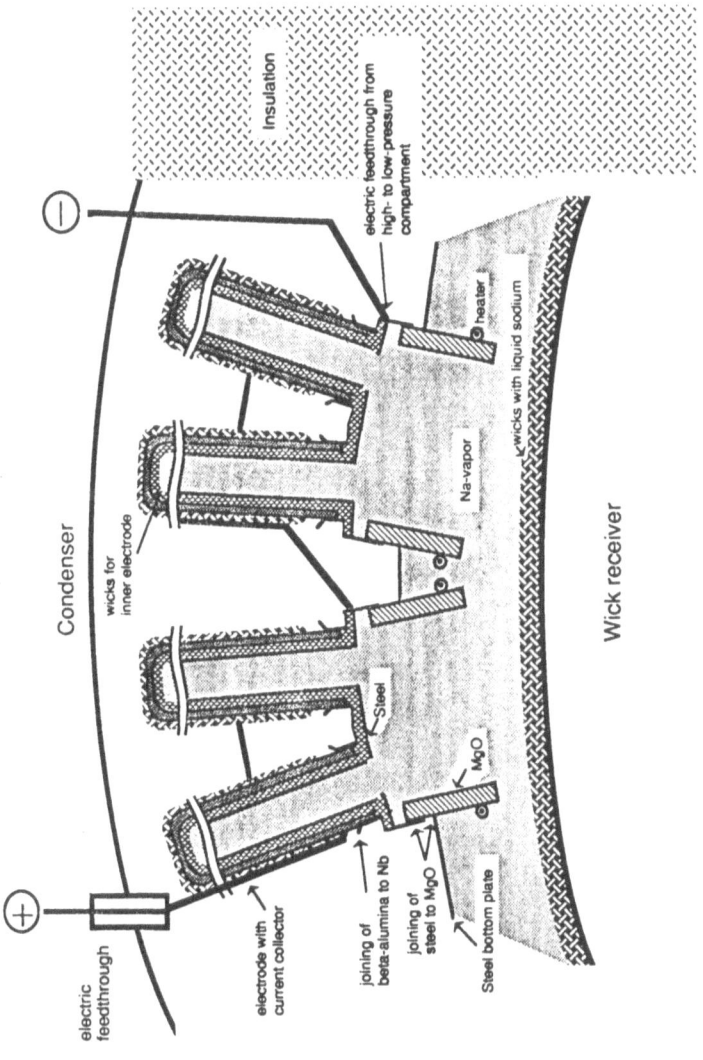

Fig. 25a: Two groups (only 2 from 3 tubes are shown) inside the module with electrical connections

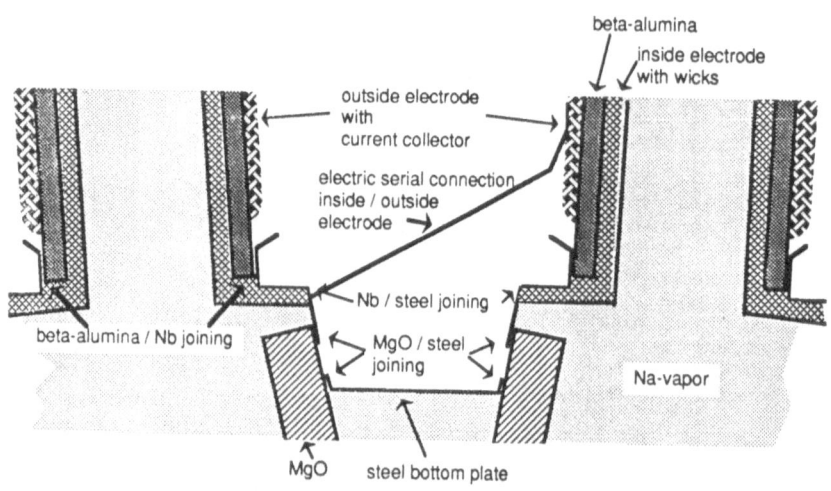

beta-alumina

inside electrode
with wicks

outside electrode
with
current collector

electric serial connection
inside / outside
electrode

Nb / steel joining

beta-alumina / Nb joining

MgO / steel
joining

Na-vapor

MgO   steel bottom plate

Fig. 25b: Details of Fig. 25a

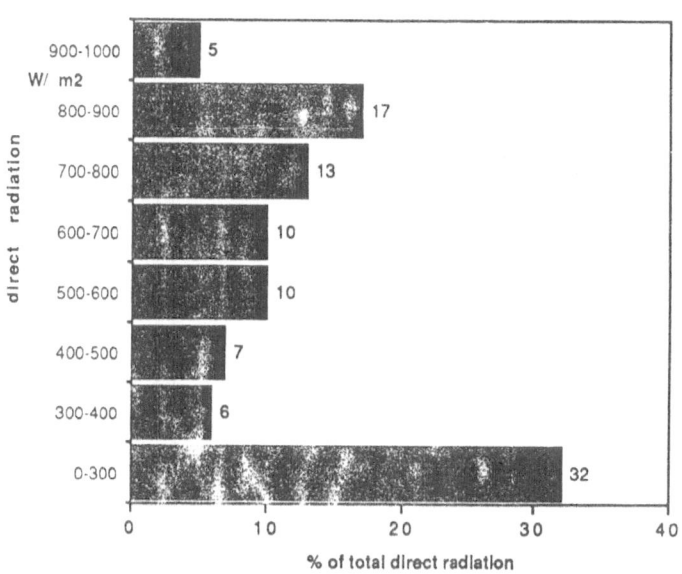

Fig. 26: Annual distribution of direct radiation in Almeria

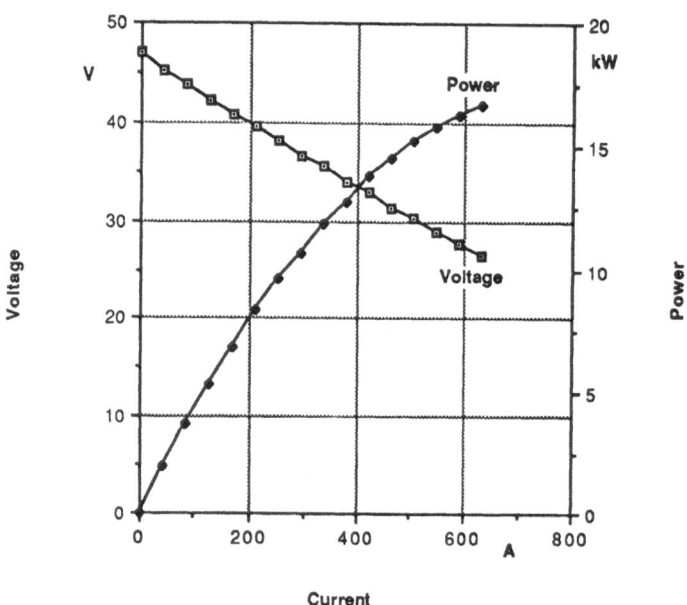

Fig. 27: Calculated voltage/current and power/current relationship
of the 10 kW-module for 1000 °C/300 °C

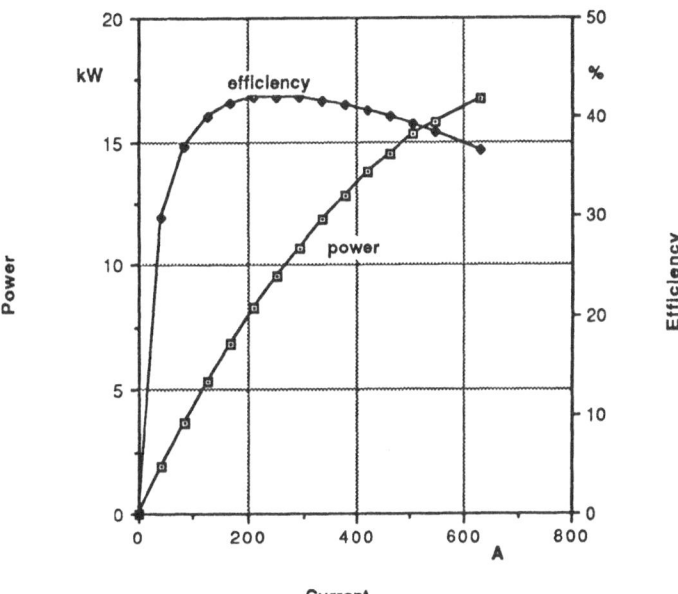

Fig. 28: Calculated efficiency and power of the 10 kW-module
for 1000 °C/300 °C

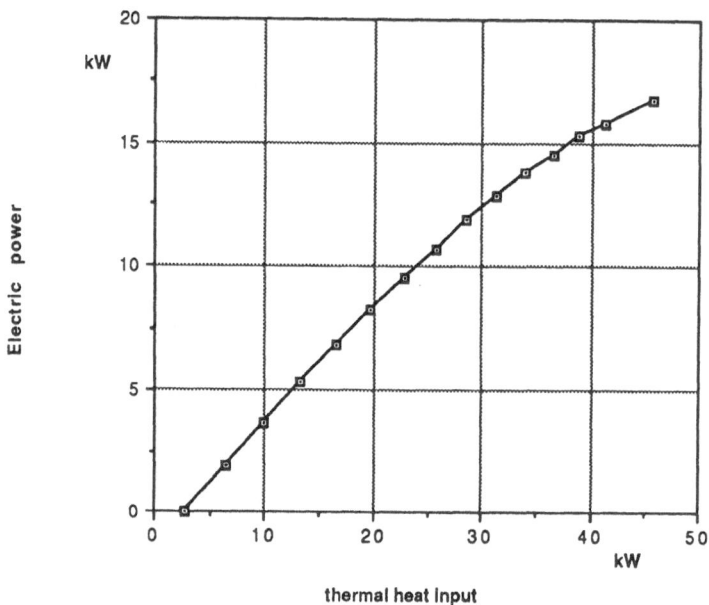

Fig. 29: Calculated electric output for the 10 kW-module at 1000 °C/300 °C in dependence of the thermal heat input.

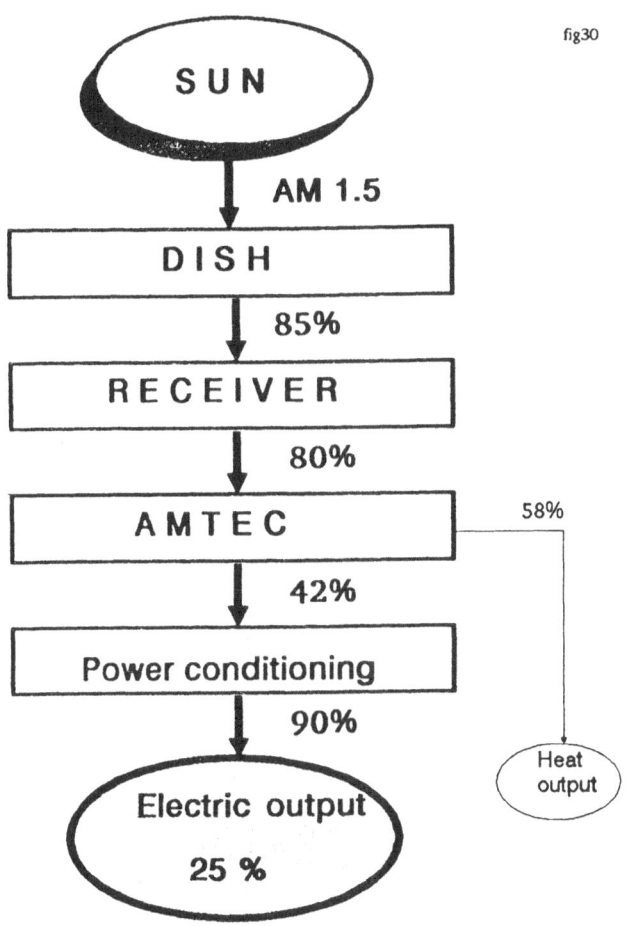

Fig. 30: Complete conversion scheme for an AMTEC/dish system

Optimization of a Selective Volumetric Foil Receiver
Detailed Heat Transfer and Flow Simulation

R. Pitz-Paal, J. Morhenne, M. Fiebig,

Ruhr-Universität Bochum,
Institut für Thermo- und Fluiddynamik

Contents

1 Introduction

Two years ago a new concept of a volumetric receiver design was proposed by the
Institute of Thermo- and Fluiddynamics at the Ruhr-University Bochum, FRG [4].
This design consists of a conventional ceramic foil receiver made of siliconized Silicon-
Carbid which is covered by a matrix of square channels made of quartz glass (see
fig. 1). The receiver is cooled by a flowing gas, which is driven through the whole
structure by a fan. Since the quartz glass structure is highly transparent to solar
radiation and only partly absorbs thermal radiation emitted from the high flux loaded
dark ceramic foil receiver, a reduction of thermal radiation heat losses is achieved.
In 1989, a mathematical model was developed which confirms the improvement in
energy conversion efficiency of about 10% compared to the case of the conventional foil
receiver even for a not yet optimized geometry and a possible gas outlet temperature
of more than $1000^{\circ}C$ [5]. This was an encouraging result since the new receiver type
would require 10% less heliostat field and aperture plane to deliver the same amount
of energy hence saving an enormous quantity of investment costs.

However, a lot of questions and uncertainties remained, so that a more detailed investigation was carried out, and the results are presented here.

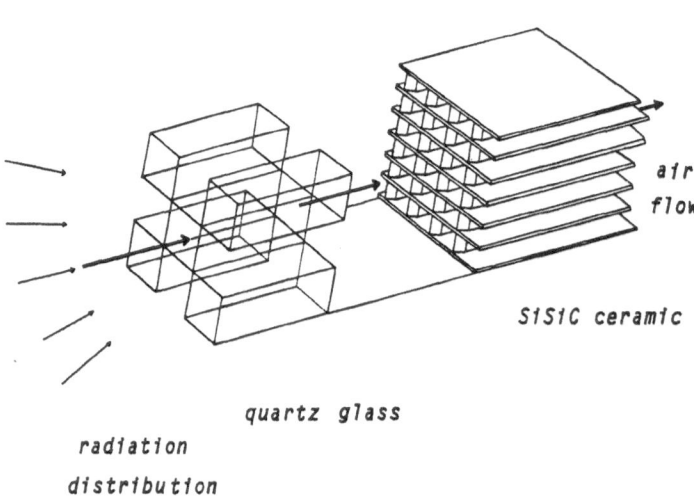

Figure 1: *Artist view of the proposed receiver design*

The Main fields of investigation were:

- **Design of the inlet region of the SiSiC ceramic foil receiver**
  The main problem of volumetric receivers is to ensure an adequate heat transfer between gas and absorber. This leads in the case of foil receivers, to very low channel cross sections in order to restrict the developing gas flow boundary layer. However, the channel diameter is limited to a minimum of 3 mm to prevent the danger of channel blocking during the manufacturing process when the whole receiver is being siliconized. Furthermore, the channel walls have finite wall strength leading to front surfaces at the inlet region which are highly loaded by solar flux. For this reason it was proposed to profile the inlet region in order to reduce the solar load and solar reflection losses (see fig. 2). Gas flow and convective heat transfer as well as solar radiation distribution will be influenced by the ridged profile. Interaction and optimal design must be found.

- **Design of the quartz glass matrix structure**
  Quartz glass channel length and channel cross section are the most significant parameters of the selective design. It must be kept in mind that in addition to the thermal efficiency, the permanent usage temperature of quartz glass and manufacturing difficulty in building the structure (i.e. melting the channels together) must also be taken into account.

- **Optimization of the gap region**
  The gap between quartz glass and ceramic part is necessary to suppress heat conduction from the hot side of the receiver to the cold side. Furthermore, the

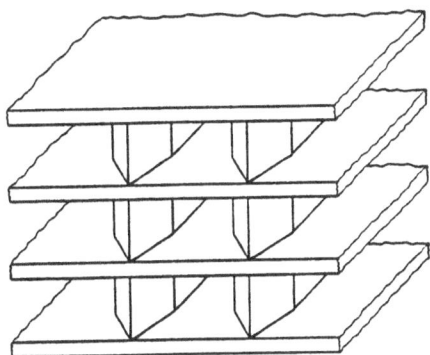

Figure 2: *SiSiC ceramic receiver section with profiled inlet region*

gas flow boundary layer, which has been developed while passing through the quartz glass channel, will level out in the gap region, so that a homogeneous inlet flow is guaranteed for the ceramic part. Flow simulation using Navier-Stokes equations have been performed in order to find quantitative results.

- **Mechanical and thermal stability of the quartz glass structure**
  Although quartz glass is well known as highly resistant to high temperatures, thermal tensions and thermoshock, the properties of matrix structures have never been investigated. Moreover, fouling by ambient dust particles which are driven with the air flow and deposited on the glass wall may induce recrystallisation and destroy the glass structure. Experimental investigations have been performed with a prototype structure of quartz glass in order learn about their stability.

The essential information were extracted from the results on the above mentioned investigations to a final design proposal for a receiver test at the Plataforma Solar de Almeria (Spain). This was transferred to construction drawings adapted to the Sulzer test equipment.

2 Receiver modelling

In order to perform a thermal analysis of the proposed receiver design a one-channel receiver model was developed. It is described in [5] but will be discussed here briefly in order to highlight some improvements that were incorporated.

2.1 One-channel receiver model

The geometry of the one-channel receiver model which calculates the axial temperature distribution is presented in fig. 3. The following assumptions are incorporated:

- concentrated solar radiation has a symmetric homogeneous distribution corresponding to the channel axis.

- wall strength of quartz glass is negligible in comparison to the cross section.

- reflection on glass wall is exclusively specular.

- mean ceramic radiation temperature can be determined by a single reference channel.

Simplification to a one-channel receiver model is possible under these assumptions as interaction between neighboring channel only took place by the following two mechanisms:

1. radial heat conduction
   Due to the assumption of homogeneous irradiation, constant temperature in the

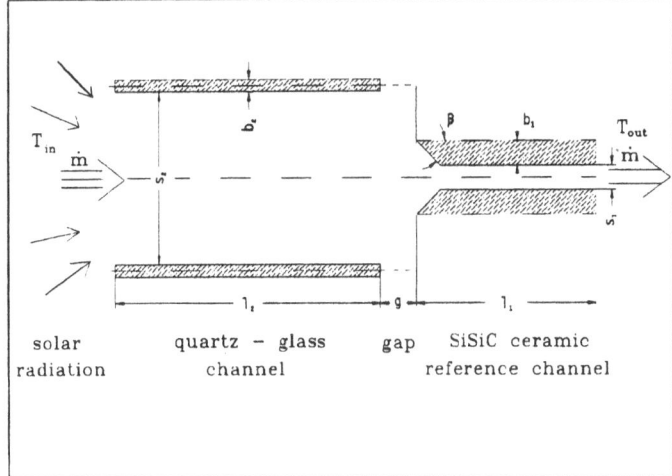

Figure 3: *The one-channel receiver model*

radial direction prevents heat transfer by conduction, so that it need not be taken into account.

2. transmittance of radiation through the quartz glass walls

Since the radiation source is assumed to be symmetric, every transmitted beam which leaves the channel is replaced by new one, which is entering at the same point where the original beam exited. Therefore, the radiation is regarded to be speculary reflected at the quartz walls with an effective reflectivity which consists of the sum of material transmissivity and reflectivity.

As nonhomogeneous radiation distribution can not be excluded in reality , the model is unable to allow global statements about the efficiency of a whole receiver under real conditions . However, it can show the potential in efficiency improvement by the quartz glass structure compared to a conventional foil receiver and is able to give limits of absorber temperatures.

As the solar flux distribution and the viewfactor matrix which are necessary as input of this calculation are determined by a Monte Carlo ray tracing method, statistical errors will influence the results due to limits in computational capacity and computational time. An estimation of this statistical error can be achieved by a comparison of the results achieved with viewfactor matrices calculated in different runs. Deviations between energy conversion efficiencies up to 1% were observed.

2.2 Improvements of the model

As described in [5] the model considers solar and thermal radiation, convection and heat conduction as heat transfer mechanisms. The heat transfer by thermal radiation has been incorporated by the enclosure method applied to wall elements having specular or diffuse reflecting surfaces (see e.g.[6]). In this method the thermal radiative heat which is transferred to a wall element can be calculated from the temperatures of all other wall elements by:

$$\dot{q}_i^{IR} = \sum_j a_{i,j} T_j^4 \qquad \text{(a)}$$

The elements $a_{i,j}$ can be calculated from specular or diffusive emissivities and from the viewfactor matrix of the problem (see [5]) This approach describes the thermal radiation being transferred in a single homogeneous wavelength band. This might be sufficient in order to describe the spectral emissivity of gray ceramic, but will result in some problems when dealing with a selective material as quartz glass (shown in fig. 4) In order to perform a more realistic description, a multiband model of the thermal radiative heat transfer is introduced. In this model it is assumed that the wavelength spectrum is subdivided into a finite number of wavelength bands $0 - \lambda_1, \lambda_1 - \lambda_2 ... \lambda_{n-1} -$

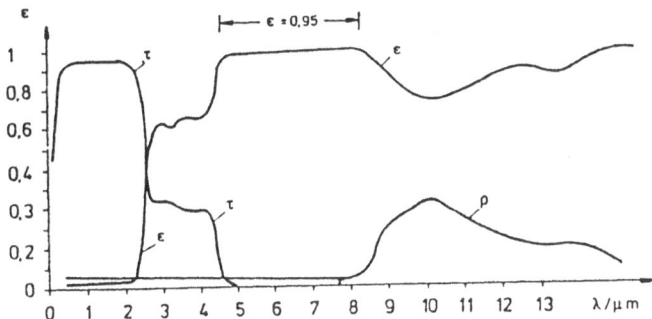

Figure 4: *Spectral emissivity of quartz glass layer (1mm)*

∞ over which the emittances of the enclosure surfaces are approximately constants. In that case thermal radiative heat transfer can be evaluated by:

$$\dot{q}_i^{IR} = \sum_{k=1}^{nband} \sum_j a_{i,j}^k E_j^k \qquad (b)$$

where E_j^k is the thermally emitted energy in the k-th band by

$$E_j^k = \epsilon_j^k \int_{\lambda_{k-1}}^{\lambda} k \frac{2\pi C_1}{\lambda^5 [\exp(C_2/\lambda T_i) - 1]} \, d\lambda \qquad (c)$$

and $a_{i,j}$ can be derived similar to the single band model as described in [5] on the basis of [6]. For every wavelength band spectral emissivity and a separate view factor matrix is necessary as we are dealing with specular reflecting surfaces. In this case (other than in the case of pure diffuse reflection/emission characteristics) the specular view factor matrix itself depends on the reflectivities of the specular surfaces and must therefore be evaluated for every wave length band separately.

In the case of quartz glass we used a two band model, splitting the wavelength band at $4.9\mu m$ and assigning the emittance of $\epsilon_1 = 0.05$ to the shortwave band and $\epsilon_2 = 0.95$ to the longwave band. The ceramic is regarded as gray over all wavelengths with $\epsilon = 0.9$ due to the lack of more detailed knowledge.

3 Optimization of the ceramic section

Optimal performance of the ceramic part of the volumetric foil receivers can be achieved by minimal channel cross sections and negligible channel wall strength in order to guarantee maximum convective heat transfer. However, these geometrical quantities are limited to a finite order of magnitude due to manufacturing constraints, creating highly loaded front surfaces. The resulting high front temperatures and solar reflective losses reduce energy conversion efficiency. Improvements can be achieved by a profiling of the front surfaces, as shown in fig. 2 which influence the convective and radiative heat transfer. Improvements are achieved by two effects:

- Solar reflective losses are reduced as the reflected radiation is partly directed into the receiver.

- The front absorber temperature is decreased as the solar load on the surface is reduced compared to perpendicular front surfaces.

However, the geometrical modification does not only influence the radiative heat transfer (solar + thermal) but also the convective cooling in the entrance region of the channels as the inlet flow is changed by the ridged profiles. Since there are no existing correlations describing the local heat transfer of such a complex geometrical shape in the literature, a three dimensional Navier-Stokes simulation of the inlet flow is performed with the PHOENICS code (Cham,London [1]).

The results of these calculations for different ridge angles of the profiled surface are incorporated in a computer code which calculates the thermal efficiency for a one channel receiver model (as described before). This model includes solar and thermal radiation, convection (by means of the given heat transfer coefficients) and conduction.

3.1 Boundary conditions

Investigations on the profiled inlet region of the matrix structure (see fig 2.) were performed under boundary conditions which are fixed by manufacturing constraints. Lower limit of the channel cross section is $3 * 3mm^2$ and of wall strength of the ceramic webs is 3mm. This wall thickness is necessary in order to achieve a self supporting structure in which ceramic rods prop thin ceramic foils of 0.8 mm thickness , so that small square channel of the aspired cross section are built. The investigated receiver design is presented in the following table. The profiled inlet region is investigated for the ridge angle of 30°, 60°, 120° and perpendicular front surfaces, here called 180°.

TABLE 1: Investigated Receiver Design

Material	SiSiC ceramic
channel cross section (d_h)	3 mm
channel length	100 mm
wall strength	0.8 / 3 mm
heat conductivity	$25 W/mK$
solar absorptivity	80 %
thermal emissivity (all wavelengths)	90%
solar flux in aperture	$1\ MW/m^2$
solar radition distribution	conic
with opening half angle	45°
heat transfer medium	air
mass flow rate	$0.7\ kg/m^2/s$
air inlet temperature	20^oC

The solar flux specification is an upper limit in central receiver systems up to now and the mass flow rate is determined by a desired outlet temperature of at least 800°C.

Temperature dependent properties of air are incorporated in the Navier-Stokes calculation for the density by:

$$\rho = \frac{348\ kgK/m^3}{T\,[K]} \qquad (d)$$

and for the kinematic viscosity by:

$$\nu\,[m^2/s] = -3.5733\,10^{-5}\ m^2/s + 1.578\,10^{-7}\ m^2/(sK)\ T\,[K] \qquad (e)$$

Specific heat capacity and Prandtl Number are assumed to be constant in the expected temperature range (20-1000 °C) ($c_p = 1.1\ KJ/Kg/K$ and $Pr = 0.7$). Heat conductivity is calculated by:

$$\lambda = \frac{\nu\,\rho\,c_p}{Pr} \qquad (f)$$

Mean gas flow velocity is (due to the given mass flow rate) $\overline{v_{in}} = 0.58\ m/s$ at the channel inlet and (increased by thermal expansion and blocking of cross section) $\overline{v_{out}} = 6.5\ m/s$ at the outlet. Reynolds numbers of the flow can be evaluated using the hydraulic diameter of the channel ($d_h = 3\,mm$) as characteristic length:

$$Re_{in} \quad = \quad \frac{\overline{v_{in}}\ d_h}{\nu(20^o C)} = 116.66 \qquad (g)$$

$$Re_{out} \quad = \quad \frac{\overline{v_{out}}\ d_h}{\nu(1000^o C)} = 112.6 \qquad (h)$$

so that laminar flow can be obviously expected.

3.2 Convective heat transfer

The convective heat transfer coefficient is determined by a Navier Stokes simulation performed with the PHOENICS code (Cham, London) under the assumptions of a con-

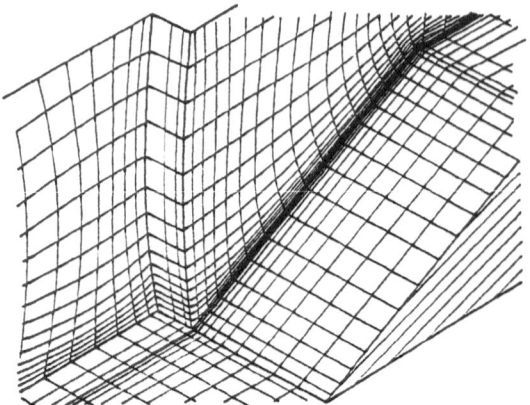

Figure 5: *Computation grid of the PHOENICS code in the inlet region of the profiled channel*

stant wall temperature of 1000°C. This code, which simulates fluid flow, heat transfer, chemical reaction and related transport phenomena, provides solutions of discretised sets of differential equations based on conservation of mass, energy and momentum. Depending on the problem, distributions of velocities, densities, temperatures, mass fractions etc. are given. Local convective heat transfer coefficients are derived from the temperature distribution which is evaluated by PHOENICS in the following way:

$$h_{konv} = \frac{\lambda(T_{wall} - T_{1.cell})}{\Delta x(T_{wall} - T_{bulk})} \tag{i}$$

were λ represents the heat conductivity of air and Δx the distance between wall and center of the cell nearest to the wall. T_{bulk} is the calometric mean temperature at the actual axial location and is derived from:

$$T_{bulk} = \frac{\int \rho(x,y)\, w(x,y)\, T(x,y)dx\, dy}{\int \rho(x,y)\, w(x,y)\, dx\, dy} \tag{j}$$

Grid and boundary conditions The calculation area is restricted to the smallest symmetric element of the channel (see fig.2), which is a fourth of a single channel implying that one half of the ceramic wall is contained in this element. The domain of the problem contains in flow direction (here called z) consists of an entrance region of 20 mm ($> 6d_h$), the profiled region and the region of square channel of 20 mm. A three dimensional boundary fitted coordinate grid, which is refined near the channel wall and in the inlet region of the point is applied to the geometry (part of the grid in the profiled region is shown in figure 5). In this first case a ridge angle of 60° is considered.

As a first step, a basic grid was created by an external FORTRAN code, which was smoothed by a Laplacian-grid-smoothing algorithm (supported by PHOENICS) in order to achieve convergence of the algorithm. Calculations were performed with a resolution of 10*10*30=3000 grid cells and 15*15*50 =11250 grid cells.

Boundary conditions of the flow calculation are:

Inlet   mass flow 0.7 $kg/(s\,m^2)$,
inlet velocity u=0, v=0, w=0.58 m/s,
inlet temperature $20^o C$

Outlet relative pressure level p=0

Wall   no-slip condition, (tangential velocity is zero at the wall)
transferred heat:

$$q_{konv} = \lambda \frac{(T_{wall} - T_{1.cell})}{\Delta x} \tag{k}$$

Grid independent solution and comparison with literature   In fig. 6 an example of the axial velocity distribution in the grid layer close to the channel center is shown. Acceleration of the flow in the inlet region is affected by the blocking effect and by thermal expansion of the heated gas. Flow separation was not observed, probably due to the low Reynolds Number.

The convective heat transfer coefficient as calculated in eqn. i) is averaged over the channel height of the profiled channel and plotted against the axial location in the receiver. The calculations were performed for two different grid resolutions (10*10*30 and 15*15*50) in order to determine uncertainties by grid discretisation. The results are presented in fig. 7. Deviation of the mean value is less than 3%. This is in an order of magnitude which is acceptable compared to all other uncertainties incorporated in the

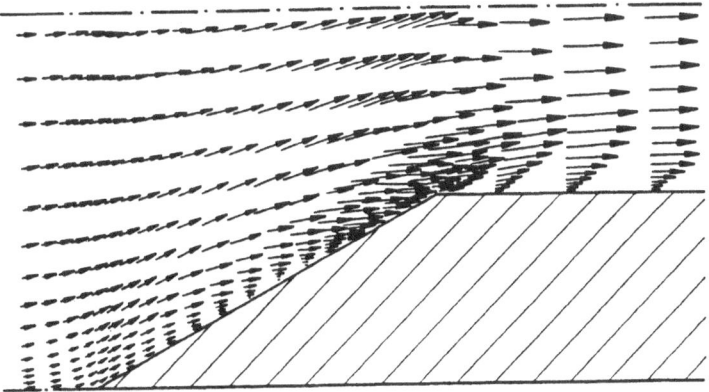

Figure 6: *Axial velocity profile in the grid layer close to the channel center*

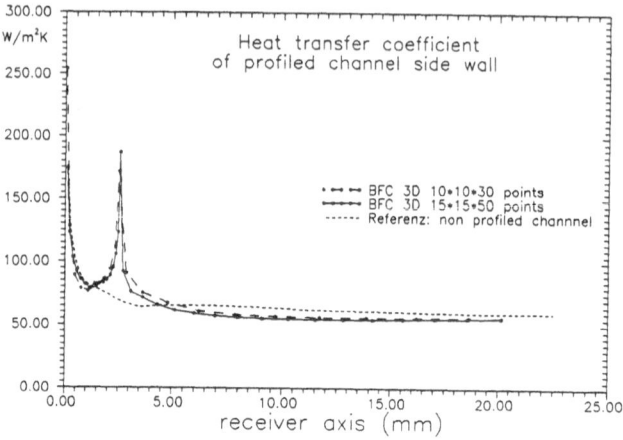

Figure7: *Axial distribution of heat transfer coefficients of two different grid resolutions*

model (e.g. material properties, assumption of constant wall temperature, application of a one-channel receiver model).

Although no comparable case of experimental or theoretical investigations of the proposed design could be found in the literature, qualitative correspondence to experimental results of Kottke [3] who investigated the flow at a flat profiled plate with finite wall strength at zero incidence could be found at low Reynolds Numbers (see fig. 8).

Effects of ridge angle In fig. 9 a) the variation of the heatt transfer coefficient is shown for different profile angles and in fig. 9 b) for the case of a perpendicular front surface. For all profiled geometries the heat transfer coefficient has a maximum at the stagnation point, decreases along the profile region and has a second maximum at the edge where the nominal channel cross section is reached.

In the case of perpendicular front surfaces a maximum value of h_{konv} is reached at the front edge. No separation of flow occurred, possibly because of the low flow Reynolds number.

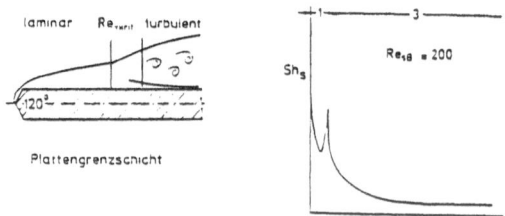

Figure8: *Dependence of Sherwood Number on axial location for Re = 200 (comparable to h_{konv} by analogy of heat and mass transfer)*

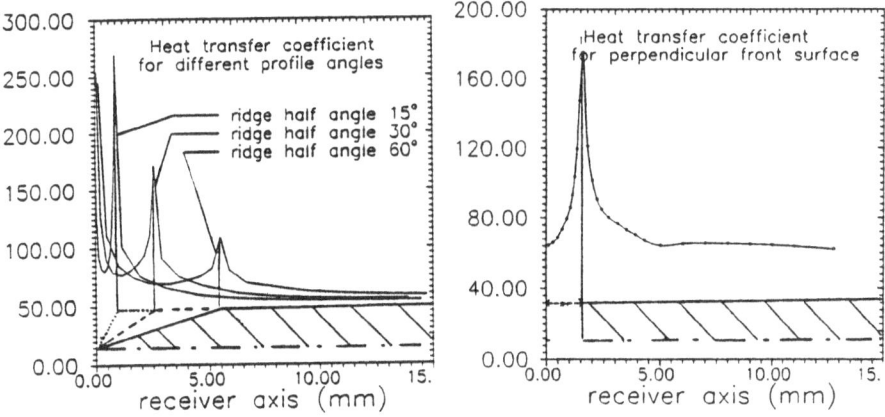

Figure 9: *heat transfer coefficient versus axial location for a) profiled front surfaces and b) perpendicular front surfaces evaluated for boundary conditions as given in Tab. 1*

Effects of the mass flow rate An example of the dependence of heat transfer coefficient on the mass flow rate is shown in fig. 10 for the case of 60° ridge angle profiling. Effects on heat transfer are only significant in the inlet region ($\approx 3d_h$) of the channel, becoming constant at fully developed flow. Values of local heat transfer coefficients for different mass flow rates are calculated by linear interpolation from these curves.

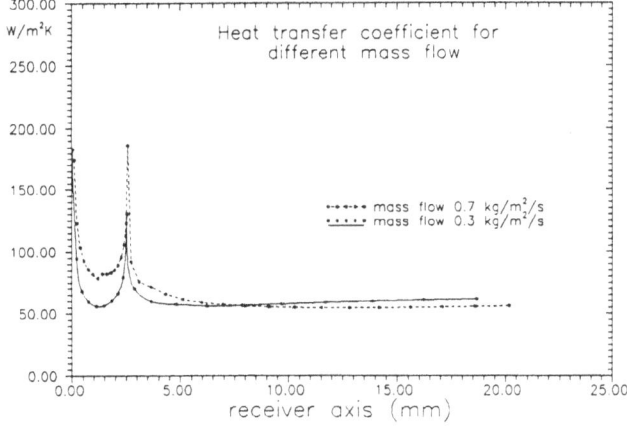

Figure 10: *Axial distribution of the heat transfer coefficient for two different mass flow rates*

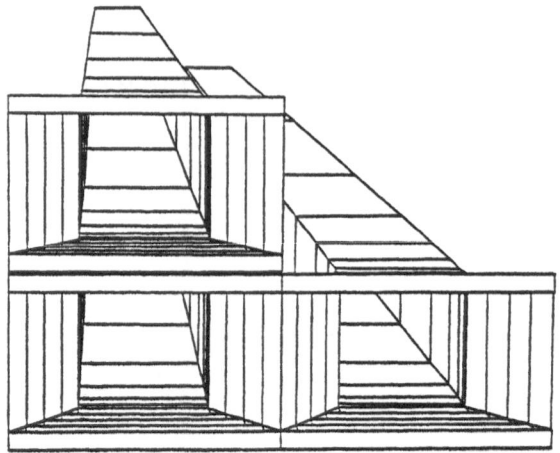

Figure 11: *Geometrical resolution of the one-channel receiver model*

3.3   Effects on thermal radiative losses

In order to analyze the effect of the profiling on thermal radiative losses of the receiver, convective heat transfer data and boundary conditions as described above were transferred to the one channel receiver model. In this model the channel is divided into 72 surface elements which are assumed to be isothermal. The geometrical discretisation is shown in fig. 11. In this figure, 3 neighboring channels are presented in order to to illustrate the shape of a single symmetric element (compare with fig. 2)

In order to illustrate the pure effect of geometry on convective and thermal radiative heat transfer, solar absorptivity is assumed to be 1.0. In fig. 12 a) the effect on temperature distribution and 12 b) on receiver efficiency are shown. Front temperature of the receiver is reduced by $77°C$ for profile angle of $30°$, whereas receiver efficiency is increased by 2.4 %.

3.4   Effects on solar radiative losses

Solar reflecting losses depend purely on the geometry and on the solar absorptivity of the material. Therefore they can be investigated totally independent of the heat transfer problem in the receiver. The effective solar absorptivity of the proposed design (fig. 2, Tab. 1) was determined by a Monte-Carlo ray tracing algorithm as a function of the material absorptivity. Pure diffuse reflection characteristic was assumed. Specular reflecting properties of the material would increase the effective absorptivity beyond that for diffuse reflection, since no reflection losses out of the channel can occur. Incident radiation is assumed to have a uniformly distributed cone profile with a half angle of $45°$. Effective solar absorptivities of three different channel profiles ($30°, 60°$ and perpendicular) are shown in fig. 13 as functions of the material absorptivity.

High improvements of effective absorptivity can be seen for materials with poor solar absorptivity.

3.5 Conclusions

Detailed numerical investigations proved that profiling of absorber front surfaces of volumetric receivers will improve energy conversion efficiency by two effects.

1. The effective solar absorptivity is increased.

2. Front temperature and thermal emissive losses are reduced.

The first point can be understood easily, as solar radiation is partly directed into the receiver. This effect is of high importance at low material absorptivities and will be increased further on if the reflectance characteristic of the absorber is of a more specular type.
The second one is determined by the interaction of different heat transfer mechanisms. Solar flux density is decreased on the tilted surfaces in comparison to perpendicular ones, the heat transfer coefficient is changed, and the thermal radiative losses of the profiled surfaces are different from either perpendicular surfaces or channel side wall elements, such that their cumulative effects can hardly be predicted without calculations. However, thermal modelling showed that energy conversion efficiency improvement of up to 2.5 % and a decrease in front temperature of $80°C$ for the case considered can be achieved by this second effect. At material absorptivities of 80%, which is a typical value for SiSiC ceramic, the increase in the effective solar absorptivity for the of $30°$

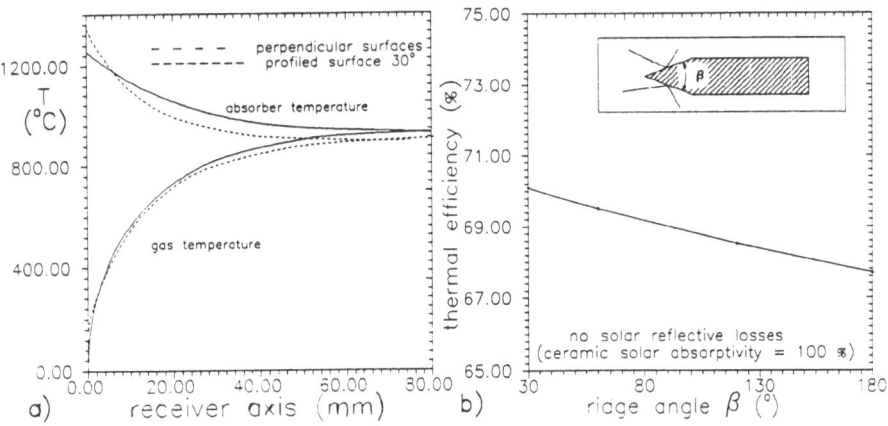

Figure 12: a) *axial temperature distribution in the receiver* b) *Receiver energy conversion efficiency versus profile angle*

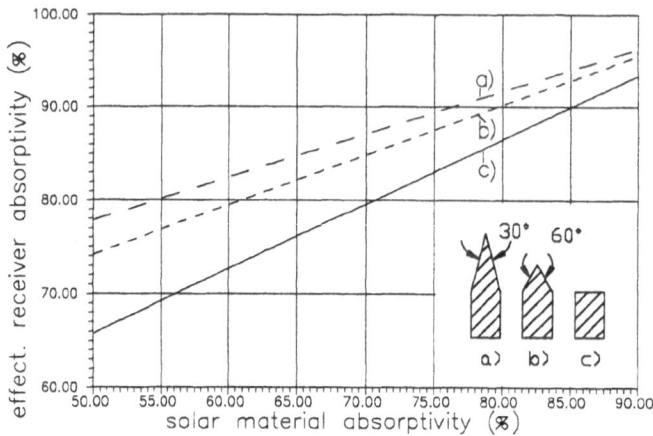

Figure 13: *Effective solar absorptivity of the receiver versus the solar material absorptivity*

profiling compared to perpendicular front surfaces is 6.4% and therefore represents the dominating effect.

The reduction in the maximum front temperature is an especially important result since it is usually a restrictive condition determining maximum solar flux density and maximum outlet temperature.

Finally, it can be remarked, that while investigating profiling of the channel inlet of volumetric foil receivers (down to ridge angles of 30°), the minimal ridge profile angle will effect highest improvement in efficiency and lowest front temperatures. This leads to the recommendation to the manufacturer to achieve the profile angle as acute as possible.

4 Optimization of quartz glass section

The quartz glass structure is mounted in front of the ceramic foil absorber and consists of square channel of quartz glass which are melted together at the edges in such a way that two channel side walls are not contacting directly. Thereby a self supporting structure is built which is able to compensate inner (thermal) and outer stresses by its grid-like geometry.

4.1 Manufacturing cost aspects

It is obvious that a minimal cross section of the matrix structure would ensure best convective cooling, whereas maximum channel length can suppress thermal emissive losses best. However, manufacturing constraints and cost estimations must be taken into account.

Manufacturing costs can be subdivided into material cost which are proportional to the fraction of channel length to channel cross section (l/s) and labor costs mainly affected by the process of welding the glass channels together. In the suggested configuration the number of melting points is proportional to $1/s^2$, which will limit the scale of the channel cross section due to cost aspects. Only if the connection process can be automated and will therefore influence the total costs only in a second order of magnitude, costs of 500 - 1000 DM/kg of manufactured quartz glass can be expected.

4.2 Typical temperature distribution

In order to find out how sensible thermal energy conversion efficiency and maximum quartz glass temperature will be affected by channel length and channel cross section, parameter investigations were performed. The channel wall strength is fixed to 1.2 mm for all calculations, which seems to be a good compromise between mechanical stability and low front surface cross section which would imply additional solar reflection losses. Inlet temperature of $20^{\circ}C$ and a mass flow rate of 0.7 kg/m^2 are fixed for all calculations. Ceramic channel geometry is assumed as described in section 3 with a ridge profile angle of 60° which seems to be in a realistic value after considering manufacturing difficulties.

A first example of a typical calculation result is presented in fig. 14 for the case of $20*20$ mm^2 glass section cross section and 80 mm channel length. In order to rate this result, the calculation results for the pure ceramic foil receiver are added to the plot. In this case the same air outlet temperature can only be reached at a lower mass flow rate which will also influence the convective heat transfer (see fig. 10). At outlet temperatures of $1000^{\circ}C$ energy conversion efficiency of 63.3 % is evaluated, whereas 79.4 % can be reached by the additional quartz glass cover.

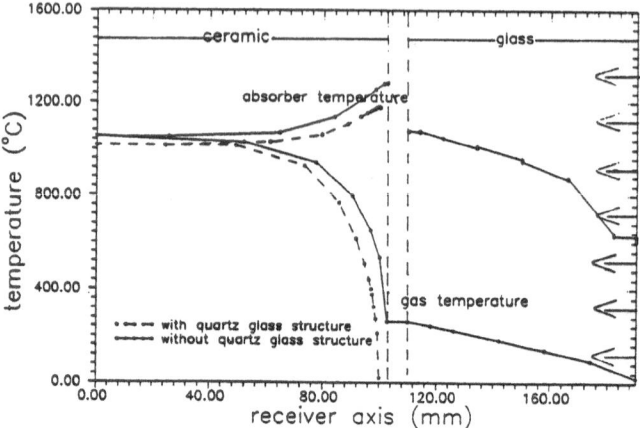

Figure 14: *Axial temperature distribution in the volumetric receiver*

The temperature distribution, as shown in fig. 14 is qualitatively the same for all calculated cases. Lowest material temperature is reached at the front side of the quartz glass matrix which is facing the heliostat field. The temperature is raised from the entrance to the inner part. The maximum quartz glass temperature is reached on the back side of the channel. This implies that the structure is not heated by direct solar radiation but by the thermal emission from the hot ceramic part. The SiSiC ceramic has its highest temperature on its front side, due to the maximum solar flux load which is occurring here. The absorber temperature drops in the inner part of the channel to the value of the outlet temperature.

4.3   Variation of geometrical parameters

Critical values of ceramic absorber temperatures could not be observed in the performed calculations under the described boundary conditions ($1MW/m^2$, $1000^{\circ}C$) , since SiSiC ceramic is able to stand temperatures up to $1400^{\circ}C$. However, maximum glass temperature can be found in the critical region , as permanent usage temperature of $1100^{\circ}C$ is given by manufacturer Heraeus. Therefore, this quantity must be considered carefully, besides the thermal efficiency.

Calculation results are shown in the figures 15 and 16. The thermal efficiency and the maximum quartz glass temperature are presented as functions of the fraction l/s for different cross sections. As expected, highest efficiencies are achieved for large values of l/s and low cross sections. It is interesting to note that the values for the 15 mm and 20 mm cross section heights only differ slightly, so that the larger geometry will be preferred due to cost considerations. The 40 mm channel height will result in maximum glass temperatures above the $1100^{\circ}C$ permanent usage temperature limit, so that this geometry can not be recommended for usage under the given boundary conditions.

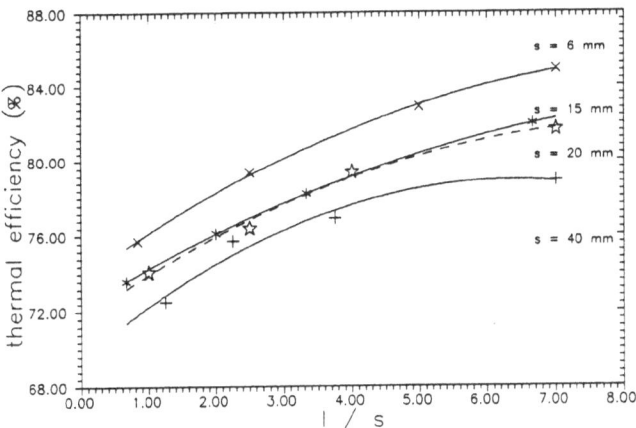

Figure 15: *Dependence of the energy conversion efficiency from the fraction of quartz glass channel length to quartz glass channel cross section*

It must be emphasized that for all cases looked at, air outlet temperatures above $1000^{\circ}C$ are achieved. For lower aspired outlet temperatures which would naturally influence the maximum glass temperature, the 40 mm channel height might be of interest because of its low cost aspect.

The 6 mm cross section shows best efficiencies and lowest glass temperatures but will need about 11 times more single channel elements as a receiver consisting of 20 mm cross section height. As long as Manufacturing costs of prototype structures will highly be influenced by labor costs, this small diameter can not be recommended. If the feasibility of quartz glass matrix structures is proved, detailed cost estimations must show whether larger manufacturing effort is worthwhile.

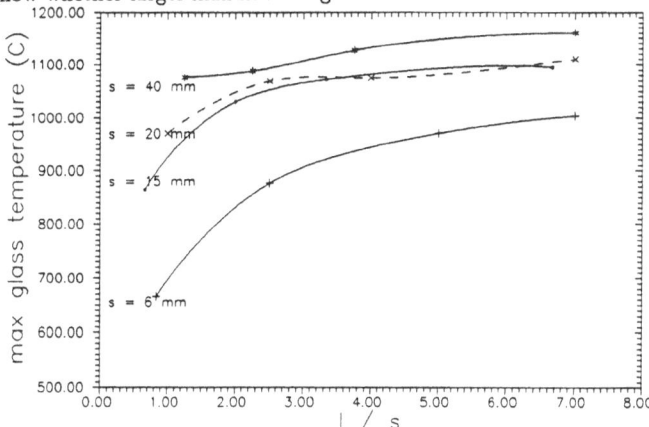

Figure 16: *Dependence of maximum quartz glass temperature on the fraction of quartz glass channel length to quartz glass channel cross section*

It must be kept in mind, that the same outlet temperature ($1000^{\circ}C$) can be achieved with the pure foil receiver only at efficiencies below 63%, so that a relative improvement in the range of 12 - 35 % will be achieved by the quartz glass structure.

4.4   Preferred quartz glass design

As described above, the prototype receiver with 20 mm cross section design is advantageous. It presents a compromise between the financial point of view and the security point of view, as maximum glass temperatures are tolerable. Concerning the channel length, a compromise between efficiency and stability of the structure must be found. As the connection can only be established at the surface front sides, longer channel will affect higher gravity forces and bending moments which have to be compensated for at the same connection points. Satisfactory results concerning stability were achieved with channel of 50 mm length in experimental investigations (see chapter 6), so that this value should be enlargeable to at least 80 mm in order to improve the efficiency. One square meter of such a structure consists of 1250 single glass channel elements and has a weight of about 21 kg. This can be compared with a glass plate of 9.5 mm thickness.

In order to decrease the maximum glass temperature on the back side a staggering of the glass structure seems possible. This can be managed by fixing shorter quartz glass channel pieces diagonally into the actual channels at their back side (see Appendix, photo 1). Convective cooling is increased due to the smaller channel cross section and the disturbance of the developed boundary layer, so that maximum glass temperatures can be decreased at the rear side. Cost aspects of this proposal need to be checked.

4.5   Thermal characterization of the design

In the previous section the effects of different geometries were discussed in for special fixed boundary conditions (1 MW/m^2, $1000^{\circ}C$) which represent an upper limit for existing and planned (e.g. Phoebus Project) central receiver systems. In order to describe the thermal behavior of the proposed design two different variations of input parameters were performed:

Variation of incident flux   In fig. 17 the dependence of the thermal efficiency on the incoming solar flux is shown for the two cases of a receiver with and without quartz glass structure. The outlet temperature is kept constant at $950^{\circ}C$. It is seen that the efficiency decreases with lower fluxes. The thermal emissive losses which depends on the absorber temperature vary only slightly with a given outlet temperature and are more or less independent of the incident flux. Efficiency improvements affected by the

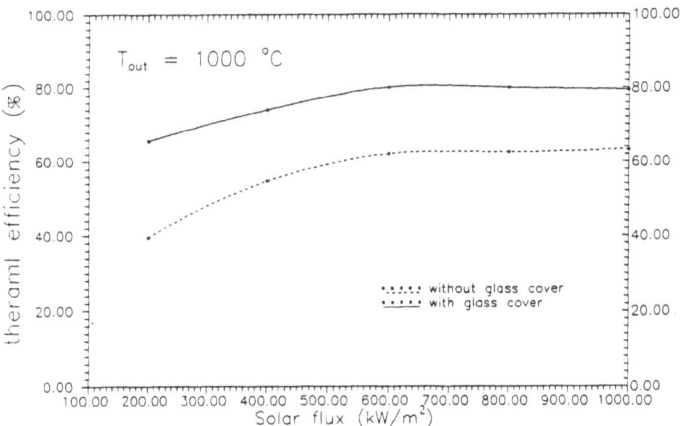

Figure 17: *Thermal receiver efficiency in dependence of incident solar flux*

quartz glass structure is greater at lower solar fluxes because the absorption of thermal radiation at (slightly) lower temperatures is better.

Variation of outlet temperature The variation of the outlet temperature was performed by adjusting the mass flow rate at an incoming flux of 800 kW/m^2. The investigated design is a quartz glass structure with 20 mm cross section and 80 mm channel length. The results are shown in fig. 18 for the cases with and without quartz glass structure.

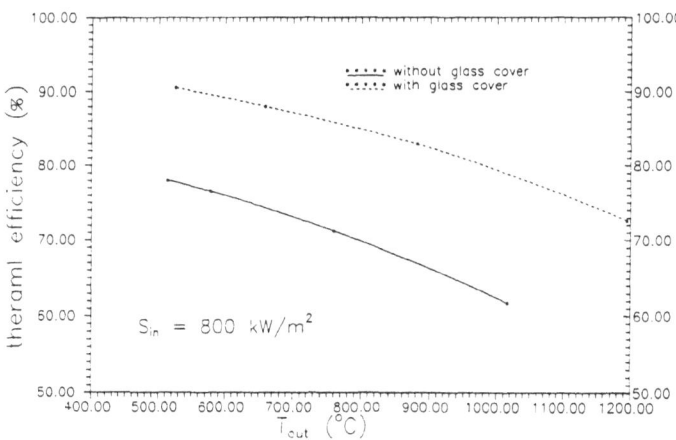

Figure 18: *Thermal receiver efficiency dependence on receiver outlet temperature*

5 Optimization of the gap region

Important fact of the proposed design of the selective volumetric receiver is that the quartz glass structure and the ceramic part are not in direct contact but separated by a gap. This is necessary because of the following reasons:

1. As highest absorber temperatures are expected in the inner part of the receiver, which will affect the improvement in efficiency compared to conventional foil receivers by a decrease of emission losses, heat conduction from the inner to the front part can be suppressed by the gap.

2. Air flow is developing in a boundary layer while passing through the quartz glass channel. If the gap distance is big enough this boundary layer can level out in the gap region in order to guarantee a homogeneous inlet flow in the ceramic channel. Nonhomogeneous inlet flow might result in inhomogeneities of the convective cooling and could affect local overheating.

In order to find out the dependence of the inhomogeneities on the gap distance, a flow simulation was performed with the PHOENICS code. Due to the complexity of the geometry (at least one quartz glass channel and various ceramic channel must be taken into account) a three dimensional treatment of the problem with a sufficiently fine resolved grid is not possible because of the huge computational requirement. Therefore a two dimensional treatment was applied to the problem and will give a first approximation of the order of magnitude of the required gap distance.

5.1 Two dimensional model

The two dimensional model of the investigated geometry is shown in fig. 19. Since relative inhomogeneities of the heat transfer in the different quartz glass channels shall be quantified but not their absolute values, the front side profiling of the ceramic channels was not taken into account. The geometrical parameters and the used boundary conditions used are presented in the following table:

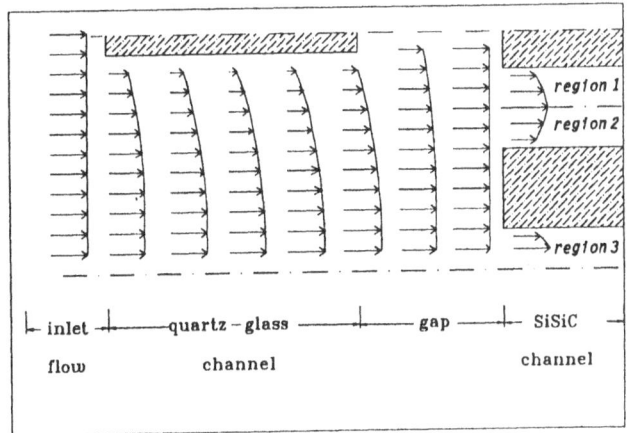

region 1

region 2

region 3

inlet flow | quartz – glass channel | gap | SiSiC channel

Figure19: *Two dimensional model of the selective volumetric receiver*

TABLE 2: 2-dim. Boundary Conditions

glass channel height	18 mm
glass channel length	80 mm
glass channel wall strength	1 mm
glass wall temperature	$820^{\circ}C$
ceramic channel height	3mm
ceramic channel length	20 mm
ceramic channel wall strength	3 mm
ceramic wall temperature	$1020^{\circ}C$
heat transfer medium	air
mass flow rate	$0.7\ kg/m^2/s$
air inlet temperature	$20^{\circ}C$
air inlet flow	homogeneous
gap distance	140-220 mm
grid resolution	34 *120 cells

Temperature dependence of material properties were incorporated as described in chapter 3. A typical flow distribution is shown in fig. 20. In order to describe the inhomogeneities of the design, the ceramic section is divided into three semi-channel-regions as is shown in fig. 19. For each of these regions the caloric mean inlet and outlet temperature is calculated according to eqn. (j). The convective heat transferred in the ceramic section is proportional to the mean caloric temperature difference between channel inlet and outlet. This may vary from one channel region to the other due to the following effects:

- The velocity distribution in the inlet region of the ceramic channel is different, so that the convective heat transfer coefficient is influenced.

- The inlet temperature distribution is different so that the driving temperature potential is changed which will in turn change the rate of the heat removed.

Therefore, it seems reasonable to regard the deviation of the transferred heat (expressed by the bulk temperature difference between ceramic channel inlet and outlet) as a value which qualifies the influence of velocity and temperature distribution of the inlet flow on the convective cooling. This percental deviation is plotted against the gap distance in fig. 21. Since this calculation is performed for a two dimensional geometry, the predicted values are not exactly transferrable to the 3 dimensional problem. However, it can be seen that the order of magnitude of this gap is a multiple of 100 mm and not of 10 mm in order to prevent overheating by nonhomogeneous inlet flow. For a large-scale receiver design there will be no problems with this gap distance as boundary effects are negligible, but in a small receiver test, the influence of the receiver boundary walls which will be highly loaded especially in the gap region could result in overheating of these side walls. Therefore, the gap distance is fixed to the relatively small value of 150 mm in the design proposal for a receiver test.

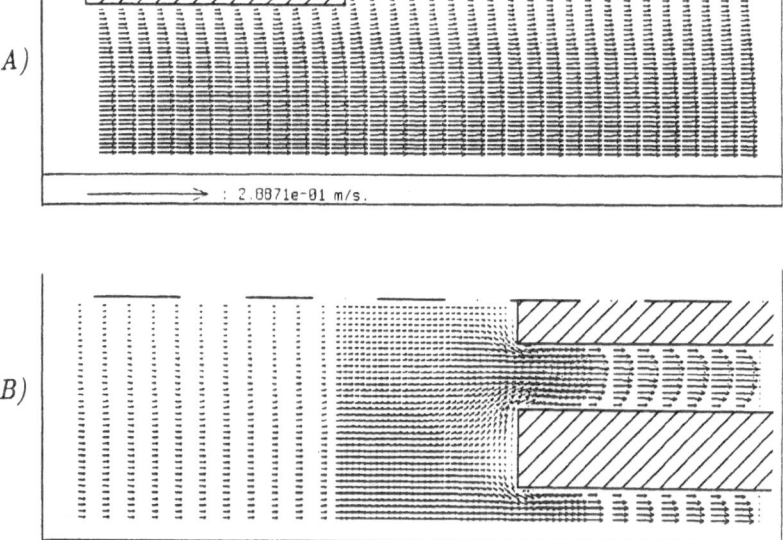

Figure 20: *Typical velocity distribution in the investigated geometry: A) outlet flow of the glass channel; B) inlet in the ceramic channel*

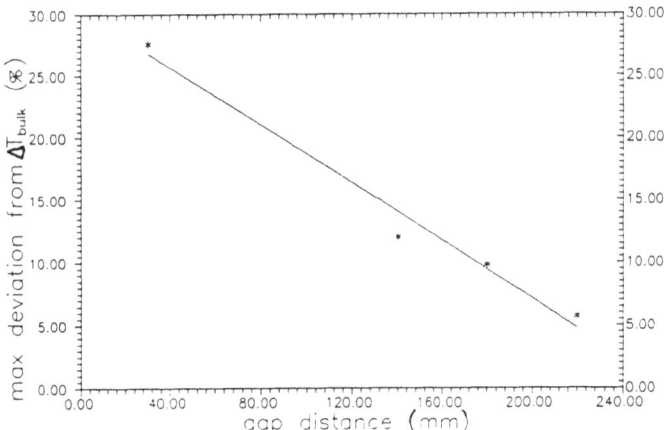

Figure 21: *Dependence of percental deviation of the transferred heat in the three channel regions on the gap distance*

6 Experimental investigations on stability of the quartz glass structure

The numerical results presented in the previous chapters show the big potential in efficiency improvement that can be achieved by the quartz glass structure. But we also learn that temperatures up to 1100° can be expected and axial temperature gradients of more than $100°C/cm$ even under steady state condition will occur. Moreover, radial temperature gradients must be taken into consideration, due to the nonhomogeneous irradiation which can not be predicted by the one-channel receiver model. Additionally, deposition of dust on the hot quartz glass might accelerate the recrystallisation process and would influence optical and mechanical properties of the receiver.

Therefore thermal load test were performed with prototype structures in the laboratory in order to prove the mechanical stability and reliability of the quartz glass structure under atmospherical conditions.

6.1 Properties of quartz glass

Relatively low values of mechanical properties like Young's modulus, torsion modulus, and stress, bending and torsion strength imply a very careful handling of the structures. Values of these quantities at ambient temperature given by the manufacturers vary strongly from piece to piece as is usual for glass-like or ceramic materials and decrease at temperatures above $800°C$. However, all physical properties of quartz glass present an extraordinarily steady temperature curve, so that phase transition from solid to liquid state can not be described precisely. The softening temperature is at approximately $1700°C$. The linear thermal expansion coefficient is a factor of 30 lower than for usual engineering materials and allow highest temperature gradients in the material. At temperatures above $1000°C$ quartz glass, like all other glass types, is tending to recrystallisation. The transition to the structurally more stable crystalline form starts at existing nuclei. The velocity of recrystallisation is strongly dependent on the temperature. At $1100°C$ it is lower by a factor of 1000 compared to $1280°C$. As the purity of quartz glass is very high, recrystallisation starts at the fouled surfaces. Acceleration is induced especially by alkali and alkali earth ions which can be found for example in perspiration. A further developed recrystallisation process will create crystobalite modification after the cooling down, which can induce cracks in the quartz glass (information from [2]).

6.2 Test set-up

As we learned from the calculations, the quartz glass structure is mainly heated on its back side by thermal radiation from the hot ceramic foil receiver. This heating

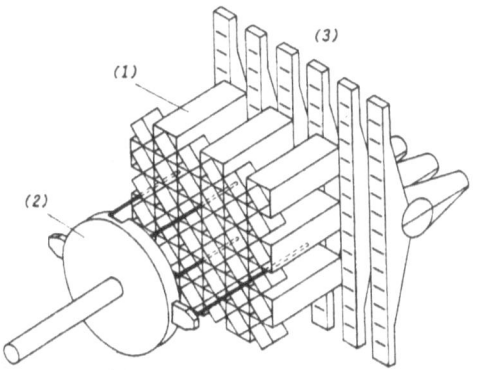

Figure22: *Propane burner (3), quartz glass structure(1) and hanging structure(2)*

will be simulated in the laboratory by a propane burner with an active flame surface area of 260 * 200 mm^2 (see fig. 22) with a power of about 30 kW which corresponds to a load of about 600 kW/m^2 It is mounted in a horizontal position and heats the glass structure from one side. Main heat transfer mechanism is, other than in the real receiver, convection. The quartz glass structure is not cooled by an additional gas flow, so that the temperature distribution will depend only on radiation and natural convection losses.

The burner is a conventional mixing gas type used for domestic heating and is made of stainless steel. It is mounted onto a turning lathe which is equipped with a ventilation system in order to remove the hot gases from the laboratory. The whole test set-up is shown in fig. 23. This setup has various advantages:

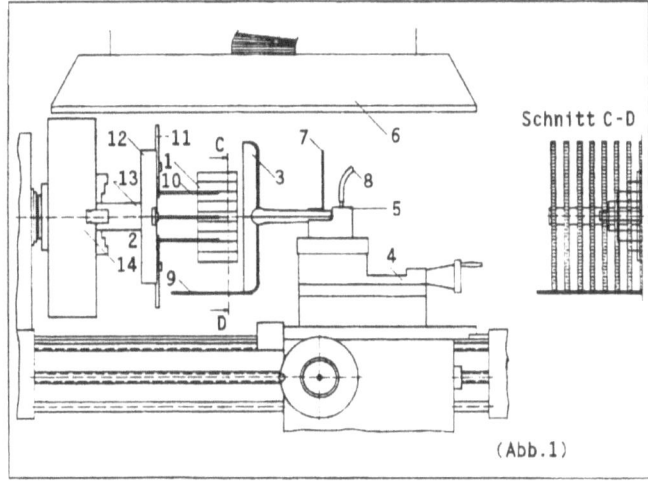

Figure23: *Test set-up of the thermal load test (Numbers are explained in the text)*

- The quartz glass structure(1) can be mounted in such a way that it is possible to rotate it during the load tests in order to level out radial temperature gradients

- The burner (3) can be mounted onto a carriage (4) by a supporting structure (5) in order to move it in axial direction. Thermal load and maximum glass temperature can be adjusted by the distance of burner matrix structure.

- Hot burning gases will be driven out of the lab by the ventilation system (6).

The fixing of the quartz glass in front of the burner is managed by 4 bars of stainless steel on which the matrix structure is propped. The position of the bars can be adjusted in a such a way that no tension will be transferred to the structure.

6.3   Investigated glass structures

Various geometries and connection techniques built of quartz glass tubes were investigated in order to determine the geometry with the best mechanical and thermal properties. Four different designs were built and tested:

A) **Manufacturer** Ruhr-University Bochum

    Design   19 glass tubes with a diameter of 15 mm, wall strength 1.5 mm, channel length 50 mm

    Connection   Melting of both front surfaces at the connection points of the channels , free cross section between channels is partly filled with extra quartz glass pieces of quartz glass bar of 0.5-1 mm diameter

    Manufacturing No cleaning of the channel before the manufacturing process, fixing of the channel during the melting process by hand.

B) **Manufacturer** Ruhr University Bochum

    Design   13 channel elements with square cross section, 15 mm channel height, 1.5mm wall strength, 50 mm channel length

    Connection   Melting at both front surfaces at the connection points, extra quartz glass bar is used

    Manufacturing Cleaning before the manufacturing process with hydro-fluoric acid, further handling with gloves, fixing of the channel elements with magnets during the melting process.

C) **Manufacturer** Heraues Quarzschmelze

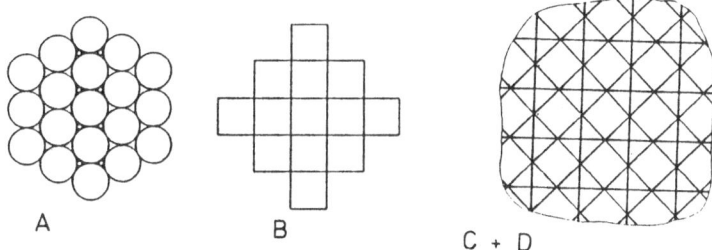

Figure 24: *Investigated designs in the thermal load tests*

Design   41 channel elements with square cross section, 1.3 mm wall strength, 50 mm channel length, at one channel end diagonally inserted square channel of 15 mm cross section, 10 mm length and 1.5 mm wall strength

Connection   Melting only at one front surface side

Manufacturing same as B)

D) **Manufacturer** Ruhr University Bochum & Heraeus Quarzschmelze

Design   same as C) but 36 channel elements

Connection   same as C) but connected at both front surface sides

Manufacturing same as B)

6.4   Temperature measurement

Glass temperature is measured with radiation pyrometer (KT 73, Heimann GmbH) which is sensible for radiation of 4.9 - 5.5 μm . The emissivity of quartz glass is practically constant in this range (see fig. 4). Absorption bands of H_2O in the range of 6.2 μm does not influence the measurement, so that hydrogen burners will not disturb temperature measurement. Limits of temperature measurement are 300 -2500 ^oC. Measurement error is given as ± 25 K by the manufacturer.

As we are not dealing with a Hydrgen-Oxygen gas burner but with a carbon containing burning gas, absorption and emission bands of CO_2 and CO influence the detection of the quartz glass radiation signal. A mathematical correction is not possible as the emitted gas radiation depends on the gas temperature, mass fraction of the carbon

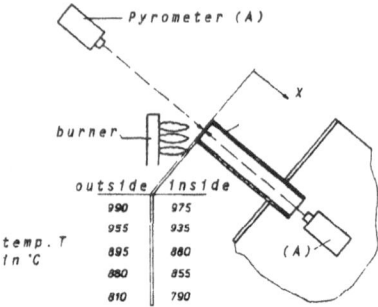

Figure 25: *Set-up of the pretest in order to quantify the influence of the gas radiation on the output signal of the pyrometer*

molecules and volume of gas between quartz glass and detector. In order to quantify the error affected by the gas radiation the following pretest was performed:(see fig 25) A glass tube of 400 mm length and 40 mm diameter with a thin bottom at one side is heated with the propane burner from the outside. The temperature of a marked point at the bottom of the Glass tube is aimed with the pyrometer from the inner and the outer side. As the inner part of the tube is totally blocked from the hot gas, but not the outside, the influence can be checked. The results are shown in fig. 25. The values are about the same, with respect to the measurement error. However, as the heat is transferred by convection, higher temperatures are expected at the outside of the glass . The measurement was repeated this time heating up the glass tube by a hydrogen burner. As no influence of burner gas on the detected signal is expected, the temperature gradient from the inner to the outer glass wall can be measured. In this case an outer surface temperature of $1200^{\circ}C$ corresponds to an inner surface temperature of $1100^{\circ}C$. It is therefore obvious that the glass temperature is underestimated in the presence of absorbing and emitting burning gas by up to $100^{\circ}C$ at a temperature level of $1200^{\circ}C$. In the following text, glass temperatures are given as measured by the pyrometer without any correction. It should be kept in mind that they represent the lower limit of the measured temperatures.

6.5   Stability tests

Axial temperature profile   The glass surface temperature profile which is shown in fig. 26 was measured along the channel axis under steady state conditions. High temperature gradients are achieved which are comparable to the predicted ones described in the previous chapters. However, in central receiver systems highest gradients will

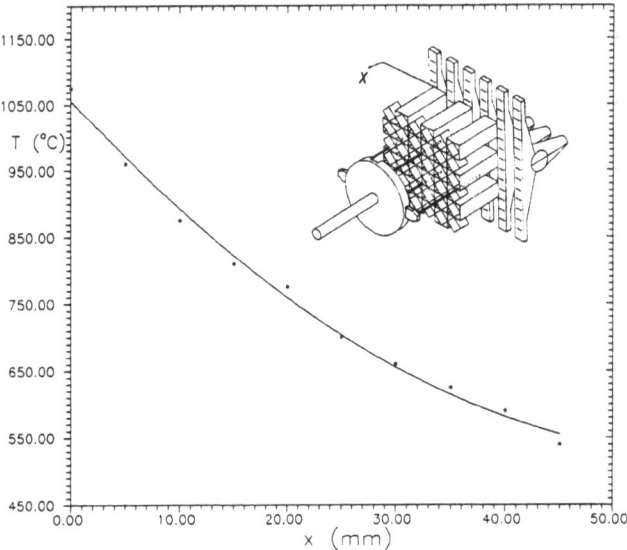

Figure 26: *Axial temperature distribution of the quartz glass structure B under steady state conditions*

occur under transient load conditions, for example when the total solar flux is switched on or off by clouds covering the sunshape. In order to achieve a maximum temperature gradient, thermoshock test were performed with all investigated structures. In these tests, water is poured on the hot glass structure in order to achieve highest local cooling rates. By this method higher temperature gradients can be reached, more than is possible when the whole structure is dipped in water, due to the local restriction of the cooling effect and the low heat conductivity of quartz glass.

Cyclic tests and results   First step of the test plan was a quick heating up to steady state conditions at maximum glass temperature above $1180^\circ C$. This condition was kept constant for a time period of 30 minutes and cooled down by ambient air afterwards by switching off the burner. With the exception of structure C, which is connected only on one front surface side, all the other structure passed the test without damage. Various glass segments broke off from the structure C at the connection points during the cooling process (Appendix, photo 3). Mechanical stability of the structure could be characterized as low, while handling and fixing it in front of the burner. Bending forces could not be compensated for by a one side connection technique.

Second step of the test was a thermoshock test as described in the previous section, which was passed by all structures (A,B,D) without any visual cracks. Structure A, which was melted together without respect to fouling by perspiration, shows strong recyristallisation at the glass surface. (white turbidness) (Appendix, photo 4).

Cyclic phases in a real receiver operation such as heating, steady state running, cooling with parallel fouling by unfiltered ambient air were simulated in the lab. For this purpose structure B was exposed to a cyclic load which was repeated four times.

First step is the simulation of fouling. For this purpose, the structure was fixed in front of car cooler for a duration of about 1000 driving kilometers. After this process, the effective fouling is much more severe than can be expected in a real receiver operation, as existing fouling will be burnt away more often (Appendix, photo 5).

In a second step the fouled structure was kept under maximum temperature for 90 minutes and cooled down by ambient air. Fouling particles burnt off nearly completely, and only a minimal number of small white recrystallisation points could be observed at the glass surface. Stability and optical properties remained unchanged.

In order to simulate fluctuating solar load in real receiver operation a sequence of 50 cycles with 45 second maximum burner power and 60 seconds cooling time were carried out. After having repeated this test program for 4 times no damage of the structure could be regarded. Enlargements of recyristallisation effects could not be noticed at the end of the test program. Multiple thermoshock tests performed after this test program were passed without any damage.

6.6 Conclusions

The results show that quartz glass matrix structures, which are connected at both front surface sides and which will not come into contact with perspiration, can stand high cyclic thermal loads even under conditions of fouling with atmospheric dust. The extent of thermoshock and fouling simulated in the lab was larger than it can be expected during the receiver operation due to the following reasons:

- Heating and cooling will be induced mainly by the thermal radiation of the hot ceramic part. Since this structure itself has a thermal inertia transient load profiles will be damped.

- Thermoshock cooling with water (rain) is not possible, as the receiver operation is usually stopped when clouds appear in the sky.

- The deposit of dust particles during receiver operation under hot condition will be suppressed by the thermophoretic force, which is directed in the opposite direction of the temperature gradient [7]. As the blower is normally not running when the receiver is out of operation, fouling is reduced during this time also.

- Larger particles, birds etc. will not induce fouling on the quartz glass because they are burnt by the intense solar flux, before they arrive at the quartz glass.

Therefore, the lab experiments give a strong indication that quartz glass structures could stand conditions of receiver operation for a longer time. However, the following questions are open:

- How will the absolute quartz glass size influence the stability of the structure?

- Due to cost aspects a volumetric receiver has to operate for years. Can this be guaranteed for quartz glass? Receiver tests durations in the order of magnitude of a month will give first informations.

- Will chemical contents and type of fouling at the receiver test location be comparable to the simulated fouling in the lab-tests?

The answers can only be given in a real receiver experiment, which should have at least the size to allow for a scaling up in modules.

7 Proposal of the final design

In order to prove the feasibility of the investigated receiver design, a receiver experiment at the Plataforma Solar de Almeria (Spain) is suggested. A proposal of the construction details of such a receiver test is added in Appendix B of this report.

The design is adapted to the Sulzer test equipment and is built up in a modular manner, so that either the whole receiver (quartz glass + ceramic) or only the ceramic part of the receiver can be tested, in order to quantify the improvement effects by the quartz glass.

The SiSic ceramic structure is built of ceramic foil of 0.8 mm wall strength which is mounted one onto the other and propped by ceramic rods of $3 * 3mm^2$ cross section. Thereby, square channel of $3 * 3mm^2$ and 92 mm channel length were created. The rods are profiled at the front side with a maximum ridge angle of 60°. The whole ceramic structure is circular with a diameter of 940 mm and consists of 9 different modules, which are connected by form closed dovetail joint. It is important that the ceramic channels be continuous throughout the whole structure including the connection domain. The size of the whole structure is fitted to the adapter piece which was built for the CARTREC receiver test, in order to reuse it. 9 holes of 85 mm depth will be drilled into the ceramic rods from the back side at different radial locations. They are used for the installation of the thermocouples used for the measurement of the material front temperature. 4 further thermocouples will be installed in holes of 10 mm depth in order to measure the rear material temperature.

The quartz glass structure consists of square channels of $20 * 20mm^2$ cross section, 80 mm channel length and 1-1.5 mm wall strength, which are melted together at the front and rear side. The structure has a circular shape with a diameter of 865 mm and will be built in one single piece.

It will be installed in a cylindric metal adapter of 250 mm length which can be flanged at the CARTREC adapter. The inside of the cylinder is coated with 50 mm white ceramic insolation material (M1260, KAPYROK), because the wall is either loaded by transmitted sunlight or by thermal radiation from the hot ceramic receiver (cavity effect). Moreover the insulation will suppress the direct contact of the quartz glass with the metal cylinder. Boreholes in the cylinder wall are necessary for the following reasons:

1. Fixing mechanism which prevents the tilting of the quartz glass structure during the installation process while the whole test bed is in a horizontal position, can be inserted and removed.

2. Fixation of the ceramic coating can be managed

3. Insertion of thermocouples which measure the ceramic insolation temperature in order to suppress overheating is possible.

Fixing of the quartz glass structure is managed by a shrouding ring with an inner diameter of 835 mm (aperture). This ring can also be fitted to the CATREC adapter, in order to guarantee testing of the pure ceramic part.

The ceramic radiation shield of the Plataforma Solar de Almeria can be fixed at this ring also. The inner side of this circular ring which will support the glass structure on a width of 20 mm will is shielded against thermal radiation load by 25 mm ceramic insolation material.

A   Photo-Documentation

B   Construction drawings

References

[1] Cham Ltd.   Bakery House, 40 High Street, Wimbledon, SW19 5AU England

[2] Heraeus   Produktinformationen 'Quarzglas und Quarzgut' zittiert nach:

                          Otto u. Thomas, Zeitschrift f. Physik 175, 1963, p 337

[3] Kottke, V.   Einfluß von Anströmprofil und Turbolenzintensität
                          auf die Umströmung längsangeströmter Platten endlicher Dicke

                          Wärme- und Stoffübertragung,10, (1977), p 159-174

[4] Pitz-Paal, R.   Studie zur Entwicklung eines volumetrischen Receivers
     Morhenne, J.   abgestzter Struktur
     Fiebig, M.
                          Investigation in the scope of SOTA AP300 1989
                          in: DLR ET-1/90 -2

[5] Pitz-Paal, R.   The Construction of a Volumetric Receiver
     Morhenne, J.   with a Staggered Structure
     Fiebig, M.
                          in M. Becker (Ed.): Solar Thermal Energy Conversion, Vol5
                          (1990/91), Springer Berlin

[6] Siegel,R   Radiation Heat Transfer
     Howell, J.R,
                          Macgraw Hill Book Company, Washington, 1981/82

[7] Talbot   Thermophoresis of particals in a heated boundary layer
     Cheng
     Schefer
     Willis   J.Fluid Mech. (1980)

Layout and Measurements
of Terminal Concentrators

U. Schöffel, R. Sizmann,

Sektion Physik
der Ludwig-Maximilians-Universität München

Abstract

A terminal (secondary) concentrator provides a second optical element in the solar radiation concentrating system. It allows the concentration ratio to better approach the thermodynamic limit of concentrating devices. For high temperature applications high flux densities are required.

OPTEC, an optimization program for terminal concentrators, developed during previous projects, has been expanded and applied to various systems ranging from a single-piece parabolic dish to a solar furnace with faceted primary concentrator in an off-axis configuration and to a heliostat field. OPTEC automatically optimizes thermal efficiency or flux by determining several parameters of single- and two-stage concentrating systems consisting of primary concentrator, terminal concentrator, and receiver. Efficiencies are calculated employing full ray tracing and Monte-Carlo integration. Optimization now is done by a genetic algorithm, which has been implemented and incorporated into OPTEC.

An experimental set-up was designed, constructed, and built to measure under laboratory conditions flux density distributions in the exit aperture of a single-stage or two-stage concentrator. A tungsten lamp was used as primary radiation source; photographic films were used as radiation detectors.

Contents

1 Introduction

This report deals with

- the layout of terminal concentrators for various configurations, and with

- measurements of local concentrations at the exit aperture of terminal concentrators.

The layout calculations have been carried out with computer program OPTEC, an optimization program for terminal concentrators, developed at Ludwig-Maximilians-Universität München in the course of previous projects [1],[2]. OPTEC is based on general ray tracing algorithms and the Monte-Carlo technique. A major change in OPTEC with respect to the previous version was the incorporation of a Genetic Algorithm as the optimization procedure, inspired by [3]. See section 4 Optimization for details.

The maximum usable heat flux at receiver temperature T_R and ambient temperature T_a is

$$\dot{Q} = I_H - S_R \left[\epsilon_R \sigma (T_R{}^4 - T_a{}^4) + U_L(T_R - T_a) + ... \right]. \tag{1}$$

$I_H = \alpha_R \, \eta_{op} \, \gamma \, \rho_H \, \eta_{cos} \, \eta_{obs} \, S_H \, E_b$ is the product of

E_b direct insolation flux density from the sun,

S_H mirror area of the heliostat field or aperture area of dish concentrator,

η_{obs} obstruction factor, i.e., the fraction of the incoming light approaching a heliostat field mirror (or dish) and not obstructed by receiver, terminal concentrator (or tower),

η_{cos} averaged cosine factor (for dish concentrator: $\eta_{cos} = 1$),

ρ_H mean reflectivity of the field or the dish

γ fraction (intercept factor) of the flux $\rho_H \, \eta_{cos} \, \eta_{obs} \, S_H \, E_b$ intercepted by the receiver,

η_{op} optical efficiency of terminal concentrator ($= 1$ without terminal concentrator),

α_R absorptivity of receiver,

ϵ_R emissivity of receiver, approaches unity for deep cavities,

S_R aperture area of receiver,

σ Stefan-Boltzmann constant ($= 5.67 \cdot 10^{-8} \mathrm{W/m^2 K^4}$).

$S_R U_L(T_R - T_a)$ is the convective heat loss with U_L as specific heat loss coefficient. Other sources of losses are neglected. The maximum thermal efficiency is:

$$\eta = \frac{\dot{Q}}{E_b S_H} = \eta_0 \ - \ \frac{L}{C} \quad \text{with} \tag{2}$$

$$\eta_0 \ = \ \alpha_R \, \eta_{op} \, \gamma \, \rho_H \, \eta_{cos} \, \eta_{obs} \tag{3}$$

$$L \ = \ \frac{\epsilon_R \sigma (T_R{}^4 - T_a{}^4) + U_L(T_R - T_a)}{E_b} \tag{4}$$

$$C \ = \ S_H / S_R \tag{5}$$

The thermodynamic upper limit of concentration is for the case of biaxially focusing systems [4]:

$$C_{\mathrm{max}} = 1/\sin^2 \theta, \tag{6}$$

where θ is the half opening angle the radiation subtends. With imaging optical systems this limit cannot be attained. With parabolic dishes alone approximately only one fourth of this limit can be attained (under the conditions of sharply defined angular extent of the incoming radiation and a receiver aperture defined by the smallest disc that accepts all the rays reflected by the parabolic dish). With an additional, nonimaging, optical element, the *terminal concentrator* (also known as secondary concentrator), in principle about 90% of the limit could be attained (i.e., for a parabolic dish of longer focal ratio combined with an ideal concentrator) [4, p. 197].

The optical efficiency η_{op} of a terminal concentrator is an intricate function of the geometry of the reflecting surfaces of the terminal concentrator, their reflectivities and surface errors. In particular, η_{op} is dependent on the spatial and angular distribution of the incident radiation. (An optical element not affecting, i.e., not at all focusing, the incident radiation has $\eta_{op} = 1$.)

While L is given as input and C can be calculated directly from design parameters, calculating η_0 requires integration over "many" rays.

Layout   Layout calculations were carried out for a number of configurations: for our laboratory experiment (a conical or a trumpet terminal concentrator in conjunction with a model of a paraboloidal primary), a planned outdoor experiment (with spherical primary), a solar furnace at Plataforma Solar de Almeria (paraboloidal primary), a study related to the PSA-CRS (heliostat field primary), and calculations in conjunction with a solar furnace to be constructed at DLR Cologne (faceted primary in off-axis configuration [5]).

For the first time the Detailed Heliostat Field Model of OPTEC was used in conjunction with automatic optimization of terminal concentrators.

Measurements of terminal concentrators   In the course of previous projects OPTEC has been developed for computer studies on the optical behaviour of terminal concentrators. One of the results was that the optical efficiency of terminal concentrators cannot be characterized independent of the spatial and angular distribution of incoming radiation that enters the entrance aperture.

During the present project one aim was to bridge the gap between the theoretical studies and the practical use of terminal concentrators in solar high flux density applications. On the one hand a partial validation of OPTEC was strived for; on the other hand the feasibility of laboratory measurements of terminal concentrators was to be investigated, where the angular distribution of the radiation entering the terminal concentrator resembles the distribution which is obtained in two-stage systems for solar applications. Such measurements require mainly the following components:

- A light source resembling the sun in certain aspects,

- a device that simulates a paraboloidal primary concentrator with symmetry of revolution,

- a terminal concentrator,

- a measuring device.

The laboratory measurements have the advantage of

- independence of weather conditions,

- reproducibility,

- prevention of thermal problems arising with high flux densities, thereby isolating the optical aspects.

During the previous project accompanying research work was done as a first step for a laboratory experiment. During the present project it was realized that several modifications were required. These modifications are mainly related to the light source (including the associated optics), and to the measurement procedure.

2 Terminal Concentrators

As was described in the Introduction a terminal concentrator (also referred to as secondary concentrator or reconcentrator) is an additional nonimaging optical element in a concentrating system. Various shapes have been proposed as terminal concentrator surfaces. Requirements are mainly: high concentration boost, high optical efficiency and easy procedure for manufacturing. In addition surface quality should be high: high reflectivity in the spectral range of interest and high specularity. High reflectivity is required not only for high optical efficiency but also to avoid excessive absorption of high flux densities by the terminal concentrator surfaces, requiring high temperature resisting coatings and/or active cooling.

2.1 Cones

Conical concentrators are the simplest biaxially focusing terminal concentrators. Conical concentrators can be manufactured much easier than trumpets or CPCs (see below).

The mathematical relationships have been developed in [1] and [2]. See section 5 Layout for the parameters of a cone that can be optimized with OPTEC.

2.2 Generalized Trumpets

Truncated hyperboloids of revolutions, so-called trumpet concentrators [4,6] are mathematically the simplest biaxially focusing terminal concentrators next to conical ones. The 'classical' trumpet concentrator (where the exit plane intersects the trumpet at the throat perpendicularly) shows nice mathematical properties [7,8], but in previous investigations [2] it proved to be of no advantage over the cone in two-stage concentrating systems. However, if the trumpet at its exit is allowed to enter the receiver plane non-perpendicularly, a new degree of freedom is introduced. Since a cone is a limiting case of a hyperboloid, an optimized generalized trumpet concentrator should always yield the same or better results than a cone.

The mathematical relationships for trumpets have been developed in [1]. See section 5 Layout for the parameters of a trumpet which can be optimized with OPTEC.

2.3 Other Terminal Concentrators

In the literature much theoretical and experimental work has been devoted to the compound parabolic concentrator (CPC), see [4]. Some scientists even use CPCs as a synonym for nonimaging concentrators in general [9]. CPCs are fourth degree surfaces. Their cross sections are formed by two parabolic pieces.

Modeling of CPCs remains to be incorporated into OPTEC in the future.

3 Primary Concentrators Used in Conjunction With Terminal Concentrators

3.1 Paraboloids

Paraboloids are mathematically the simplest imaging concentrators used as a first stage in a two-stage concentrating system. However, paraboloids are not as easy to manufacture as spheroids (see below).

The mathematics regarding paraboloids can be found in many textbooks. The most important property of a paraboloid is: light rays entering a paraboloid parallel to the axis of the paraboloid meet in exactly one point, the focus. However, for sunlight, since it has a finite angular extent, an exact focus is not formed (see Introduction).

3.2 Spheroids

Spheroids are much easier to manufacture than paraboloids. Mathematically spheroids are nearly as simple as paraboloids. In textbooks the geometry of spheroids is covered. Unlike paraboloids spheroids do not show an exact focus for incoming parallel light. The spot is the larger the smaller the focal ratio (focal length / aperture diameter). Here the focal length is taken as the half of the radius of the sphere with same curvature as the spheroid. At that point paraxial rays will approximately meet after reflection by the spheroid.

3.3 Detailed Heliostat Field Model and Faceted Primary

The Detailed Heliostat Field Model (DHFM) implemented in OPTEC is described in the previous report [2]. Here we want to point out that the DHFM can be used to model a faceted primary that is used, e.g., for a solar furnace. The DHFM provides faceted heliostats with flat, spherical or paraboloidal curvature. The facets of a primary can be specified either by curved facets of one single heliostat or by many single-faceted heliostats the centers of which have to be specified explicitly by three-dimensional vectors.

4 Optimization for Layout

The physical problem is to design an optical system consisting of a primary and a terminal concentrator for maximum thermal efficiency η. For this an optimization calculation is necessary. The function to be optimized is of the form (see Introduction):

$$\eta(\mathbf{p}) = \eta_0(\mathbf{p}) - \frac{1}{C(\mathbf{p})} L, \tag{7}$$

where the geometrical concentration ratio $C(\mathbf{p})$ can be determined directly from the geometrical parameters $\mathbf{p}$ of receiver, primary and terminal concentrator. In general $\eta_0(\mathbf{p})$ is an integral over an at least four-dimensional domain [2]. Within the concentrator multiple reflections can occur; rays from some directions might be lost by leaving the terminal concentrator through the entrance aperture after one or several reflections. Therefore a detailed ray tracing through primary and terminal concentrator is necessary.

4.1 Objective Function as Monte-Carlo Integral

Because of practicability reasons Monte-Carlo integration (see for example [10]) has been used for evaluation of the efficiency $\eta(\mathbf{p})$. The error estimate of the numerical integral is in proportion to $1/\sqrt{N}$ and therefore controllable by the number of rays N. The integral value for one parameter vector $\mathbf{p}$ appears to be 'noised'[1], if in every evaluation a different random number sequence is used. If there is an optimization algorithm which eliminates noise, systematic errors of a deterministic integral evaluation are also avoided. Precautions have to be taken to ensure that the noise is widely independent of $\mathbf{p}$. This is a precondition for identifying optimal parameters of the 'noised' function as optimal parameters of the (unknown) exact function. This is achieved by first evaluating the function for a given parameter vector $\mathbf{p}$ N_0 times. This yields an average $\langle\eta\rangle_{N_0}(\mathbf{p})$ for the efficiency and a standard deviation for η:

$$\sigma_{\eta,N_0}(\mathbf{p}) = \sqrt{\langle\eta^2(\mathbf{p})\rangle_{N_0} - \langle\eta(\mathbf{p})\rangle_{N_0}^2} \tag{8}$$

as well as an estimated standard deviation of the *mean* $\langle\eta\rangle$ of η[11]

$$\sigma_{\langle\eta\rangle,N_0}(\mathbf{p}) = \frac{\sigma_{\eta,N_0}(\mathbf{p})}{\sqrt{N_0}}. \tag{9}$$

Then further evaluations are done up to a total parameter-*dependent* number $N(\mathbf{p})$ so that the mean is less than a set parameter-*independent* tolerance $\sigma_{\mathbf{set}}$:

$$\sigma_{\langle\eta\rangle,N}(\mathbf{p}) = \frac{\sigma_{\eta,N}(\mathbf{p})}{\sqrt{N}} \leq \sigma_{\mathbf{set}}. \tag{10}$$

The choice of $\sigma_{\mathbf{set}}$ is guided by the requirement for a sufficient accuracy of η at minimum computer time. The choice of N_0 is guided by competing requirements: on the one hand, in the interesting parameter regions predominantly

$$\sigma_{\langle\eta\rangle,N_0}(\mathbf{p}) > \sigma_{\mathbf{set}}; \tag{11}$$

on the other hand, N_0 is as large as possible to ensure a good estimate of $\sigma_{\langle\eta\rangle,N_0}(\mathbf{p})$.

4.2 Optimization with Genetic Algorithm

The optimization algorithm used now is a Genetic Algorithm inspired by [3]. (The previously employed stochastic gradient method proved to be too limited to be applicable to the present problems[2].) Optimizing with Genetic Algorithms requires mapping of the parameter vector to a bit string, which is considered as a (haploid) chromosome of an individual; its fitness is the (rescaled) objective function value. For the present case this mapping was performed by discretizing the parameter ranges of each dimension (using 6 or 7 bits); these bits are then concatenated. The Genetic Algorithm repeatedly processes a population of bit strings that are subjected to the following mechanisms: selection according to fitness, random crossing-over between pairs of selected "parents", and occasionally mutation, **Fig. 1**.

The advantages of a Genetic Algorithm in the present context are:

- a possibly multimodal function can be optimized with high reliability without implementation of a super-strategy,

---

[1]superimposed by noise

[2]In optimization with more than two free parameters several numerical problems arose [2].

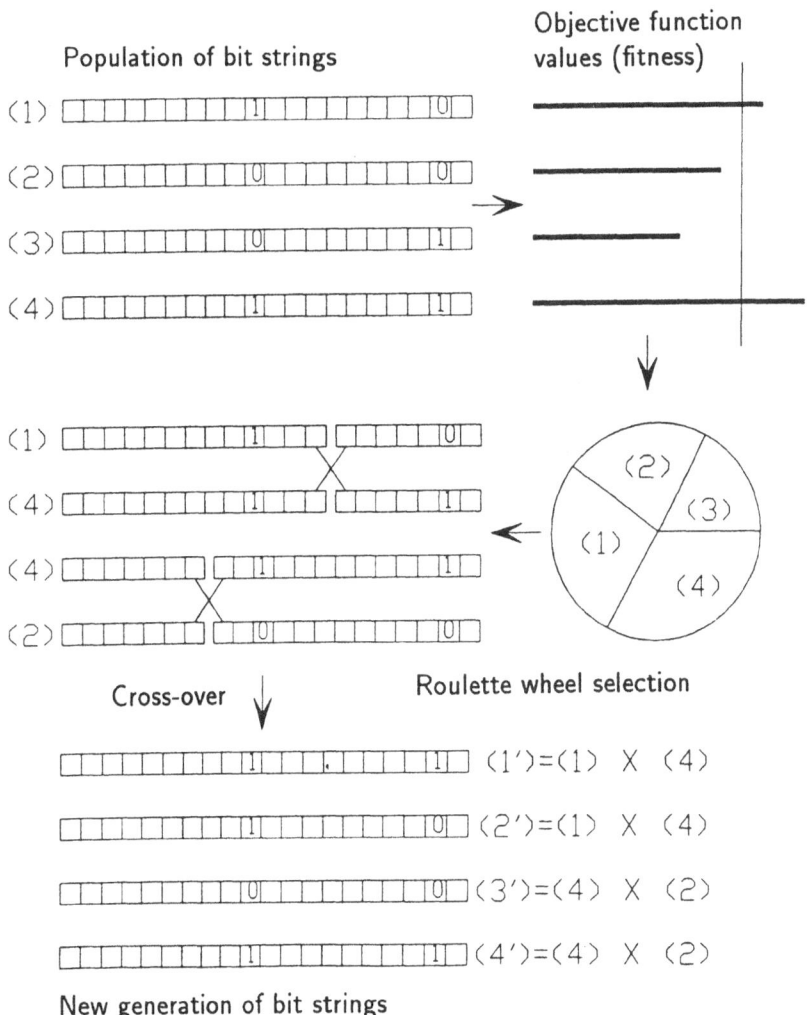

Figure 1: Simplified schematic of the procedural steps in the Genetic Algorithm used in OPTEC.

- 'noised' objective functions (here: evaluation of the objective function for a given parameter vector several times using different random number sequences yields different values distributed randomly about the exact function values) are treated, in principal, like smooth objective functions.

The disadvantages of a Genetic Algorithm in the present context are:

- a discretization of the parameter space is necessary beforehand (one will use step widths of the order of, e.g., manufacturing tolerances; a very fine discretization would enlarge the search space too much),

- an optimal coding is not known beforehand; therefore a (sub-optimum) standard coding is used [3],

- several parameters of the Genetic Algorithm itself (like mutation rate or rescaling factor) have to be chosen beforehand without knowledge of a (possibly) optimal setting.

4.3 Procedure for Calculating Detailed Results of an Optimized Configuration

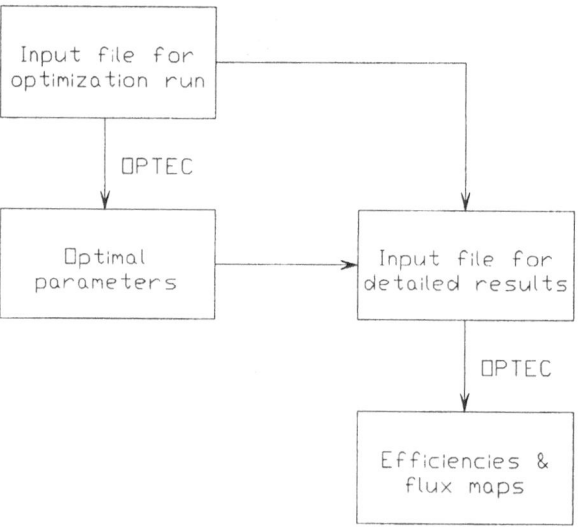

Figure 2: Schematic of the steps in determining an optimized configuration.

The steps to be taken in determining an optimized configuration are shown in **Fig. 2**. The first step in determining an optimized configuration is to prepare an input file for OPTEC specifying an optimization run. For that one has to specify

- the values of mandatory fixed parameters,

- specific values for optionally free parameters that are to be held fixed,

- a range along with a discretization width for each parameter that is to be optimized during this specific run.

The results of this run are the optimal values calculated for the parameters that were specified as free parameters.

The second step is to prepare an input file for OPTEC specifying a non-optimizing run. The optimal parameter values obtained from the optimization run have to be inserted as fixed values now. The results of this OPTEC run are averaged efficiencies and efficiency factors, statistics of the rays processed during ray tracing, as well as optional maps of absorbed flux densities at the receiver surface, and the terminal concentrator inner and outer surfaces.

Problems Optimization becomes difficult if the thermal efficiency of even an optimized configuration is approximately zero. This occurs inevitably in those cases where the specified insolation is not sufficient to achieve the specified receiver temperature. This difficulty is to be kept in mind when using OPTEC for solar furnaces that are to be used at very high temperatures (a problem that typically arises for receiver temperatures above ca. 2000°C).

Another problem with the Genetic Algorithm arises, if the search space is huge and the objective function (thermal efficiency) is greater than zero only within a small fraction of this search space. In such cases it is possible that instead of a global a local optimum is found. An example is the optimization of all geometrical parameters of a generalized trumpet: the entrance aperture radius of the trumpet and the entrance aperture radius of the cone that is the tangent at the exit of the trumpet are strongly interrelated. One has to be careful when setting up the input file for the optimization run. A different parameter set should be chosen to avoid this problem.

5 Layout for Various Configurations

Layout calculations have been carried out for various configurations without terminal concentrators as well as with cones and trumpets as terminal concentrators using OPTEC.

5.1 Laboratory Experiment

The configurations labeled laboratory experiment are those which section 6 Measurements refers to. These configurations consist of a parabolic primary (represented by a parabolic profile covered with highly reflective silvered foil) and a conical or a trumpet concentrator.

Schematic

A schematic of the laboratory experiment is shown in **Fig. 8**.

Conditions

In **Tab. 1** the conditions are shown which have been chosen to optimize configurations of the laboratory experiment. The parameters for insolation level and receiver temperature are necessary to obtain an optimal configuration that is in some sense representative for an outdoor case. In the laboratory only the optical performance was investigated. For that no high flux levels were needed. In fact the flux density levels used in our experiments were very low (on the order of Watt per square meter or lower).

parabolic primary	
reflectivity	0.92
surface errors	0
radius	42 cm
focal length	75.6 cm *
receiver	
temperature	1000°C
absorptivity	1
convection coefficient	16 W/m²/K
terminal concentrator	
reflectivity	0.95
meteo	
ambient temperature	20°C
insolation	900 W/m²
pillbox distribution	1.37° rim angle

Table 1: Fixed conditions of the optimization for the laboratory experiment. * This number was obtained previously from a preliminary optimization.

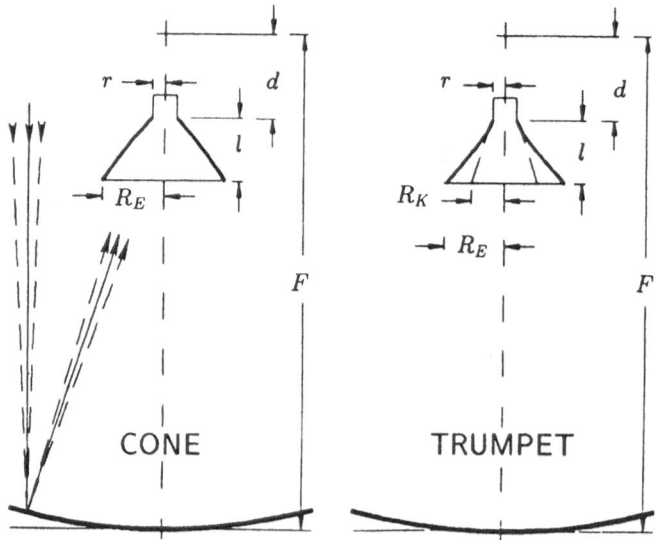

Figure 3: Dish configurations for optimizing (b) and (c) with conical terminal concentrator (left). For optimizing (d) and (e) with hyperboloid of revolution (trumpet concentrator, right).

		without t.c.	cone	trumpet
r	exit radius	19 mm	12 mm	12 mm
d	displacement	1.5 mm	18 mm	13 mm
l	length	—	77 mm	86 mm
R_E	entrance radius	—	70 mm	71 mm
R_K	radius of tangent cone entrance	—	—	59 mm
η_{op}		[1.00]	0.94	0.96
η		0.47	0.68	0.70

Table 2: Optimal values of the free parameters for the laboratory experiment; calculated terminal concentrator optical efficiency η_{op} and thermal efficiency η.

Results

Tab. 2 shows results for the terminal concentrator obtained by running OPTEC to find optimal parameters for entrance and exit aperture radii as well as the length of conical and trumpet concentrator. In the case of the trumpet the entrance aperture radius of the tangent cone (see **Fig. 3**) has also been determined.

These optimal parameter values then were used for the laboratory experiment. Because of practical reasons the actual dimensions of the manufactured terminal concentrators deviate. Also the displacement values were not used (in fact the nominal focal length was used instead). In section 6 Measurements the calculated and measured local concentrations are shown for this configuration as a function of the radial absorber position.

5.2 Terminal Concentrator Solar Furnace at LMU

Schematic

At Ludwig-Maximilians-Universität, Munich, a small-scale terminal concentrator solar furnace is planned. In **Fig. 4** a schematic is shown. A heliostat with an area of $2\,\text{m} \times 2\,\text{m}$ redirects sunlight onto a spherical concentrator which in turn concentrates the light onto a receiver. In front of the receiver a terminal concentrator can be mounted.

Conditions

The conditions for the optimization runs concerning the planned solar furnace are listed in **Tab. 3**

Results

In **Tab. 4** results are listed for the conditions given in **Tab. 3**. Optimal configurations for conical concentrators are shown for two distinct parameter domains: i) with a receiver radius smaller than that without terminal concentrator, and ii) with a receiver radius larger than that without a terminal concentrator. In case i) the flux density is increased (note the increased stagnation temperature). In case ii) the thermal efficiency is increased by increasing the receiver area at approximately no change in average flux density (same stagnation temperature).

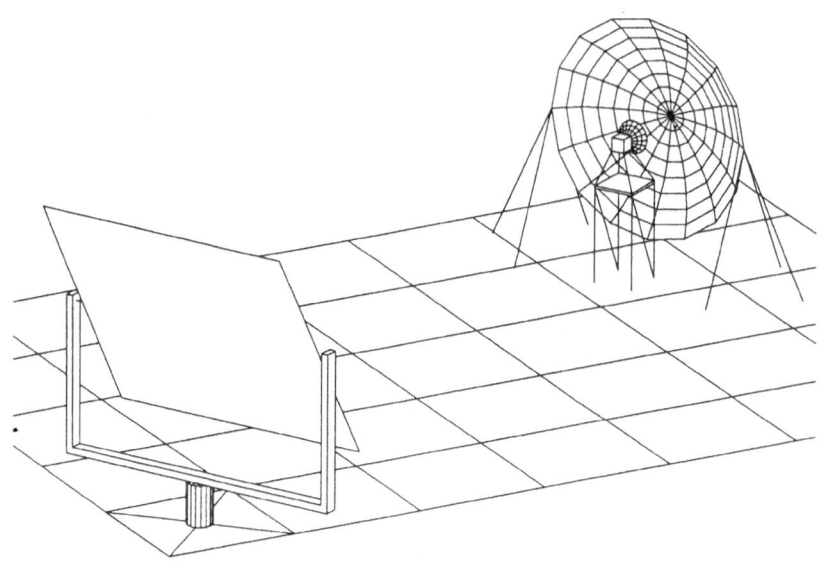

Figure 4: Schematic of the planned solar furnace at LMU Munich.

Optimization calculations carried out with generalized trumpet concentrators yielded no better results.

5.3   PSA Solar Furnace

Four heliostats redirect sunlight onto a faceted parabolic primary, approximately 11 m in diameter.

One of the first planned applications is the melting of ZrO_2 (at a temperature of approximately 2700°C), having an absorptivity of 0.31 (yellow modification) [18].

The dish has been simply modeled with a single-piece paraboloid. Gaps between the heliostats and partial shading of the dish by a shutter have been taken into account by specifying an appropriate mirror use factor according to [18]. The receiver has been modeled simply with a cavity having an absorptivity of 0.31.

Conditions

The conditions for the optimization run concerning the planned solar furnace are listed in **Tab. 5**. Reflectivity of the heliostats is 0.9, reflectivity of the primary is 0.92. Since the

spherical primary *	
reflectivity	0.81 (including heliostat)
surface errors	1 mrad
radius	0.775 m (outer), 0.05 m (inner)
focal length	0.698 m
receiver	
temperature	1200°C
absorptivity	1
convection coefficient	16 W/m^2/K
terminal concentrator	
reflectivity	0.92
meteo	
ambient temperature	20°C
insolation	700 W/m^2
brightness distribution	LBL standard solar scan [12]

Table 3: Fixed conditions of the optimization for the planned LMU solar furnace.
* A single-piece spheroid is available at LMU.

		without t.c.	cone	cone
r	radius exit	11.5 mm	9.48 mm	13.2 mm
d	displacement	36 mm	48.4 mm	62 mm
l	length	—	84.5 mm	83 mm
R_E	radius entrance	—	82.0 mm	96 mm
η_{op}		[1.00]	0.45	0.55
η		0.22	0.30	0.32
	stagnation temperature	1757°C	2050°C	1789°C

Table 4: Optimal values of the free parameters for the planned LMU solar furnace.

heliostat cannot be modeled in OPTEC separately, an effective primary reflectivity has been used, which is taken just as the product of both reflectivities.

paraboloidal primary	
reflectivity	0.83 (0.9 × 0.92)
surface errors	1.1 mrad
radius (outer | inner)	5.5 m | 0.78 m
focal length	7.45 m (planned) †
mirror use factor	0.816 *
receiver	ZrO_2 sample
radius	0.5 cm (sample radius) and 1.0 cm
temperature	2700°C
absorptivity	0.31 (yellow ZrO_2)
convection coefficient	16 $W/m^2/K$
terminal concentrator	
reflectivity	0.95
meteo	
ambient temperature	20°C
insolation	960 W/m^2
Gaussian sunshape	3.78 mrad (standard deviation)

Table 5: Fixed conditions of the optimization for one of the planned PSA Solar Furnace experiment.
* This factor is due to gaps between the heliostats as well as to the shutter blades.
† For reference also the focal length has been used as a free parameter.

Results

Calculation of efficiencies without a terminal concentrator yielded the result that the desired sample temperature could not be attained. However, nothing was known concerning the actual process of melting (change of absorptivity due to increasing or decreasing impurities in ZrO_2 or due to changes in surface structure).

Then optimization calculations were carried out for conical concentrators, where the exit aperture was fixed to once or twice the value of the sample radius, respectively. In **Tab. 6** results are shown for the conditions listed in **Tab. 5**.

The results in **Tab. 6** show that under the above-mentioned conditions the desired sample temperature can be attained with conical terminal concentrators if the exit aperture is not significantly larger than the sample (radius 0.5 cm) or if a larger focal length of the primary is possible. However, even in these cases the stagnation temperatures calculated by OPTEC are only several degrees higher than 2700°C.

5.4 PSA SSPS-CRS

The SSPS-CRS, Plataforma Solar de Almeria (PSA), Spain, originally consists of 93 heliostats, distributed in a north field aiming at the center of a receiver located 43 m above ground; each heliostat has a reflective area of 39.3 m^2 with reflectivity 0.911.

r	radius exit	[0.5 cm]	[0.5 cm]	[1.0 cm]	[1.0 cm]
F	focal length	[7.45 m]	9.49 m	[7.45 m]	13.51 m
d	displacement	0 cm	3 cm	−4.3 cm	−2.8 cm
R_E	concentrator radius entrance	1.15 m	0.67 m	0.52 m	1.10 m
l	concentrator length	1.51 m	1.41 m	0.65 m	2.76 m
η_{op}		0.0244	0.0340	0.0915	0.0828
η		* 0.0002	* 0.0002	−0.0001	* 0.0000

Table 6: Optimal values of the free parameters for the PSA solar furnace. In brackets: Fixed values.

* The small, but *positive* values indicate, that the desired temperature can be attained.

Conditions

From the construction report [13] of the SSPS-CRS the following parameters were taken for the design point at equinox noon: Gaussian sunshape with one-dimensional standard deviation (including primary shape deviations) 3.1 mrad; direct insolation 920 W/m²; absorptivity and emissivity of receiver 1.0. For the present investigations additional conditions were chosen (complete list see **Tab. 7**).

primary	93 rectangular heliostats, spherical shape
reflectivity	0.911
surface errors	(any errors included in sunshape)
receiver	(inclined by 28°)
temperature	1000°C
absorptivity	1
convection coefficient	16 W/m²/K
terminal concentrator	
reflectivity	0.9
meteo	
ambient temperature	25°C
atmospheric attenuation factor	0.99
insolation	920 W/m², equinox noon
Gaussian sunshape	3.1 mrad (degraded sun)

Table 7: Conditions for optimization of PSA SSPS-CRS, partially taken from [13].

The receiver is modeled as a cavity with a circular (instead of octagonal) entrance aperture, inclined by 28° (manually calculated from the extent of the field). Heliostats are modeled as single-piece rectangles (6.3 m × 6.25 m) with spherical curvature (four different focal lengths) that are individually tracked (This is similar to the on-axis alignment of computer code HELIOS [14], used in [13] for flux density calculations). Cones and trumpets have been considered as terminal concentrators.

Results

Tab. 8 shows results for the above-mentioned configuration and conditions. All parameters of the terminal concentrators were optimized simultaneously (see **Fig. 3**). No changes to the heliostat field or to the tower height have been made.

Case	Optimal parameters / [m]				Efficiencies and efficiency factors						
	r	R_E	l	R_K	η_{obs}	η_{cos}	γ	η_{op}	η_0	L/C	η
(a)	0.99	—	—	—	1.0	0.95	0.90	[1.0]	0.77	0.15	0.62
(b)	0.83	4.45	5.25	—	1.0	0.95	0.99	0.93	0.79	0.11	0.68
	0.96	2.2	2.05	—	1.0	0.95	0.99	0.97	0.82	0.14	0.68
(c)	0.83	4.15	4.2	2.8	1.0	0.95	0.99	0.95	0.80	0.10	0.70

Table 8: Results of optimization calculations for one- (a) and two-stage CRS configurations with conical (b, two different parameter domains) and trumpet concentrator (c), respectively (see text). In brackets: given or fixed values. $\alpha_R = 1.0$, $\rho_H = 0.911$, atmospheric attenuation 0.99.

5.5 DLR Solar Furnace

At DLR, Köln a solar furnace is planned inspired by the high flux solar furnace (HFSF) at SERI, Colorado. A faceted primary is used in an off-axis configuration, i.e., the focal spot is outside the region heliostat—primary. The off-axis configuration allows placement of a laboratory around the focal zone. However, flux distributions in the focal plane are necessarily non-symmetric. Consulting has been undertaken by professor Lorin L. Vant-Hull. Regarding flux maps U. Schöffel has run computer calculations with computer program OPTEC using the built-in Detailed Heliostat Field Model for a number of design parameters.

To answer the question whether an optional terminal concentrator would bias strongly the decision towards one of two candidate configurations additional optimizations for conical terminal concentrators were carried out [5]. Only the results for these additional investigations are presented here.

Schematic

A plan of the solar furnace under design at DLR, Köln is shown in **Fig. 5**. **Fig. 6** shows the hexagonal facet structure of the primary concentrator. The spherically shaped hexagons have been modeled in the present investigations by area-equivalent rectangles with spherical curvature, located at the centers of the hexagons.

Conditions

Since many important data for the design study were not available at the time of the study, some even numbers were taken which allow rescaling of flux maps as soon as precise data are available. However, when optimizing (a terminal concentrator), the end use (e.g., desired melting temperature of a target with given absorptivity) as well as solar brightness distribution and mirror surface errors have to be taken into account. Therefore assumptions were made which are listed in **Tab. 9**. A low beam insolation has been chosen to comprise lower (but still unknown) beam insolation as well as non-ideal materials properties.

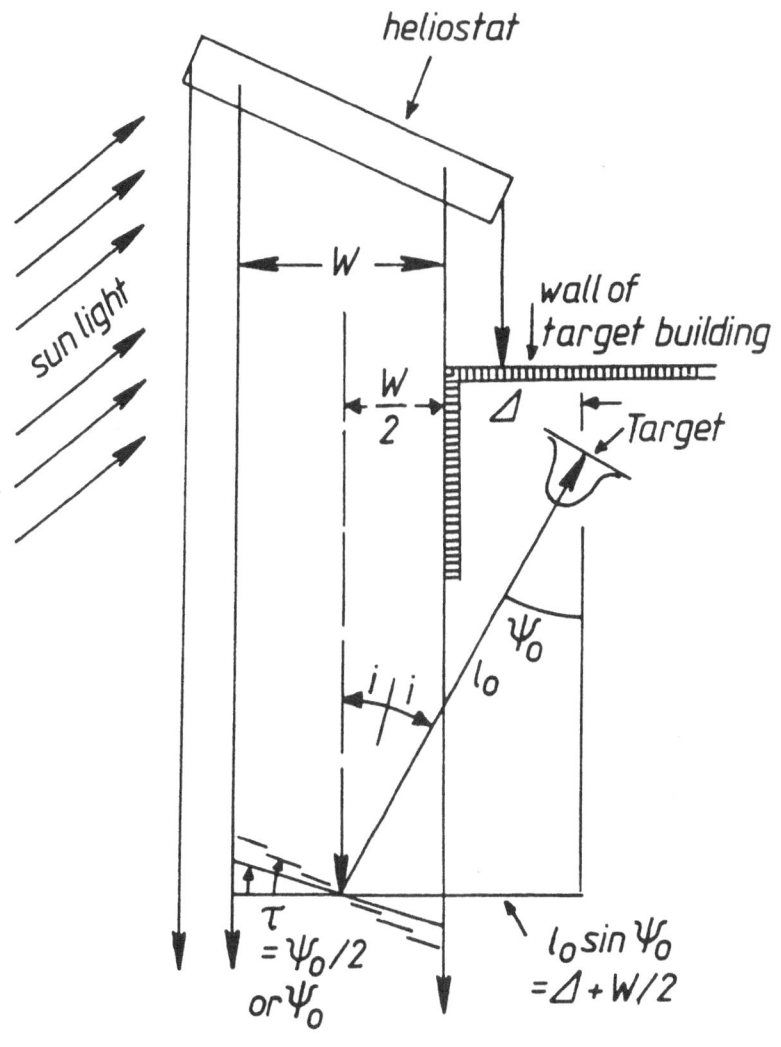

Figure 5: Plan of the solar furnace under design at DLR Köln (from [5]).

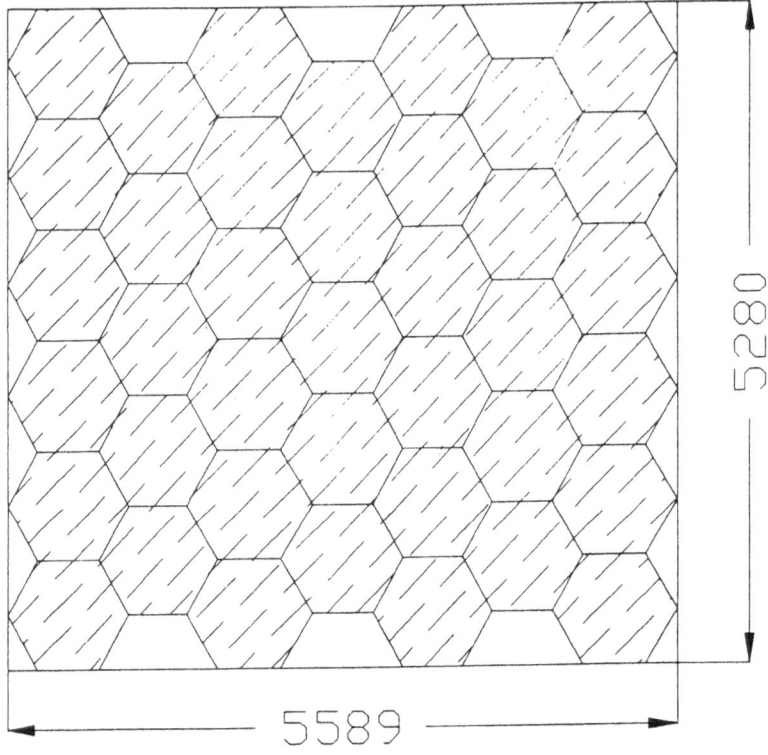

Figure 6: Facet structure of the solar furnace under design at DLR Köln (from [5]).

primary	39 spherical facets each having its own focal length, located on a structure parallel to the target plane.
reflectivity	1
surface errors	1 mrad
diameter	≈ 5.3 m
focal length	8.3 m | 7.3 m for respective off-axis angles 30° | 34°
receiver	
temperature	2500°C
absorptivity	1
convection coefficient	16 W/m^2/K
terminal concentrator	
reflectivity	0.9
meteo	
ambient temperature	20°C
insolation	500 W/m^2
brightness distribution	LBL standard solar scan [12]

Table 9: Fixed conditions of the optimization for the DLR solar furnace.

Results

Tab. 10 shows results for the conical terminal concentrator obtained by running OPTEC to find optimal parameters for entrance and exit aperture radii as well as the length of the cone. Since no forces would prevent the terminal concentrator from becoming very large, a restriction for its length (20 cm) were chosen. **Fig. 7** shows absorbed flux densities vs. horizontal radial absorber position for the two candidate configurations with optimized conical terminal concentrator. Note the horizontally asymmetric flux density distribution due to off-axis configuration.

		without t.c.	cone	without t.c.	cone
F	focal length	[8.3 m]	[8.3 m]	[7.3 m]	[7.3 m]
r	radius exit	2.6 mm	1.7 cm	1 mm	1.7 cm
d	displacement	[0 cm]	[0 cm]	[0 cm]	[0 cm]
l	length	—	17.3 cm	—	[20 cm]
R_E	radius entrance	—	8.0 cm	—	9.7 cm
η_{op}		[1.00]	0.70	[1.00]	0.69
η		< 0	0.32	< 0	0.29

Table 10: Optimal values of the free parameters as result of the optimization for the DLR solar furnace. Additionally the optical efficiency of the terminal concentrator and the overall thermal efficiency are shown. In brackets: fixed or given values.

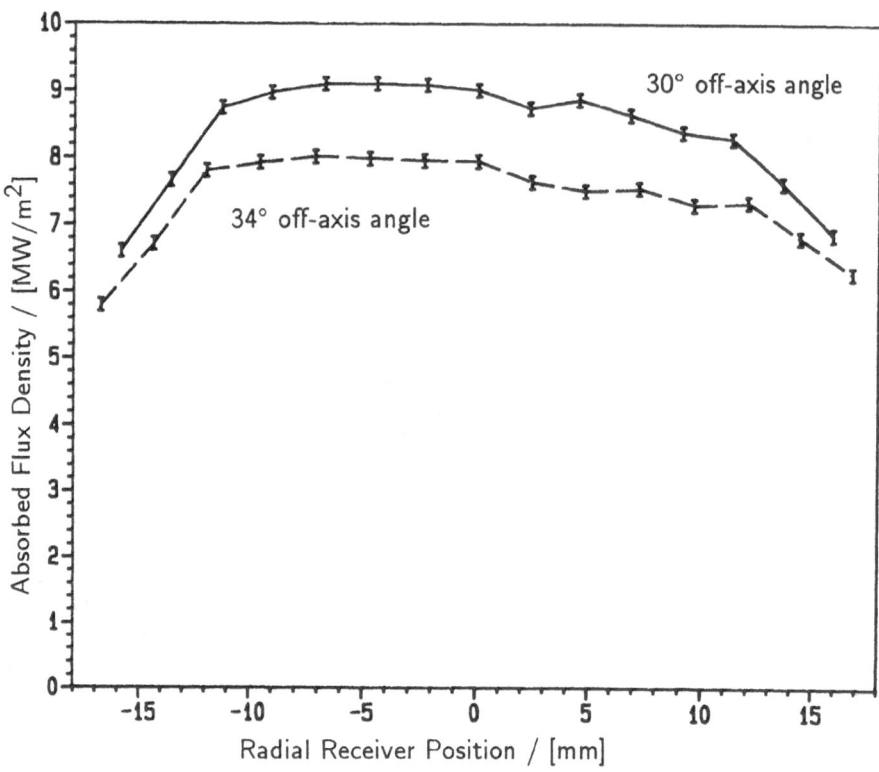

Figure 7: Flux density vs. horizontal radial absorber position in the exit aperture plane for two candidate configurations with off-axis angles 30° (focal length 8.3 m) and 34° (focal length 7.3 m).

6 Measurements of Terminal Concentrators

In the experimental phase of the present project, one aim was to bridge the gap between the theoretical studies and the practical use of terminal concentrators in solar high flux density applications. On the one hand a partial validation of OPTEC was strived for; on the other hand we wished to investigate the feasibility of laboratory measurements of terminal concentrators, where the angular distribution of the radiation entering the terminal concentrator resembles the distribution which occurs in two-stage systems for solar applications.

Ideally a laboratory experimental set-up would be a test bed for various terminal concentrators, consisting of a (variable) primary concentrator and mounts for terminal concentrators. Several problems are encountered when trying to design such a set-up: First, the different scales of the primary and receiver lead to large set-ups while at the focal zone the distances are quite small. Second, when using a complete biaxially focusing primary, quite different levels of flux densities have to compared because of the intended concentration. Third, some properties of sunlight have to be simulated (spectral distribution, angular characteristics, local uniformity across the primary).

6.1 Measuring Process

Our objective was to carry out laboratory measurements of the optical efficiency of terminal concentrator in a geometrical configuration comparable to a two-stage system typical for solar applications in an outdoor configuration. The initial principle ideas in implementing this objective were:

- The primary concentrator, which is assumed to have symmetry of revolution, is reduced from three to two dimensions, i.e., by using a two-dimensional meridional section,

- a light source is provided with defined angular extent, 'seen' from each point of the primary concentrator in the same direction,

- to avoid speckle phenomena [4] a tungsten lamp (rather than a laser) is used,

- the light is scanned along the primary using a redirecting mirror on a positioning device at a rate proportional to the weight of of the represented areas,

- recording of flux densities is carried out photographically; locally simultaneously at all positions in the focal plane; integrating over time,

- the recordings are symmetrized, thereby obtaining the radial flux density distribution to be expected for a full three-dimensional primary concentrator.

Detailed investigations showed that it was necessary to take separate photographs for different radii (see below), instead of a continuous time-integrated scan. This required an additional numerical superposition of the different digitized recordings.

6.2   Schematic of Experimental Setup

A schematic diagram of the experimental set-up is shown in **Fig. 8**. The principles of the set-up are:

- The beam from a stabilized thermal lamp (tungsten lamp) is expanded, collimated, redirected by the scanning mirror, and focused onto the primary concentrator. By moving a carriage with a redirecting mirror and a convergent lens each point along the diameter of the primary concentrator consecutively 'sees' a radiation source with defined angular extent in the same direction. This simulates some ray properties of the sun. The angular extension chosen is 3–4 times larger than that of the degraded sun because of practicability reasons (extensions of the terminal concentrator).

- A mounting allows measurement of different terminal concentrator models. Two conical and a trumpet terminal concentrator were manufactured.

The parallel expansion was achieved by a combination of a microscope objective, a spatial filter, and a convergent lens. Defined attenuations of the light beam can be obtained by a chopper with variable opening ratio. A shutter enables exposures with defined time intervals. Details of the experiment can be found in [15].

6.3   Investigations

Photographic film angular dependence   To find the angular dependence of the absorptivity characteristics of the plane photographic films exposures were taken with the film tilted at various angles of incidence between 0° and 75°. Higher angles of incidence could not be set up because of the protruding rim of the mounting for the films. The data taken within the above-mentioned range can be described within the error tolerance by Fresnel's formulas for reflection. To enable a more accurate fit for the parameters in the Fresnel formulas more, and more accurate, measurements would be desirable, especially at angles beyond 75°.

Time scan   Because of several reasons the scanning approach leads to exposure intervals of photographic recordings which are long (up to 30 seconds) compared to usual usage of photography. These long intervals are necessary because of properties of the photographic films that are available, light source brightness, mechanical limitations of the positioning device, and consequences of high dynamic ranges to be covered by the measuring films at the receiver plane due to light concentration.

Reciprocity failure effect   Inspired by [16] reciprocity failure of photographic films was investigated. Reciprocity failure is the effect that equal exposures (flux density times time) do not yield equal optical densities if times are 'long' with respect to an intrinsic time constant of the film at ambient temperature: A grain partially activated by one quantum of light recombines if not hit by a second quantum within this time interval. A computer modeling was done concerning the behaviour of photographic films exposed for long time intervals to continuous weak flux densities and intermittent higher flux densities, respectively. As a result of these investigations, automatic scanning of the primary was finally given up since the error induced by reciprocity failure would have rendered the resulting data difficult to interpret. Therefore the number of exposures had to be increased to cover the primary at

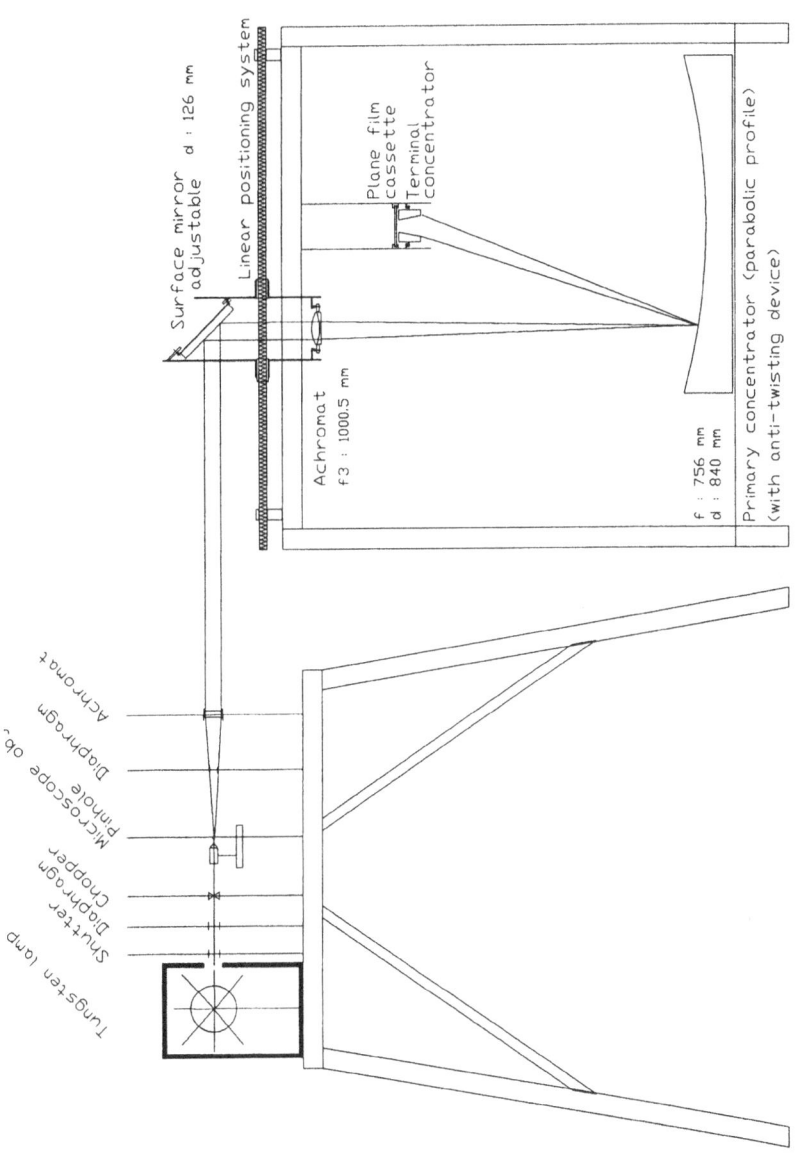

Figure 8: Schematic of the experimental set-up for the laboratory experiment.

a sufficient number of radii. Instead of one, now up to eight exposures were taken. This discretization of course, also introduces an error that is not easy to estimate.

Reference exposures Due to manual developing of exposed photographs equal deposited energy quantities do not yield identical optical densities. Therefore every photograph has to carry a reference field along with the proper measuring field. The reference field was provided by a recording of a gray scale put into direct contact with the film and exposed to parallel light. The gray scale was prepared once and was used throughout all recordings thereafter. Although the gray scale not only absorbs but also scatters the light, calibration properties of the reference field, which is irradiated from various angles, are retained.

Evaluation After digitizing of the developed photographs with a microdensitometer the data obtained from the measuring field were compared to the data of the associated reference field. This compensates for variations of the parameters of the developing process insofar as developing is locally uniform all over the exposed film. A phenomenological density function (Hurter-Driffield curve [17]) depending on exposure was fitted to the data of the reference field. Inference from the optical densities in the measuring field to exposure values was done by interpolation using the phenomenological fit.

6.4 Results

Altogether three terminal concentrators were manufactured: two cones and one trumpet. The trumpet and one of the conical concentrators were turned on a lathe from aluminum, then polished and silver coated. The other conical concentrator was made by galvanic deposition of copper onto a highly polished kernel, that had been turned on a lathe from V2A steel, then polished and silvered. The latter one (made by galvanic deposition) showed the best optical surface properties of all three.

Experimental results are presented for the second (galvanically deposited) conical concentrator. The dimensions were taken primarily from optimization. Because of practicability reasons slight changes have been undertaken (**Tab. 11**). Comparisons with calculated data shown below are based on the data and dimensions actually realized in the experiment.

Fig. 9 and **Fig. 10** show evaluated local concentrations without and with conical terminal concentrator, respectively. Also calculated results are shown, where the angular absorptivity characteristic of the photographic film has been taken as a property of the receiver. The measured data in both figures are obtained after recording of eight photographs, each taken for one radial position along the parabolic profile, then digitizing, fitting of the density curve to the reference field, symmetrization, and superposition. Note that the local resolution of the calculated data is not as fine as of the measured data, so the differences between them at the wing are probably due to the linear interpolation of the calculated data.

Fig. 11 and **Fig. 12**, which show the symmetrized measured contributions from the individual radial primary positions, give an insight why the the agreement of the measured with the calculated data for the one- and two-stage systems are so different. **Fig. 11** shows that the contributions from different primary positions to the superimposed local concentration are smooth in the case of the one-stage system, whereas **Fig. 12** shows peaked contributions in the case of the two-stage system due to the additional concentration by the terminal concentrator. So one could expect that a finer local resolution of the primary positions would yield a better agreement.

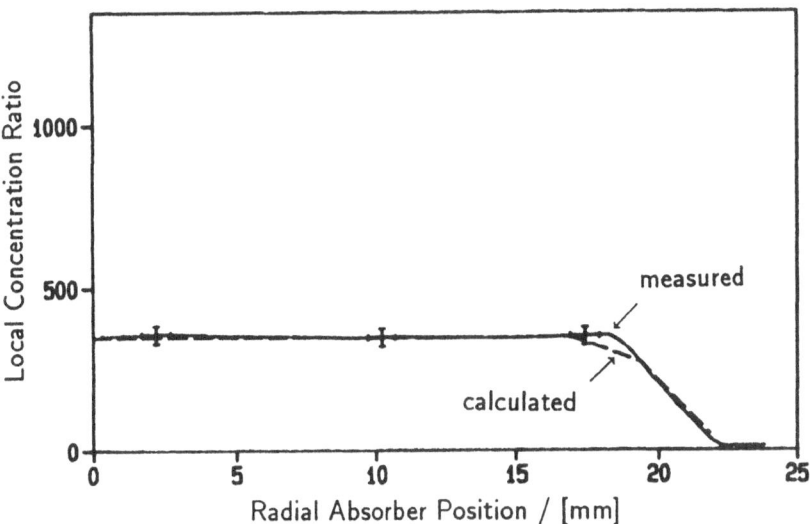

Figure 9: Local concentrations vs. radial absorber position. Measured (lines) and calculated (dashes) concentrations *without* terminal concentrator.

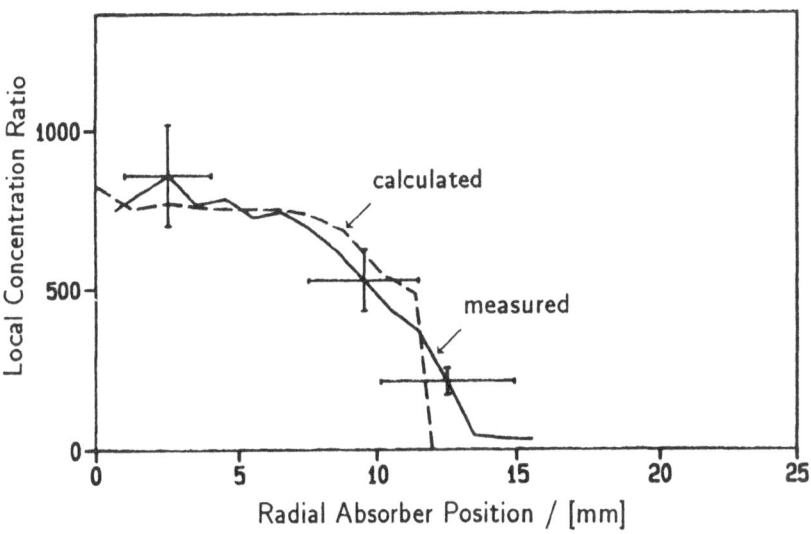

Figure 10: Local concentration vs. radial absorber position. Measured (lines) and calculated (dashes) concentrations with terminal concentrator. For three selected positions also the estimated margins of error have been drawn.

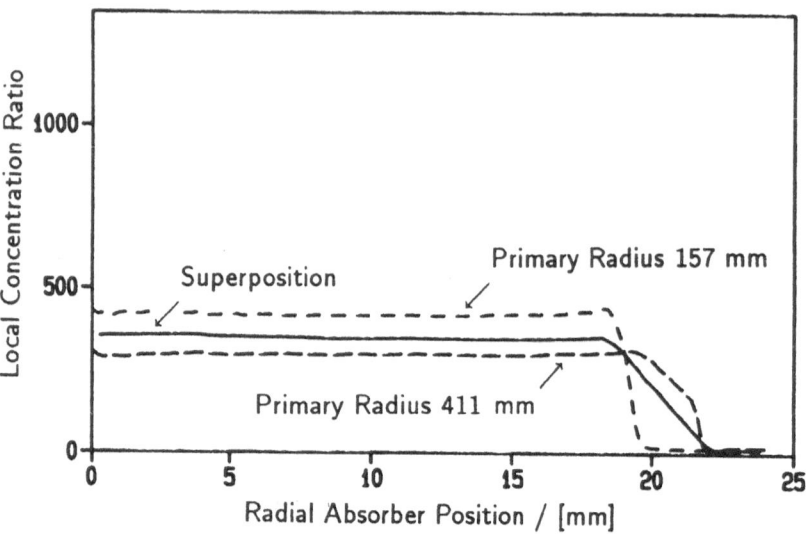

Figure 11: Measured local concentration vs. radial absorber position without terminal concentrator: contributions of 3 selected primary concentrator radial positions and the superposition for all 8 primary concentrator radial positions.

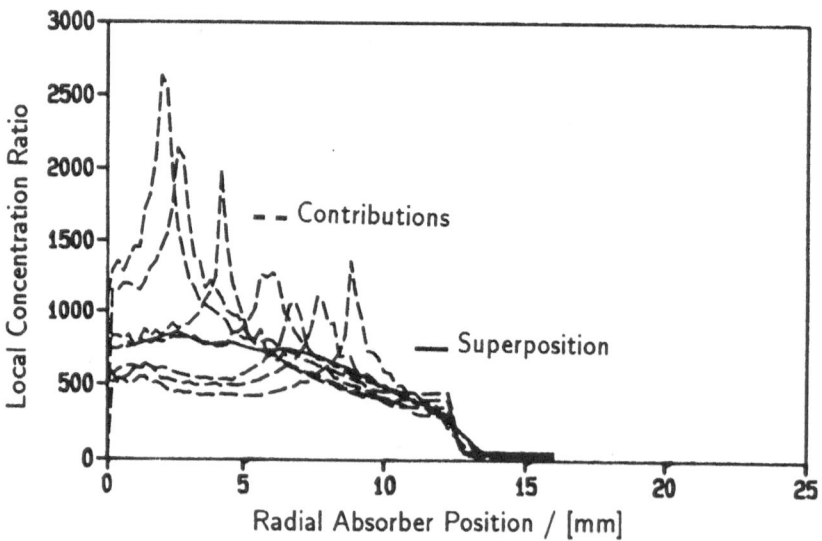

Figure 12: Measured local concentration vs. radial absorber position with conical terminal concentrator: contributions and superposition for all 8 primary concentrator radial positions.

	used for optimization	in the experiment
terminal concentrator		
radius of exit aperture	12 mm	12.4 mm
displacement from focal plane	18 mm	0 mm
length of cone	77 mm	77.7 mm
radius of entrance aperture	70 mm	69.8 mm
reflectivity	0.95	0.92
primary concentrator		
inner radius	0 mm	90 mm
reflectivity	0.92	0.95
half opening angle of light	1.37°	1.43°
absorber absorptivity characteristics	uniform (deep cavity)	Fresnel formula (film)

Table 11: Dimensions of the configuration in the laboratory experiment. Middle column: as used in optimization, right column: actually realized.

Still one remark has to be made about the non-zero concentrations outside of the exit aperture of the two-stage system in **Fig. 12**. At these positions no flux density is to be expected. However, since rays leave the exit aperture of the terminal concentrator (and hit the absorber plane, i.e., the film) under large incidence angles, obviously leakage occurs. This might be due to not totally close contact between the film and the terminal concentrator, so that rays are partially reflected between them until they are completely absorbed.

6.5   Assessment of the Measuring Method

During this project the following advantages and drawbacks of photographic recordings as a means to measure flux densities at exit apertures of terminal concentrators were found (+ means advantage, − means drawback):

+ easily available,

+ low cost,

+ locally simultaneous recording of flux density distributions,

+ manual handling possible (but time consuming),

− limited dynamical range, implying strong constraints for the layout of the experiment,

− reciprocity failure,

− angular dependence of film (especially at angles above approximately 60° due to reflective properties of dielectrics). This is a problem when measuring flux densities with terminal concentrators where rays enter the receiver plane at large incidence angles).

− on each photograph recording of a reference field is needed which is exposed to the same developing process as the proper measurement field.

− identification of positions from the developed negative is difficult,

- limited dynamic range to cover large ranges of flux densities caused by concentration effects,

- automation is difficult.

From the disadvantages it follows that the accuracy of the results is not very high. In view of all the difficulties mentioned above the results can actually be considered to be good.

6.6 Alternative Experimental Setups

The problems faced with photographic measurements of flux densities with terminal concentrators led us to consider alternative measuring methods. There are several aspects to this:

- light generation,

- scanning of primary concentrator,

- measurement method.

Light Generation Light from the sun is incoherent. Therefore, when laser light is used to simulate the sun, it has to be made incoherent by e.g., a rotating ground glass. However, this means a considerable attenuation, so that the laser has to have high power. Thermal light sources (as in our experiment) can be used instead. However, then a higher divergence angle has to be accepted. While laser light has a Gaussian profile, thermal light sources show a uniform profile. To emulate actual sun profiles continuous density filters could be helpful.

When reflecting surfaces are used, one has to keep in mind that their reflectances are wavelength-dependent. Laboratory measurements which are to characterize concentrators for use in solar applications have to be weighted appropriately.

Scanning of Primary Concentrator A large range of primaries with symmetry of revolution could be emulated by a small mirror that can be positioned along one diameter of the virtual primary and in the direction of the axis of the primary. In addition the mirror has to be properly canted.

Enhanced Photographic Recordings First, an enhanced version of our method of measuring is presented. Many problems were introduced by reciprocity failure of available photographic films. While desensitizing (prevention of occurrence of optical density by weak infrared radiation due to ambient conditions during storage) is needed in commercially available films, reciprocity failure as a consequence of that should be eliminated during measurements. This could be achieved at least partially by cooling the photographic films during recording by liquid nitrogen.

Instead of exposing the reference field to a gray scale, each subfield of the reference field could be illuminated individually and directly. This would prevent degradation of the gray scale brought into contact with the photographic films. A helpful device would be a developing machine to enable short time developing which produces a higher dynamic range of the photograph.

Electronic Measurements A totally different measurement method would use photo-diodes or photodiode arrays (CCD) as are used in video cameras. However, CCDs still are expensive devices that are available only for small areas. Also CCDs will show angular dependence that is most pronounced at large angles of incidence (information about this phenomenon were not available when this project commenced). If a special treatment of the surface of CCDs is not possible to eliminate the angular dependence, a hemispherical detector with a small entrance aperture (diaphragm) could be developed. Such a detector would ensure that rays entering the entrance aperture even under large angles of incidence would hit the detector elements perpendicular. As a disadvantage a two-dimensional scanning of the terminal concentrator exit aperture is necessary. Of course a time integration during scanning of the primary has to be carried out in addition. Thus a three-dimensional scan is necessary: one direction for the primary concentrator, and two directions for the detector.

Anticipated advantages and drawbacks are

+ automation is possible,

+ higher reproducibility,

+ larger dynamic ranges provided,

− a hemispherical CCD detector has to be developed,

− an appropriate micropositioning device has to be designed,

− high cost.

7   Conclusions

Layout of terminal concentrators OPTEC has been proved to be a useful tool in designing thermally optimized terminal concentrator configurations for a variety of systems. These configurations consist of a primary concentrator:

- single-piece spherical concentrator or

- single-piece paraboloidal concentrator or

- heliostat field or faceted dish with flat, spherical or paraboloidal facets,

and an optional terminal concentrator:

- conical concentrator or

- generalized trumpet concentrator.

Surface errors can be modeled by circular Gaussian distributions of local surface normals. In addition to uniform ('pillbox') and Gaussian sunshapes, a linearly interpolated data table can now be specified. This mechanism allows a much more realistic representation of the angular brightness distribution of the sun. Now also a table look-up mechanism is provided for receivers with absorptivities having angular dependence.

Implementation of a genetic algorithm and its incorporation into OPTEC enabled optimization of multi-parameter objective functions. However, in certain cases alternative sets of free (optimizable) parameters should be used in the future to decrease computing time and to enhance the probability of reaching global optima. This will be achieved by a general mechanism allowing the user to define relationships between the parameters and to define special objective functions, which could be cost functions.

OPTEC has also been useful in calculating absorbed flux density distributions on receiver surfaces as well as on terminal concentrator surfaces.

Measurements of terminal concentrators Three principle problems of designing a laboratory experimental set-up for the measurement of concentrating systems were encountered:

1. the largely different scales of primary concentrator and receiver, along with the requirement for accurate adjustments,

2. providing of a light source simulating the angular and local properties of sun light, and

3. measuring flux density distributions, which are caused by light entering the detector surface from the half-space.

An experimental set-up was designed, constructed, and built for measurements of local flux density concentrations in the exit aperture of one-stage and two-stage concentrators. The primary concentrator is a reflective parabolic profile representing a three-dimensional paraboloidal mirror. Two conical and one trumpet concentrator were designed, manufactured, and tested in the set-up. By using a thermal lamp and appropriate optics, an angular uniform (pillbox) brightness distribution was realized. Flux densities were recorded by photographic films. By time-consuming post-processing the local flux density in the exit aperture was obtained.

The requirement for sun-like light prevents irradiation of the complete primary simultaneously. Thorough analysis of the low irradiance reciprocity failure of photographic films led to several changes of the original experimental approach.

Photographical radiometry as an approach to measuring flux density distributions in the exit aperture of concentrating systems caused several difficulties:

- Automatic scanning of the primary leads to slowly varying flux densities in the concentrator exit aperture, over a period which is long compared with the time constant of decay of partially activated photographic grains (low irradiance reciprocity failure). Under such conditions experimental data would be hard to interpret.

- While scanning the primary, at each point in the aperture a widely varying flux density is observed, which cannot be recorded very accurately by a photographic film because of its limited dynamic range.

- Highly exact manual execution of exposures in an unlit room is needed to avoid radiation scatter and to achieve good reproducibility. The same applies to the developing of the recordings in the dark room.

- Digitizing of the local optical transmission of the developed photographs by a micro-densitometer generates a large amount of data, which are required to calculate the symmetrized flux density distribution for a three-dimensional primary concentrator with symmetry of revolution.

Because of the first two points automatic scanning was given up, and several individual recordings had to be taken instead. Because of time limitations, only for one of the conical concentrators could the complete process be carried out thoroughly, after several preliminary experiments. In view of all these problems the results presented can be considered to be quite good. However, as a general approach to measuring flux density distributions in concentrating systems irradiated by radiation resembling the sun, photographic recordings are not be recommended, especially if a general test bed for terminal concentrators is to be designed.

OPTEC flux density calculations appear to be consistent with measured flux density distributions in the laboratory experiment.

8 Nomenclature

Letters

Symbol	Description
C	concentration ratio
E_b	insolation
I_H	radiation flux in the solar range
N	number of rays
$\dot{Q}$	heat flux
r	radius (esp. of exit aperture of terminal concentrator)
R_E	radius of the entrance aperture of terminal concentrator
S	area
T	absolute temperature
U_L	convective loss coefficient

Greek Symbols

Symbol	Description
α	absorptivity
ϵ	emissivity
γ	intercept factor
η	efficiency
ρ	reflectivity
σ	parameter of Gaussian distribution

Subscripts

Symbol	Description
a	ambient
cos	cosine effect
H	heliostat field or dish
K	tangent cone entrance
max	maximum
obs	obstruction
op	optical
R	receiver
T	terminal concentrator

References

[1] Schöffel, U., Sizmann, R., 1989. Terminal Concentrators. In: Becker, M., Funken, K.-H., Schneider, G. (ed.) *Solar Thermal Utilization, Vol. 4: Final Reports 1988*, Springer-Verlag Heidelberg (1991).

[2] Schöffel, U., Sizmann, R., 1989. Optimization of Terminal Concentrators. In: Becker, M., Funken, K.-H., Schneider, G. (ed.) *Solar Thermal Utilization, Vol. 5: Final Reports 1989*, Springer-Verlag Heidelberg (1991).

[3] David E. Goldberg, *Genetic Algorithms in Search, Optimization, and Machine Learning*. Reading, Mass. (1989).

[4] Welford, W.T., Winston, R., *High Collection Nonimaging Optics*, Academic Press. New York (1989).

[5] Vant-Hull, L., Schöffel, U., *High Flux Density Solar Furnace Design Study for DLR Köln*, unpublished report. München (1991).

[6] O'Gallagher, J., Winston, R., Test of a "Trumpet" Secondary Concentrator with a Paraboloidal Dish Primary, *Solar Energy* **36**, 37 (1986).

[7] Winston, R., Welford, W. T., Geometrical Vector Flux and Some New Nonimaging Concentrators. *J. Opt. Soc. Am.* **69**(4) (1979), p. 532–536.

[8] Winston, R., Welford, W. T., Ideal Flux Concentrators As Shapes That Do Not Disturb the Geometrical Vector Flux Field: A New Derivation of the Compound Parabolic Concentrator. *J. Opt. Soc. Am.* **69**(4) (1979), pp. 536–539.

[9] Winston, R., Nonimaging Optics, *Scientific American* (March 1991).

[10] Hammersley, J.M., Handscomb, D.C., *Monte Carlo Methods*. London (1967).

[11] Gould, H., Tobochnik, J., *An Introduction to Computer Simulation Methods. Applications to Physical Systems. Part 2: Simulations with Random Processes*, Chapter 10, (corrected reprint), Addison-Wesley Publishing Company, Reading, Mass. (1988)

[12] Bendt, P., Rabl, A., *Effect of Circumsolar Radiation on Performance of Focusing Collectors*, Tab. 4-1, p. 21. SERI/TR-34-093 (1980).

[13] Becker, M., Ellgering, H., and Stahl, D., 1983. *Construction Experience Report for the Central Receiver System (CRS) of the International Energy Agency (IEA) Small Solar Power Systems (SSPS) Project*. IEA-SSPS Operating Agent DFVLR, Köln, 1983.

[14] Biggs, F., Vittitoe, C.N., *The Helios Model for the Optical Behavior of Reflecting Solar Concentrators, SAND76-0347* (1979).

[15] Dehalt, O., *Vermessung der Strahlungsflußdichte für Konzentratorsysteme*, Diplomarbeit Universität München, Lehrstuhl Prof. Sizmann. München (1991).

[16] Chang, C.T., Bjorkstam. J.L., Effect of Nonuniform Irradiance, and Irradiance Fluctuations, upon the Response of Photographic Film. *J. Opt. Soc. Am.* **12**, 1495 (1975).

[17] Green, A.E.S., McPeters, R.D., New Analytic Expression of Photographic Characteristic Curves. *Applied Optics* **14**, 271 (1975).

[18] Müller, R., Plataforma Solar de Almeria. Personal communication (November 1990).

Investigation of a Secondary Concentrator Model and of High Temperature Resistant Materials

R. Anton, G. Lensch, W. Rudolph,

N.U. Tech GmbH,
Neumünster

Zusammenfassung

Diese Studie (Vertragsnr. 5-395-4405) wurde im Rahmen des SOTA Programms AP 300 für die DLR durchgeführt.

Basierend auf den Ergebnissen der beiden letztjährigen Projekte [1] wurden weitere Testbeschichtungen zur Erstellung einer temperaturbeständigen Spiegelschicht durchgeführt. Dabei wurde TiN mit einer Schutzschicht gegen Oberflächenoxidation und mit einer Sperrschicht gegenüber dem Substrat zur Unterbindung von Diffusion bei erhöhter Temperatur versehen. Es ließen sich so Temperaturen bis 400°C erreichen, ohne daß es zu Schäden an der Oberfläche kam, die den Reflexionsgrad merkbar gesenkt hätten.

Durch dielektrische Beschichtung der Goldschicht ließ sich diese gegen mechanische Einflüsse schützen, und gleichzeitig konnte der Reflexionsgrad um 500nm herum deutlich erhöht werden. Bei der Temperung dieser Schichten wurde bis 480°C keine Abnahme des Reflexionsgrads gegenüber den Meßwerten bei Raumtemperatur festgestellt.

Ferner wurde ein doppelt linearisiertes Trumpetmodell mit geometrisch 3facher Verstärkung gebaut und in verschiedenen Ebenen mittels des verbesserten Meßaufbaus optisch vermessen. Dabei ergab sich ein Strahlungsdurchfluß durch die Austrittsapertur, der dem Reflexionsgrad des Spiegelmaterials entsprach.

Summary

This study was performed within the SOTA program AP 300 for the DLR under contract no. 5-395-4405.

Based on the results of the last year's studies [1] further deposition tests were accomplished to create a reflector layer, resistant to high temperature. Surface oxidation of the TiN - layer was prevented with a protecting layer, whereas diffusion of substrate material at high temperature was eliminated by a barrier layer. In this way temperatures up to 400°C were applied without surface damage, which would decrease the specular reflection distinctly.

Deposition of dielectric layers on top of the gold layer protected it against mechanical wear, and also increased the specular reflection around 500nm. Annealing of the multilayer system up to 480°C caused no degradation of reflectivity compared with the values at room temperature.

Further, a double linearized Trumpet concentrator model was built with a geometrical concentration factor of three, and different sectional planes were measured with an improved measurement arrangement. At the output aperture we found a radiation flux, which corresponded with the reflectivity of the used mirror material.

Concept of the study

The procedure of testing the materials and models was as follows:

Reflecting layers:

-   Based on last year's results we selected protection and barrier layers against oxidation and diffusion, respectively.
-   The layer systems were investigated in tempering tests. Results were used to gradually improve the layer design consecutively
-   Evaluation and reporting

Concentrators:

-   The measurement arrangement for the models was modified to measure the whole area of the exit aperture, and was automatically controlled.
-   Measurements were done on the last year's CPC - model
-   Construction and measurement of an improved Trumpet model
-   Evaluation and reporting

Contents

1. Introduction

This study has been performed within the SOTA program AP 300 for the DLR under contract no. 5-395-4405, and is a continuation of the last year's investigations to find materials resistant to high temperatures with high reflection and low absorption, which may be used in a secondary concentrator, where great variations in temperature and radiation density occur.

This requires special properties of the material, which can only partly be realized by conventional reflector materials. The last year's study has shown that polished ceramic needs a deposition to rise the specular reflection, suitable multilayer systems were intensively investigated, based on the experience that the reflectivity and temperature stability of unprotected TiN is limited.

The second part of the study describes the newly built concentrator model with a greater geometrical concentration. Both, the new model and the last year's CPC-model were investigated with the changed measurement arrangement, which simulates the incidence of light at the input aperture of the model near to the real incidence. The measurement was controlled by a computer program and the intensity distribution of the whole area of the exit aperture was detected.

2. Investigation of materials for reflecting layers

2.1 Material selection

Our last year's study has shown that the specular reflection of a TiN - layer is quite large but not as high as that of gold. On the other hand the resistance against mechanical wear is much greater than of gold. However, at high temperatures the surface of TiN oxidized, and consequently, the reflectivity decreased. So a protection layer was suggested. A good candidate for this was assumed to be gold.

In the project of this year this possibility was investigated. Further, different barrier layers between the glass substrate and the TiN - layer were tested in order to increase the temperature at which surface damage occurres. Several combinations of metal and ceramic coatings were tested.

The first aim was to build up a simple layer system to minimize production costs. So we tested a single gold layer as well as a gold layer with a thin chromium intermediate layer on glass, which should increase the adhesive strength.

A gold mirror shows a very high reflectivity for the infrared (IR), and the material is chemically very stable, but it is also relatively soft and needs protection against mechanical damage.

Mirrors for the total solar spectrum from 350nm to 2500nm should also exhibit a high reflectivity in the visible part of the spectrum, as the maximum intensity is at around 500nm. There, the reflectivity of gold reaches values between 40 and 50 % (for normal incidence of light) and steeply increases up to 95 % for longer wavelengths (see Fig.3). Coating the gold with dielectrical layers may protect the surface against mechanical wear, while this provides also the possibility of increasing the specular reflection for wavelengths below 500nm by interference, if one selects suitable materials and optical thicknesses of the layers. Fig.11 shows the calculated reflectivity for two and four dielectrical layers on a thick nontransparent gold layer. Depositions were produced according to the theoretical calculation nd tested with regard to the reflectivity and thermal stability.

2.2 Production and tests

All films were produced by magnetron sputtering in a physical vapor deposition (PVD) - apparatus (ALCATEL SCM 650) owned by our company [1]. As substrate materials we used glass (BK7) or polished ceramic discs for tests at temperatures above 500°C. Because of the porosity of the ceramic material reflectivities of coatings were generally by about 10% lower than on glass (see Fig.3).

We measured the specular reflection at the gonioreflectometer [1] at an angle of incidence of 15°. The high temperature behaviour was investigated by annealing the samples at constant temperatures for 5 hours at each step, increasing from room temperature to 450°C. After each annealing step, the spectral distribution of the reflectivity was recorded. One sample was also tested at 300°C and a duration of 300 hours.

Several ceramic samples like (Al_2O_3, Al_2TiO_5, hard porcelain) of last year's study were annealed at up to 1000°C.

2.3 Results

The annealing experiments showed that the reflectivity of gold or gold plus chromium on glass degrades at temperatures above 250°C. Already after the first annealing step at 280°C, the degradation of reflectivity appeared for the gold layer (see Fig.1). Fig.2 shows the degradation of the reflectivity, when a thin Cr layer was used to increase the adhesive strength between the gold film and the substrate. The degradation started at lower temperature, and at a temperature of 400°C the reflectivity was decreased to only a few per cent in the visible part of the spectrum, while the reflectivity of the single gold layer was still acceptable.

The results for the double layer system TiN / Au are presented in Fig.4 for a glass substrate and in Fig.5 for steatite ceramic. On glass, the degradation began below 380°C and the TiN reacted with the substrate. Using the steatite substrate the reflectivity did not change at this temperature. But at 480°C the reflectivity decreased partly by about 10%. Further increase of the temperature changed the layer structure with the result that the surface became ocher colored and the reflectivity decreased rapidly. At 1000°C the reflectivity reached the values of polished steatite (nearly 5%), as if there was no coating at all.

The experiences of these tests first suggested the use of a barrier layer between the substrate and TiN layer. Figs.6 to 9 show the results of the annealing tests, starting at a lower temperature than before. The degradation of the reflectivity by increasing the temperature was reduced compared with the case without a barrier layer. The best results were obtained with a Ti - layer at temperatures up to 400°C (Fig.7). Beyond this temperature, the reflectivity decreased quickly. Using an Al_2O_3 - AlN double layer or a TiO_2 - layer as a barrier an extension of the stability regime up to 440°C resulted with only a small degradation of reflectivity. At 480°C, however the degradation became stronger, especially for the Al_2O_3 - AlN layer. At higher temperatures, Al and Ti alloyed, while at the surface, the gold film desintegrated into large crystals. Fig.10 shows a SEM - photography of the annealed sample.

In contrast to these results at higher temperatures, annealing at 300°C for 300 hours led to only a small degradation of reflectivity at the visible part of the spectrum, but there were no changes at the IR - region (Fig.9).

Fig.12 to Fig.15 show the results of the annealing tests of the layer systems with the dielectrical films. One can see the increase of reflectivity around 600nm, but the absolute values are smaller than the theoretical values. Depending on the type of barrier layer the degradation of reflectivity began at different temperatures.

With a TiO_2 - layer, temperatures up to 480°C can be applied without degradation of reflectivity. Up to 440°C even a slight increase of reflectivity was observed after annealing especially in the visible and near IR (see Fig.12). At 520°C the whole layer system stripped off the glass substrate because of stress. With Ti as a barrier layer degradation started at a temperature of 300°C (Fig.13). Fig.14 shows the result for a similar layer system on a steatite substrate. The strong degradation of reflectivity at 350°C is conspicuous. At this temperature most of the dielectric coating peeled off.

As the length of the optical path varies with the angle of incidence, the reflectivity depends on angle. Fig.15 shows the reflectivity as a function of wavelength for several discrete angles of incidence between 15° and 70°. When the angle increases, the minimum in the reflectivity shifts to shorter wavelengths. At an angle of incidence of 70°, the degradation of reflectivity at the minimum is significant, so that the reflecting white light becomes green. This effect must be taken into account when practical applications are considered. In the case of not normal incidence, for which the system was designed, the thicknesses of the dielectrical layers must be changed corresponding to the actual angle of incidence.

The ceramic samples Al_2O_3, Al_2TiO_5 and hard porcelain were annealed at more than 1000°C, in order to obtain a definite reaction of the surface. The color of the Al_2O_3 and Al_2TiO_5 samples became lighter, but was not steady. The color of the sample of hard porcelain remained white. At 1000°C, the glaze plasticized which excludes the use at such high temperatures. Perhaps, there are other glazes with higher fusing temperature. To get a smooth surface, the glazing of the surface is superior to expensive polishing of the ceramic surface. In general, depositions on hard porcelain appear glossy with a wavy surface. As a conclusion it appears to be possible to produce mirrors to high temperature resistant with relatively high efficiency at reasonable costs. Best results are obtained for smooth substrates like glass. For substrates of ceramic material, the reflectivity is generally by about 10% lower.

3. Model for secondary concentrator

3.1 New Trumpet model

Our previous models were quite big. With a relatively low geometrical concentration of two, only a small part of the incident light impinging near the edge of the input aperture was reflected at the mirror. This year, a smaller model with a greater geometrical concentration was built. The special linearization, as used before, was kept [6].

For linearizing the hyperbola a curve with a concentration of four was selected, resulting in a relatively small number of segments, which facilitates production. The first segment, standing vertically on the plane of the exit aperture, and being a cause of multiple reflections, was omitted. This was simulated in the calculations by choosing a great length for this section, so that ommission would actually shorten the whole length of the model. The lengths of the other segments increase with increasing distance to the focal plane (= exit aperture), as the hyperbola approximates to a straight line at large distances to the focal point. This also facilitates production.

The Trumpet concentrator model was built with dimensions as follows:

input aperture	50.0 mm
exit aperture	28.8 mm
max. inner diameter	80.0 mm
total length	60.3 mm
number of hyperbola segments	6

The circle was devided into twelve linear segments, which appears to be sufficient for this size. The coated substrates were cemented at circular carriers, which were connected by vertical pins. The completed model is shown in Fig.21.

A model with ceramical mirrors was desired, as these are expected to be thermally stable. Because the surface roughness of the real secondary concentrator had to be scaled down for the model, a very fine polishing of the surface was necessary, which resulted in high costs for the purchase of the material. As an example, we bought polished ceramical discs with a diameter of 25 mm, tested last year, for DM 250 a piece. To build a model many plates are necessary, which are larger than actually needed, in order to be able to cut the edges off by

LASER machining. After that, the edges must be finished to connect the pieces without any gap. This requires a special fixture to save the polished surface. Even so, part of the reflectivity is lost by scattering effects.

Taking into account all these aspects, we selected glass as substrate material as a compromise. Coating was done after cutting and polishing the edges. A construction with glazed ceramical thin film substrates would result in more intensity losses because of the waviness and roughness of the available substrates.

Because the high temperature mirrors were not finished at the time of production of the model, we used aluminium instead. Thus, direct comparison with last year's models was possible, as this was also made from Al.

3.2 Test device and measurement

The measurement arrangement of the models were changed in that way that the incident light was not focused into a point as before, but the light formed a cone with minimal diameter at the plane of the input aperture, and which expanded beyond that plane. We simulated an extended light source in order to optimize this behaviour by shining a HeNe LASER beam onto a scattering disc. The scattered light was collimated with a Fresnel lens to the input aperture of the model with the ratio 1 : 1. Fig.17 shows a scheme of the arrangement, Fig.18 is a photograph of the general view, and Fig.19 shows the angular distribution of the transmitted light, measured at a great distance behind the aperture. Large scattering angles occur relatively often.

The measurements were automated. After the exchange of the manual displacement unit by stepper motor drive, an area of 50 mm x 50 mm or a circle of diameter 50 mm could be scanned, controlled by a computer program. The test data were saved in the computer memory to enable the evaluation at a later time.

Level diagrams of the measured area or of the concentration relatively to the incident radiation, as well as several cross sectional planes of these diagrams could be plotted on the screen. Further, the integral of radiation flux was calculated for the measured area, but only for the new Trumpet model, because the diameter of its reference was 50 mm and detected completely. For the CPC - model the diameter of the reference was 60 mm, so that the edge of the area could not be covered.

The measurements were done at different distances behind the exit aperture, and, for the Trumpet model, also before the aperture by taking off the last circular carrier (see Fig.20).

3.3 Results of measurement

From the test data we calculated the light distribution in the different levels and cross sections for the reference and the planes parallel to the exit aperture. Figs.22 to 27 show the results for the CPC - model. Fig.22 and Fig.23 show the diagrams of the reference with diameter 60mm, Figs.24 and 25 the diagrams of the model at the exit aperture. The cross sections look similar to the curves of last year's measurements. Just above the mirror planes the intensity shows maxima, but at the center the intensity decreases to a local minimum. One can distinguish six local maxima at the level diagram near the linear mirror segments. For a concentrator with an ideal form, a circularly shaped maximum near the edge of the aperture area should be expected.

Behind the exit aperture the intensity distribution expands. The light reflected at the mirrors near the exit aperture crosses the optical axis at a short distance behind the exit aperture. So the intensity at the optical axis first increased and then decreased for increasing distance from the aperture. Figs.26 and 27 show the level diagram and cross sections for a distance of 10 mm behind the aperture. Now, the maximum of intensity is at the center of the area, and the form of the cross section looks approximately like the reference curves, but they are slimmer than these with a local concentration of 1.3 at the center. Near to the aperture one finds a plane with a central area exhibiting a nearly constant intensity distribution.

Figs.28 to 36 show the diagrams for our new Trumpet model. We selected the reference with a diameter of 40 mm corresponding to the results of the model. The reference with 50 mm diameter was selected to calculate the radiation flux. We chose the results for the planes at 13.2 mm, 0 mm and -10 mm (see the scheme in Fig.20). The intensity distribution is similar to the reference with a maximum of intensity at the center. The cross sections of the measurement at the exit aperture (0 mm) show the strongest increase at the edge of the measured area (Fig.30), while the reference and the measurement at 10 mm behind the aperture show Gaussian like distributions. Figs.33 and 34 present the curves of concentration relatively to the reference. The concentration factor reaches values from 1.5 to 1.7, decreasing to one side as a consequence of not optimum adjustment.

The curves for the plane at 13.2 mm, measured after removing two circular carriers near the exit aperture, diverge from the curves for the other planes in that way that the intensity is relatively high near the edge of the measured area, recognizable from the curves of concentration (see Figs.35 and 36). Near the edge the concentration reaches 1.4, while it decreases to 1.3 at the center of the area. One finds twelve local maxima of intensity at the level diagram near to the edges.

In addition to these diagrams, we integrated the radiation flux for all measured planes of the Trumpet model. The values are listed in Fig.37. At the exit aperture we got a radiation efficiency of nearly 0.899. This corresponds to the degree of reflection for the aluminum coating at a wavelength of 633 nm (vertical incidence). The efficiency becomes greater in the inner part of the model because of the larger area which allows that more of the radiation is passing the area directly without any reflection. At a location far behind the model we could not detect the whole radiation distribution, as the diameter of the measurement area was only 40 mm. So the measured efficiency became smaller, although the theoretical value must be constant.

4. Conclusion

The tested multilayer systems with barrier and protection layers turned out to be stable up to a temperature of 400°C without a significant degradation of reflectivity. Gold layers alone on a glass substrate are not practicable so far, especially with a chromium interlayer, which was thought to increase the adhesive strength. Also, layer systems with additions of aluminum and titanium are critical at high temperatures.

Dielectrical layers on top of the gold layer provide protection against mechanical wear and, simultaneously, the reflectivity was increased, especially with regard to the visible part of the spectrum. With suitable selection of the barrier layer and of the coating parameters, temperatures up to 480°C can be reached without degradation of the reflectivity. Only at higher temperatures the layer system was damaged because of internal stresses.

In order to obtain stability at even higher temperatures than 480°C, further investigations are needed to optimize the reflection of materials. Moreover, the thicknesses of the dielectrical films should be adjusted corresponding to the angular distribution of the reflectivity.

With our new measurement arrangement for the concentrator models, the whole exit aperture was measured. Both models showed the expected distribution similar to the curves measured last year with maximum intensity at the edge, and low intensity at the center of the area for the CPC - model, but with maximum intensity at the center of the exit aperture for the Trumpet. For the Trumpet model, an efficiency of nearly 0.9 was calculated at the exit aperture, corresponding to the reflectivity of the used material. The shortening of the model and the special linearization contributed to this result, because the proportion of multiple reflections was reduced thereby.

As a consequence of scaling down the size of the model, the roughness of the reflecting surfaces must be very small. With glass as substrate material we got an optimum reflectivity with minimal scattering losses. In reality, the surface roughness can be samewhat larger, dependent on the size of the concentrator, to obtain similar results for the efficiency. Possibly, ceramic material with a surface glaze is advantageous, provided the fusing temperature is high enough.

Further investigations should be done on the high temperature stability of coated mirrors. Especially, a long duration test under real conditions is necessary, perhaps with a parabolic dish, and cooling to limit the temperature rise of the device to below 400°C as is necessary at present.

5. Nomenclature

I   intensity

c   concentration factor

x,y   length parameter

$\propto$ angle of incidence

λ wavelength

ς degree of reflection

Φ flux

6. References

[1] Lensch,G.   Investigation and Selection of Materials Resistant to
Lippert,P.   Temperatures and Radiation to Construct a Metallic/
Rudolph,W.   Ceramic Secondary Concentrator as well as Measure-
ments at Premodels
and
Anton,R.   Investigation of Hard Coating and Heat Mirrors for
Lensch,G.   Simultaneous Energy Conservation in a Photovoltaic/
Rudolph,W.   Solarthermic Hybrid System or for Use in a Secondary
Ueth,R.   Reflector
In: Solar Thermal Energy Utilization: Vol.5, (1989),
Springer Verlag Berlin Heidelberg New York

[2] Bock,J.P.   Temperaturwechselbeanspruchung keram. Werkstoffe
durch Luftabschreckung
Ber. Dtsch. Keram. Ges., 51 (1974), pp.252-255

[3] Karlsson,B.   Optical Properties of Transparent Heat Mirrors Based on
Ribbing,C.G.   Thin Films of TiN, ZrN and HfN
SPIE , Vol.324 (1982), pp.52-57

[4] Rabl,A.   Optical and Thermal Properties of CPC`s
Solar Energy, 18 (1976), GB, pp.497-511

[5] Winston,R.   New Concentr. for the Generation of Very High
O`Gallagher,J.   Temperatures from Solar Energy
Univers. of Chicago, Final Techn. Report, Grand No.
DE-FG 02-79ET-00089 US-Depart. of Energy

[6] Rudolph,W.   Materialien und Modelle für Sekundärkonzentratoren in
Solarenergieanlagen
Diplomarbeit FH - Lübeck (Phys. Technik) 1989

7. Supplement

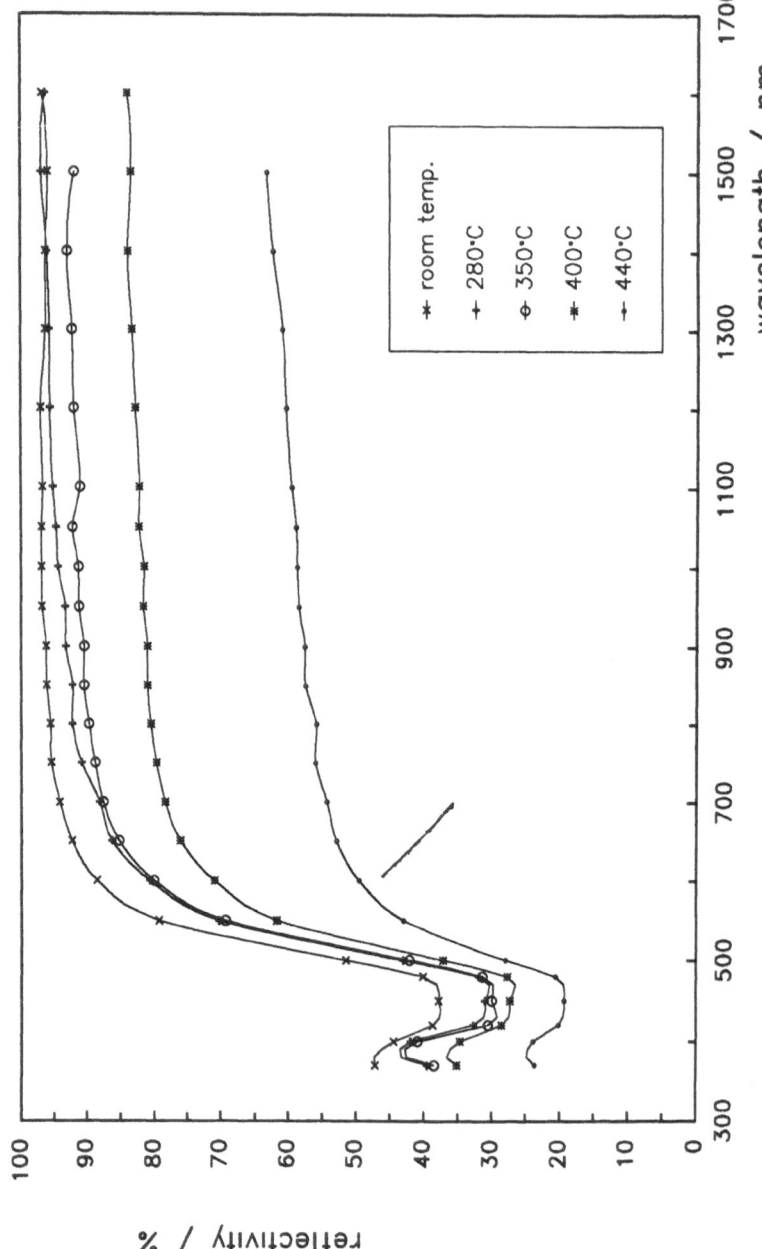

Fig.1   Reflectivity of a RF– sputtered Au– film (80nm) on glass measured at room temperature and after annealing at higher temperatures (5 hours each temp.)

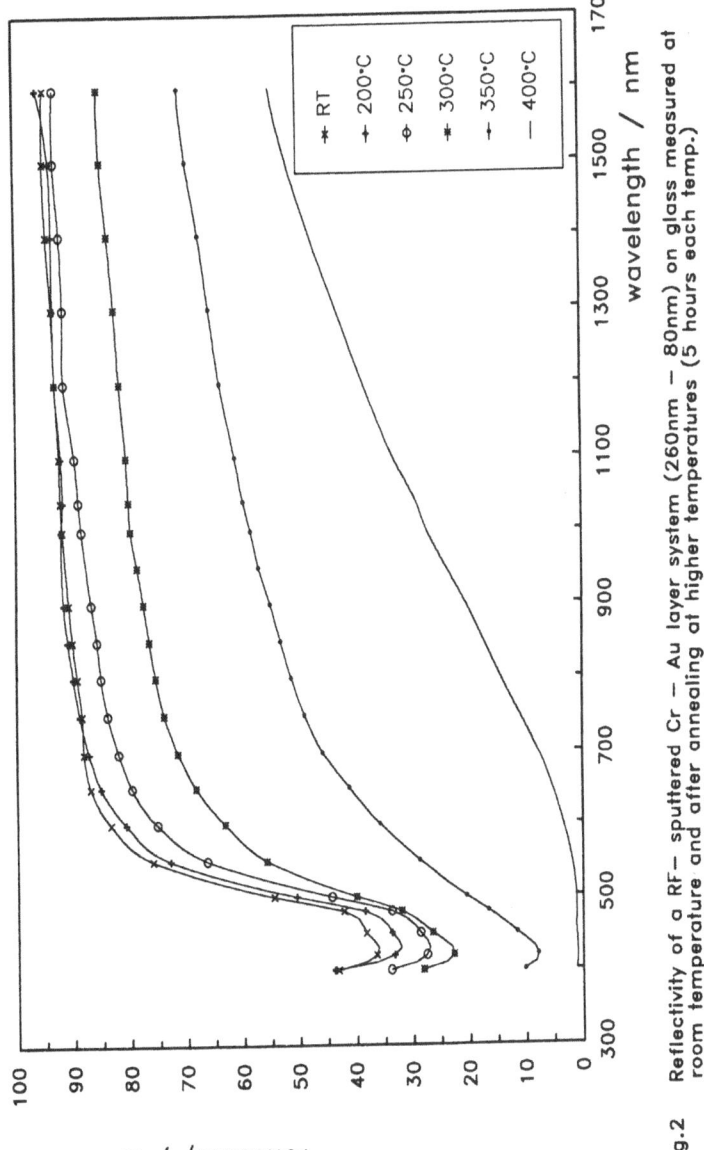

Fig.2   Reflectivity of a RF— sputtered Cr — Au layer system (260nm — 80nm) on glass measured at
room temperature and after annealing at higher temperatures (5 hours each temp.)

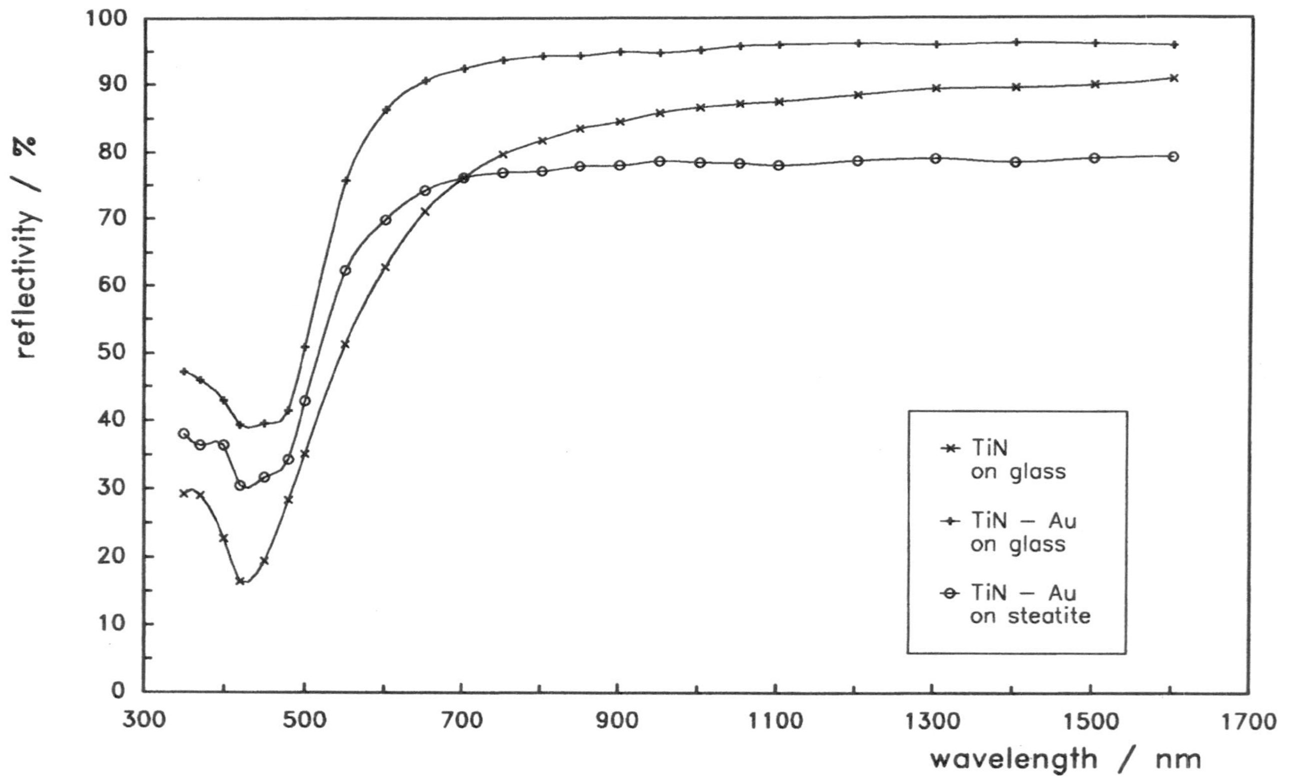

Fig.3 Reflectivity of a RF- sputtered TiN film (500nm) and a TiN (500nm) – Au (80nm) layer system on glass or steatite substrate

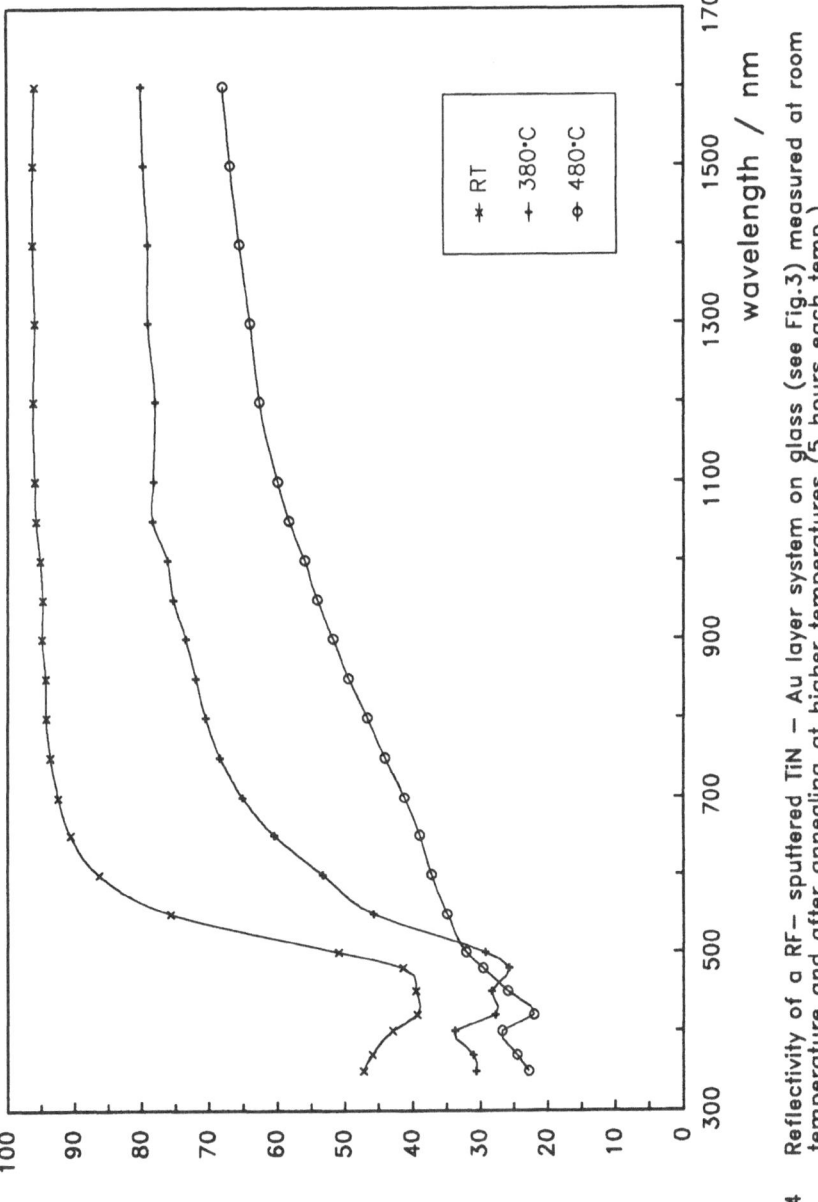

Fig.4   Reflectivity of a RF– sputtered TiN – Au layer system on glass (see Fig.3) measured at room temperature and after annealing at higher temperatures (5 hours each temp.)

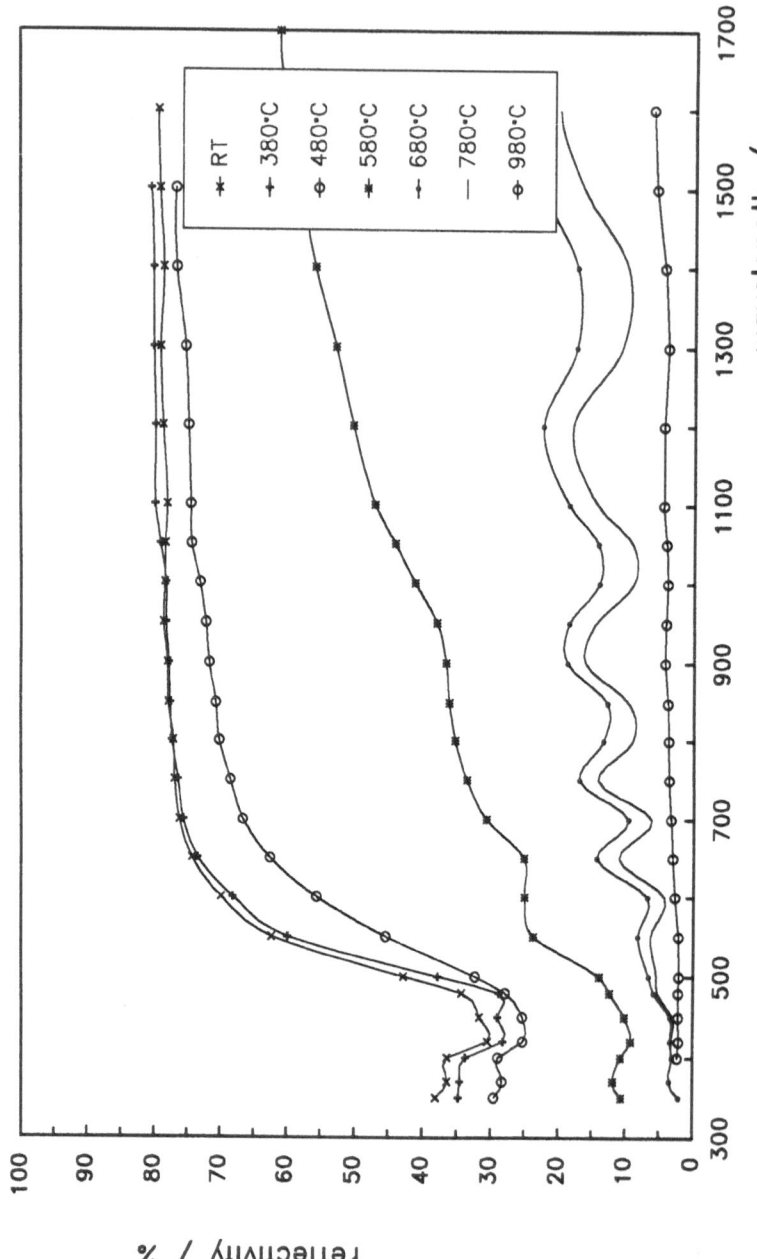

Fig.5 Reflectivity of a RF- sputtered TiN − Au layer system on steatite (see Fig.3) measured at room temperature and after annealing at higher temperatures (5 hours each temp.)

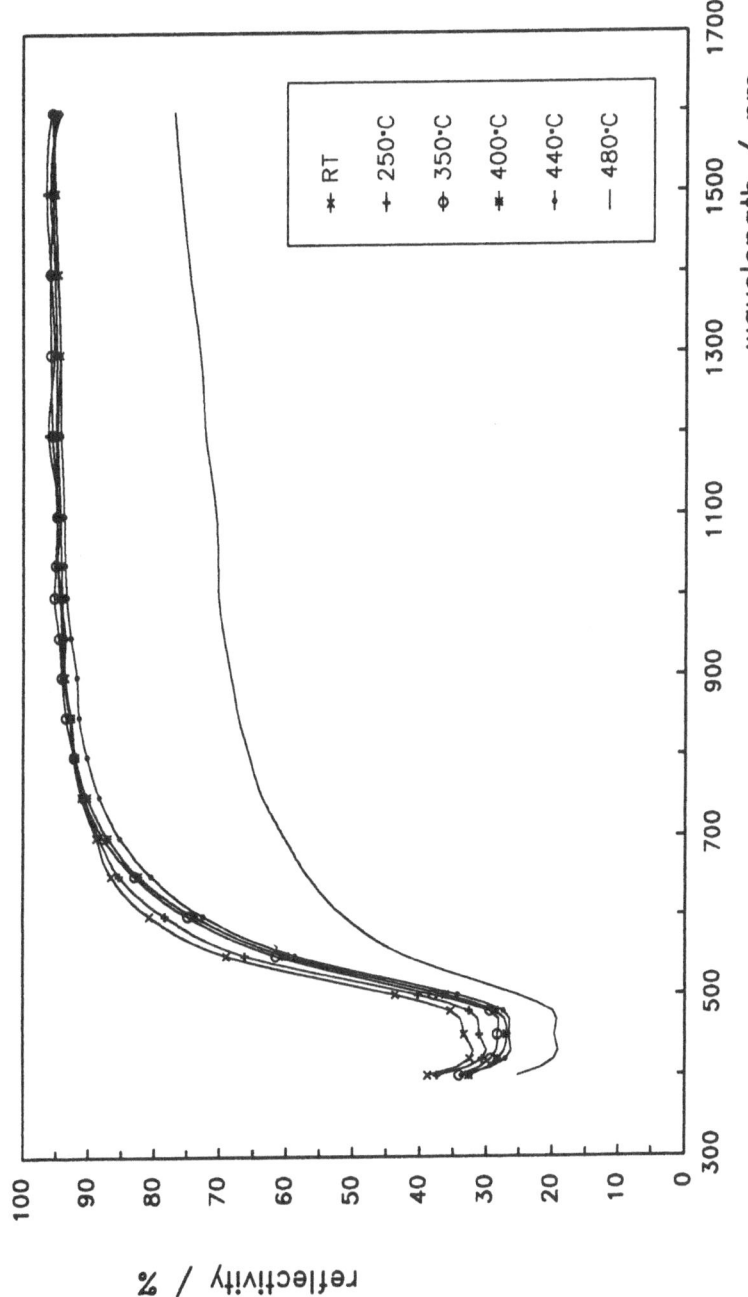

Fig.6 Same as Fig.4 but with a TiO2 barrier layer (250nm)

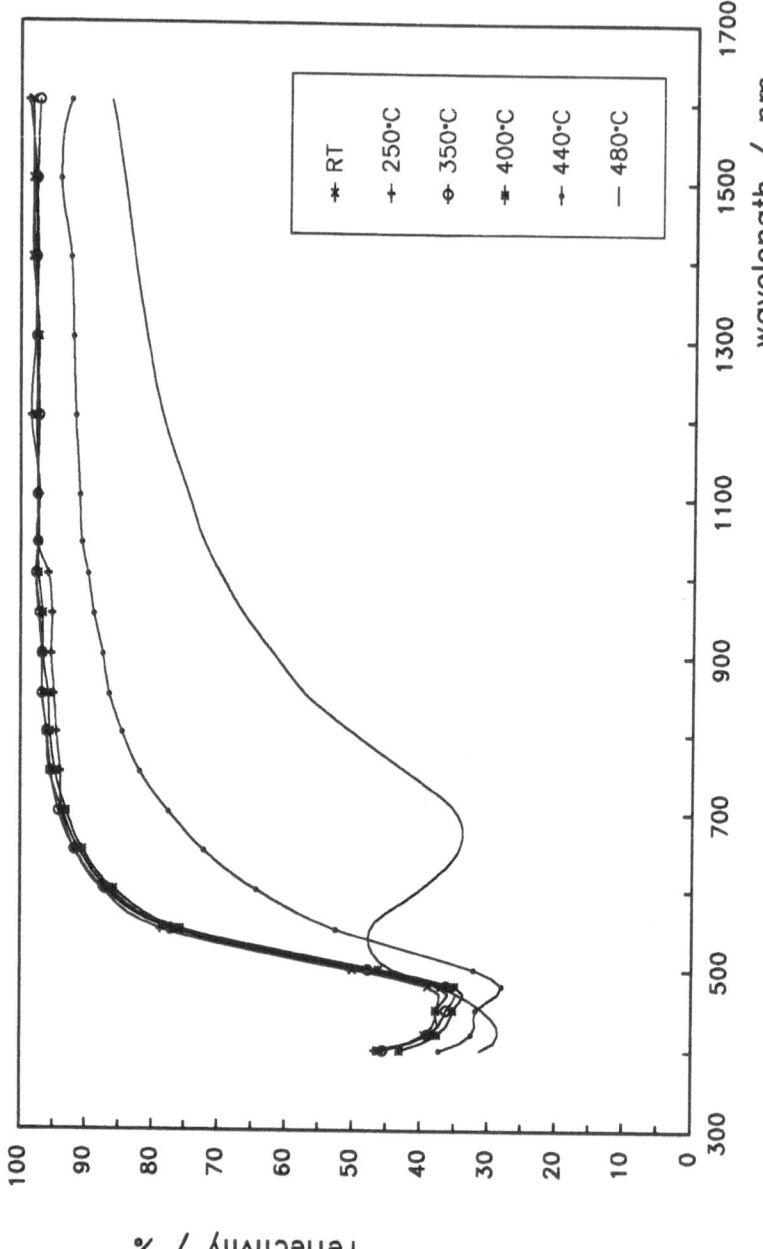

Fig.7 Same as Fig.4 but with a Ti barrier layer (60nm)

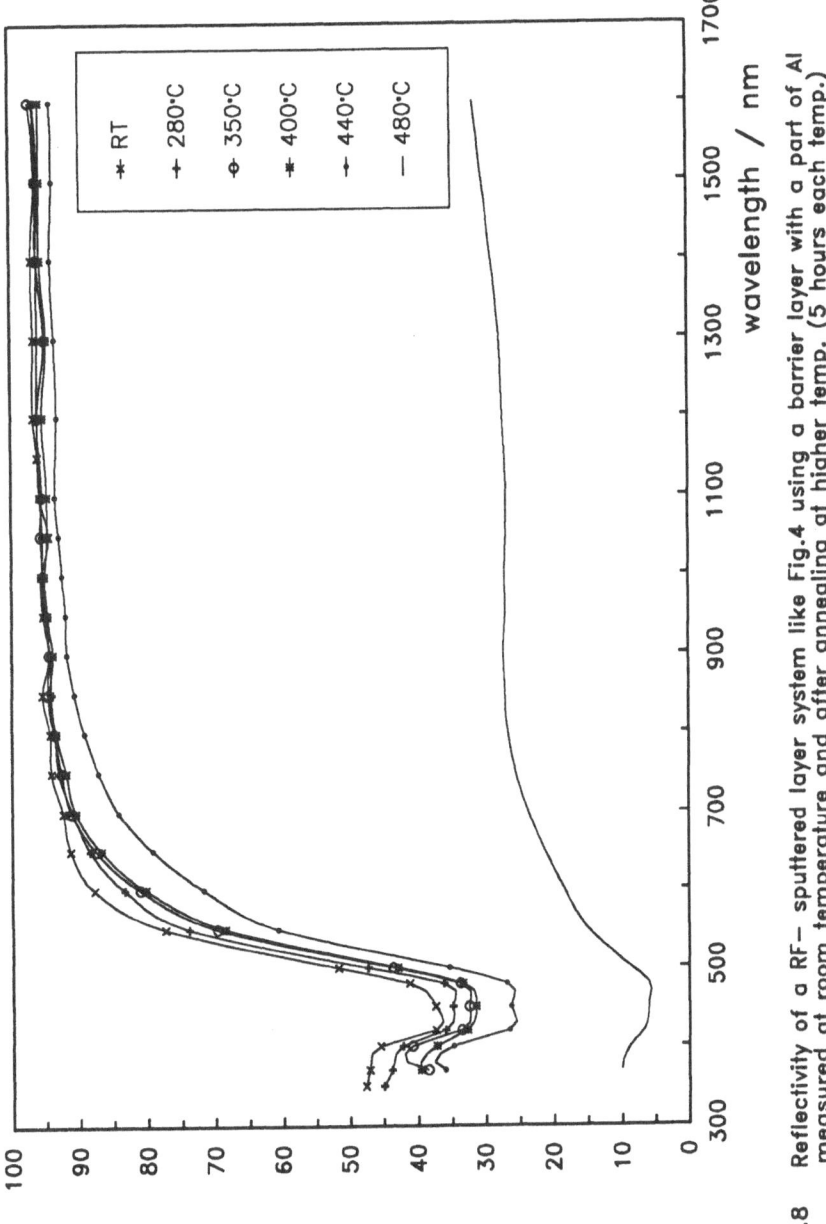

Fig.8   Reflectivity of a RF- sputtered layer system like Fig.4 using a barrier layer with a part of Al measured at room temperature and after annealing at higher temp. (5 hours each temp.)

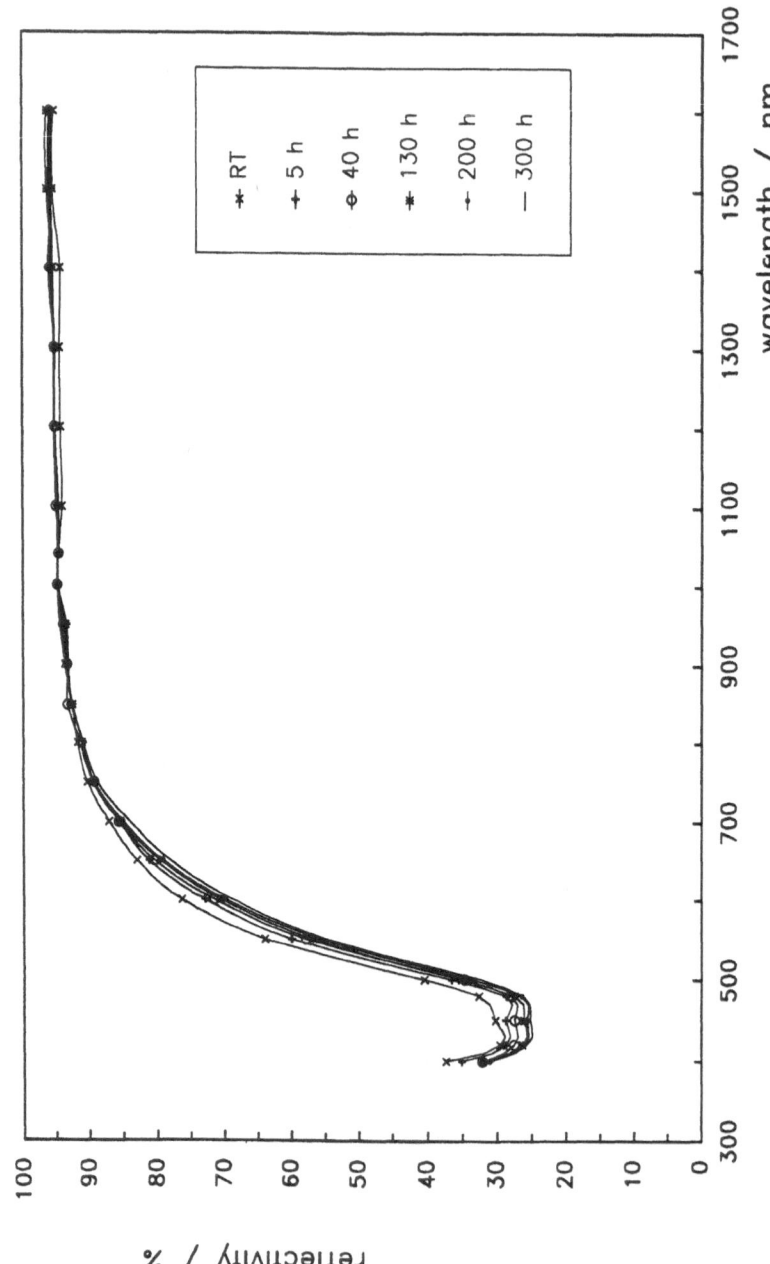

Fig.9 Long time test with the layer system also used at Fig.6
Temperature 300°C

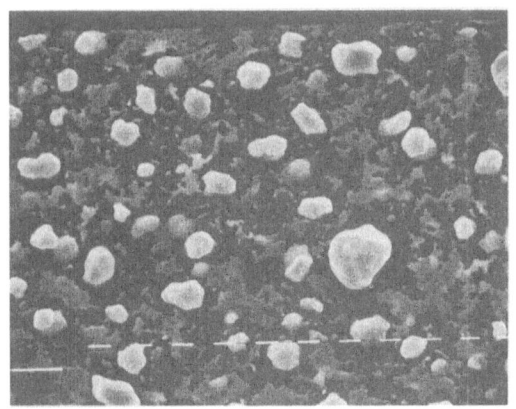

Fig.10   SEM – photography of the sample of Fig.8 after
         annealing at 480°C
         one graduation mark corresponds to one μm

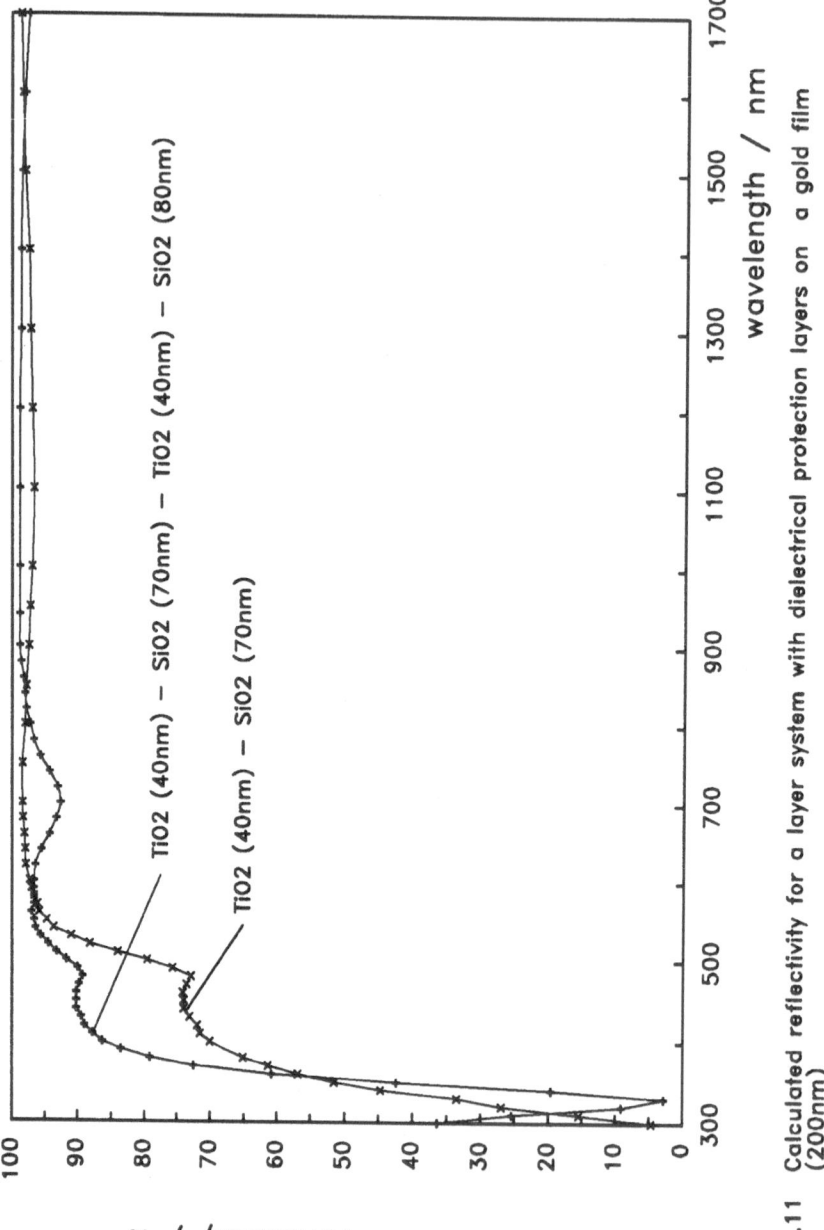

Fig.11 Calculated reflectivity for a layer system with dielectrical protection layers on a gold film (200nm)

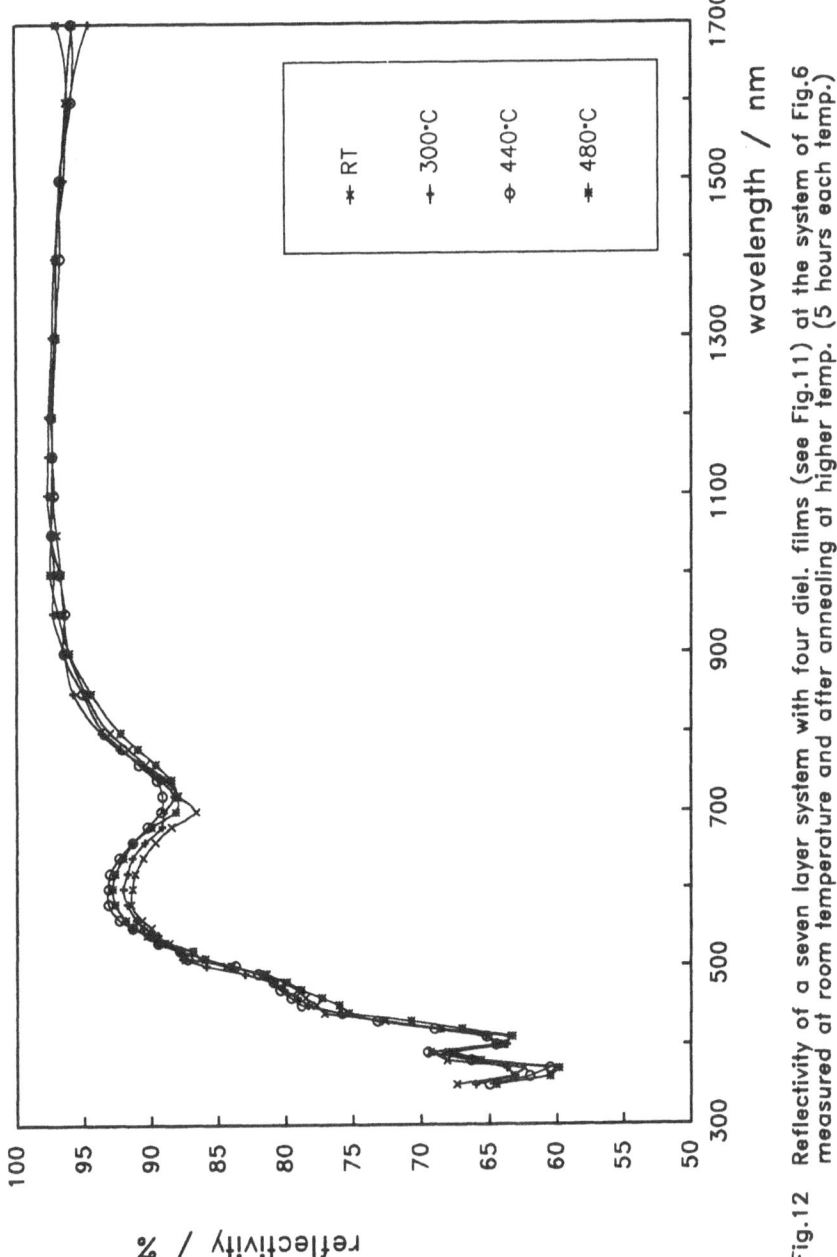

Fig.12 Reflectivity of a seven layer system with four diel. films (see Fig.11) at the system of Fig.6 measured at room temperature and after annealing at higher temp. (5 hours each temp.)

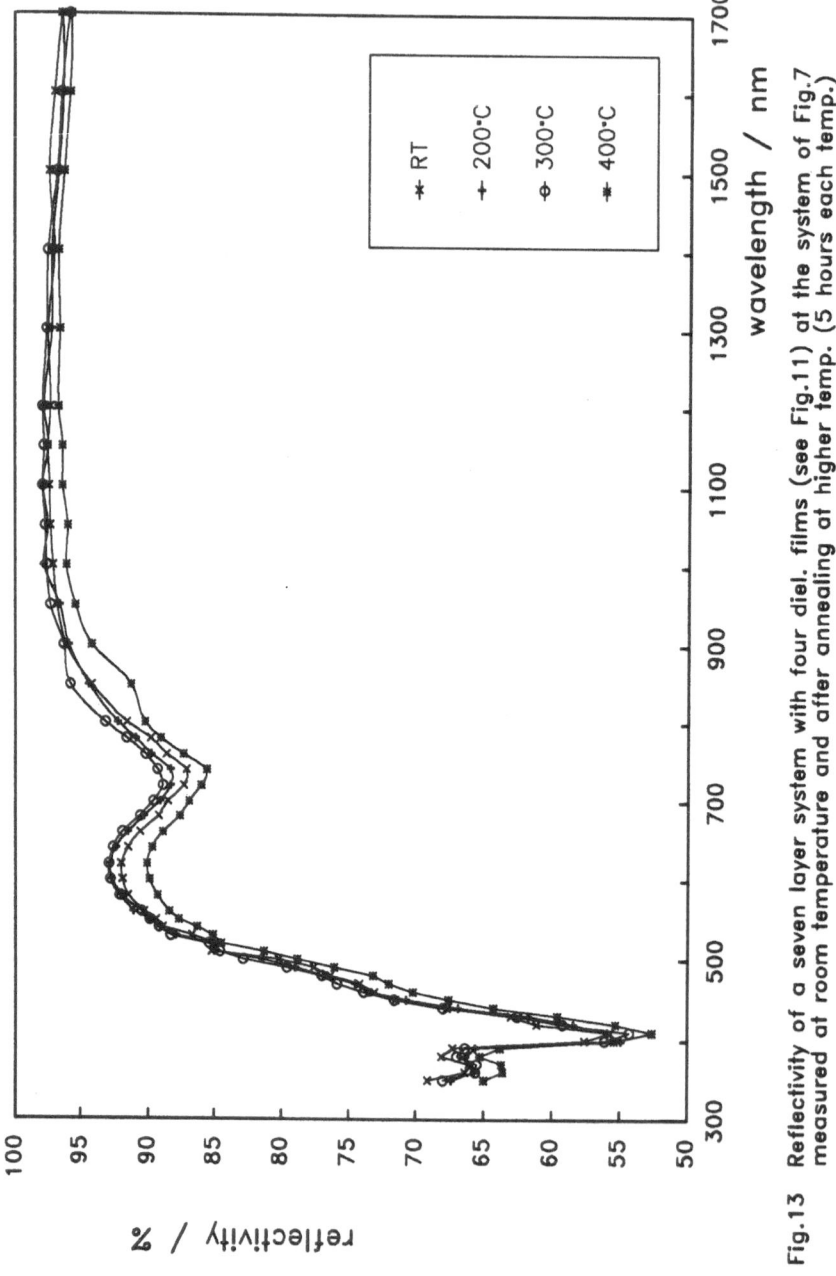

Fig.13 Reflectivity of a seven layer system with four diel. films (see Fig.11) at the system of Fig.7 measured at room temperature and after annealing at higher temp. (5 hours each temp.)

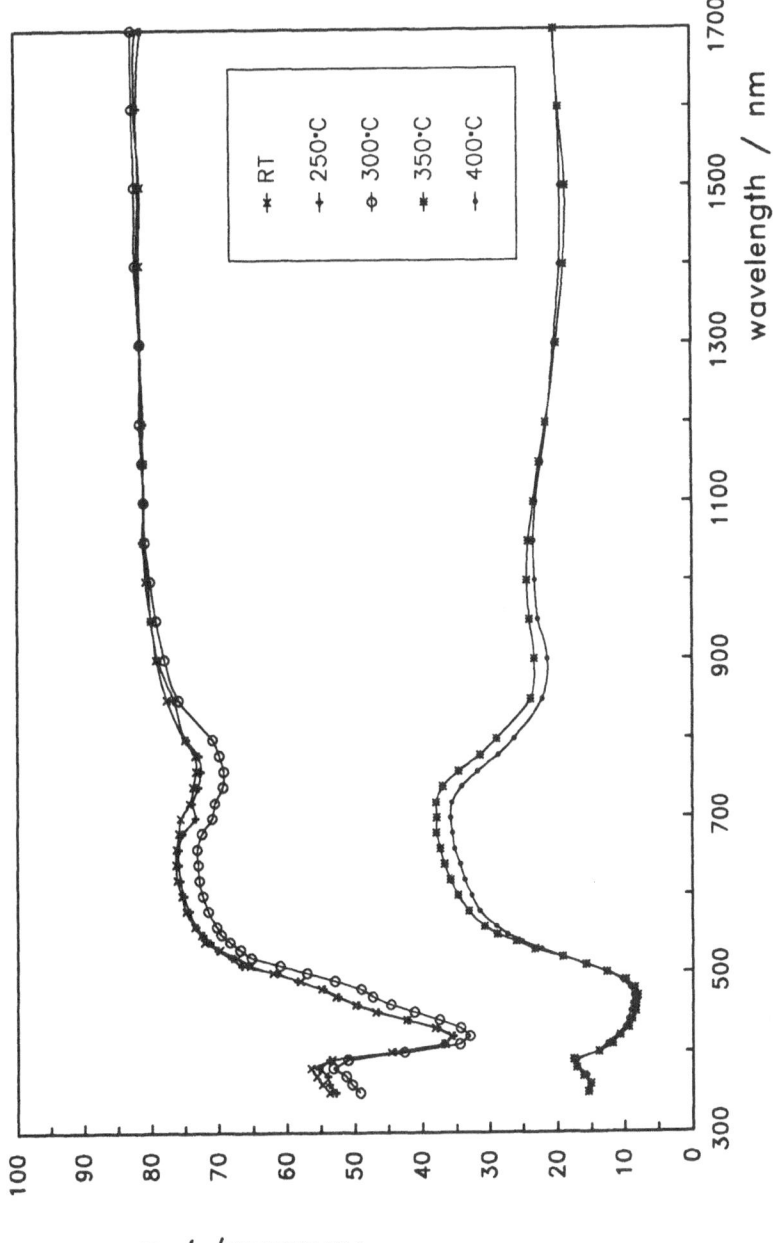

Fig.14 Same as Fig.13 but on a steatite substrate

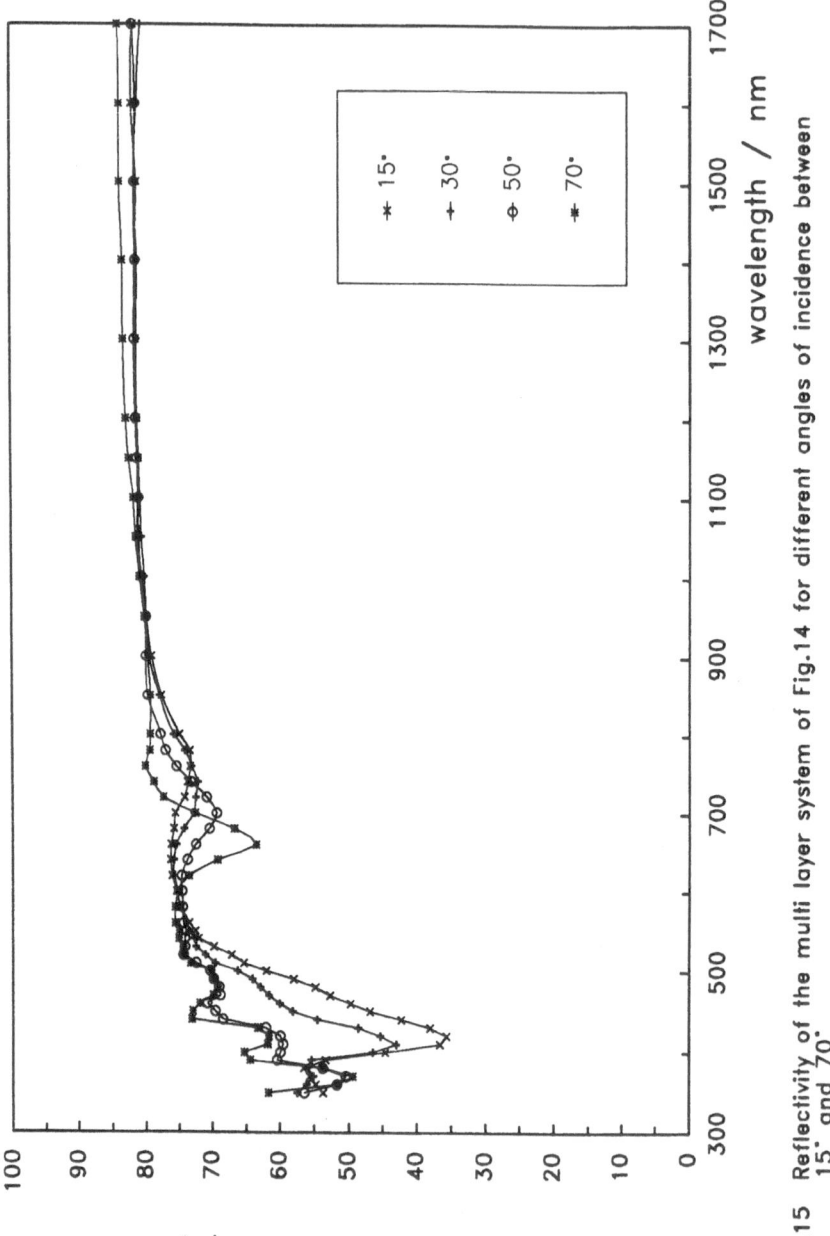

Fig.15 Reflectivity of the multi layer system of Fig.14 for different angles of incidence between 15˚ and 70˚

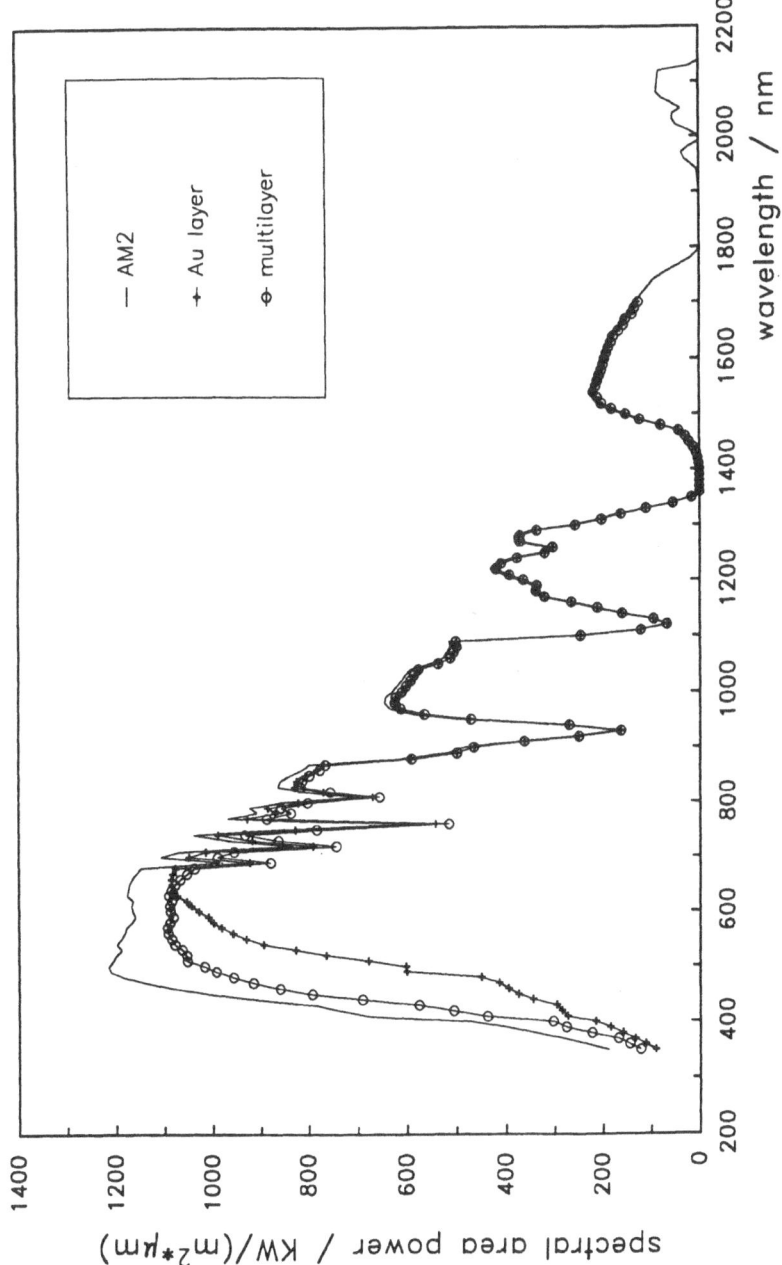

Fig.16 Spectral area power of solar spectrum AM2, reflected by a gold layer (see Fig.6) and by the best multilayer system (see Fig.12)

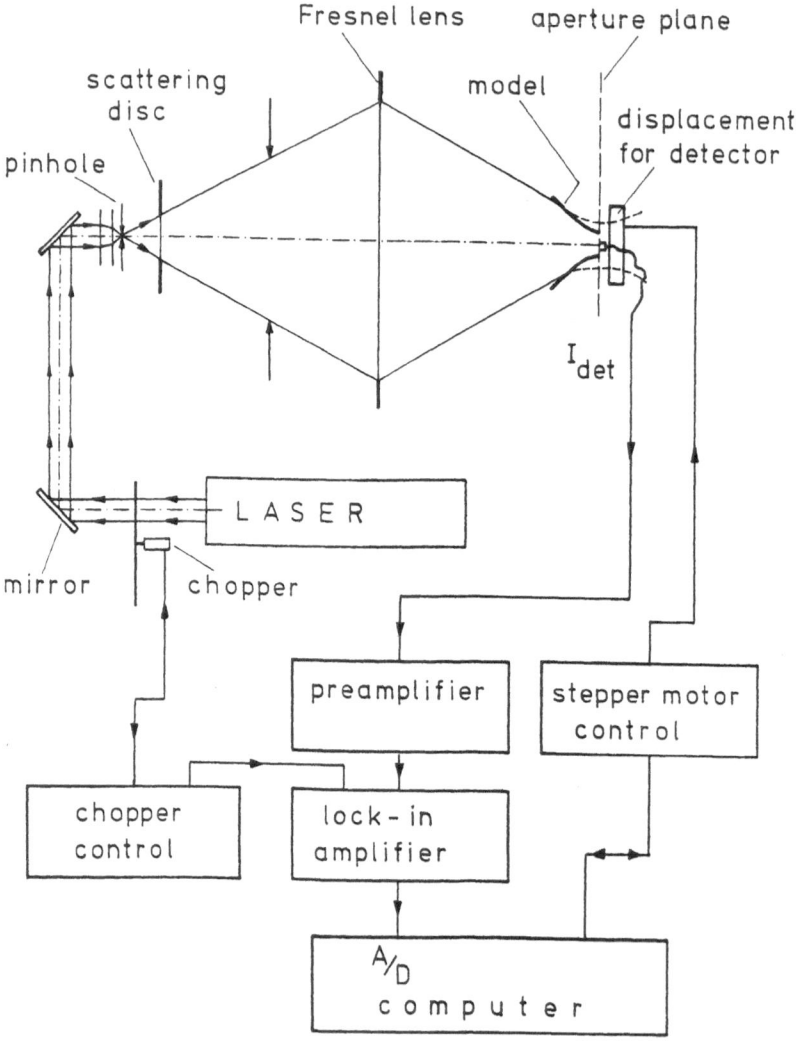

Fig.17   scheme of the measurement arrangement for
          the models

Fig.18 Photography of the measurement arrangement for the models

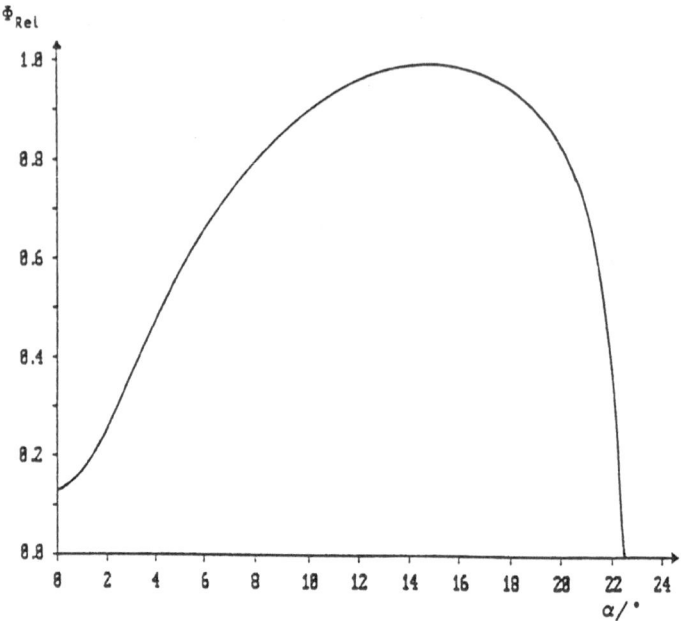

Fig.19   Distribution of angle of incidence, passing the input
           aperture

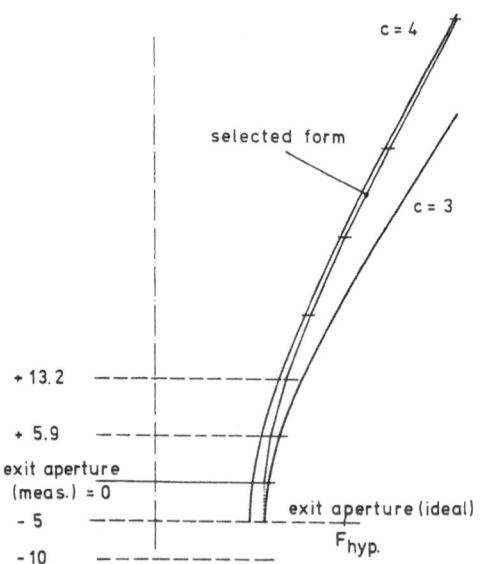

Fig.20   Scheme of the new Trumpet model with the position
           of the measuring planes

Fig.21 Photography of the new Trumpet model. View into the model

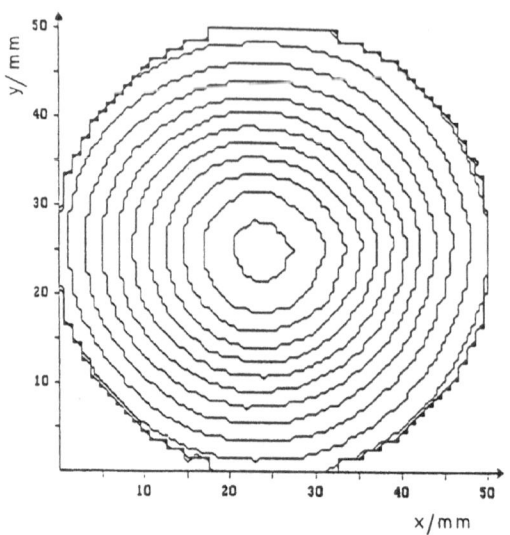

Fig.22 Level diagram of the reference with diameter 60mm
(CPC — model)

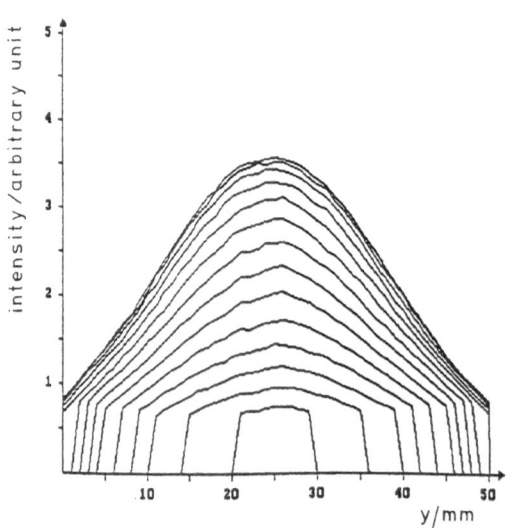

Fig.23 Cross sectional curves of the reference, corresponding
to Fig.22. 2mm distance between the curves

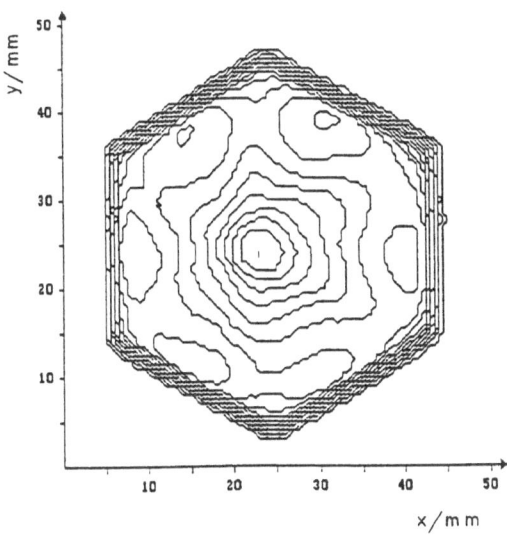

Fig.24 Level diagram of the CPC — model, measured at the
exit aperture of the model

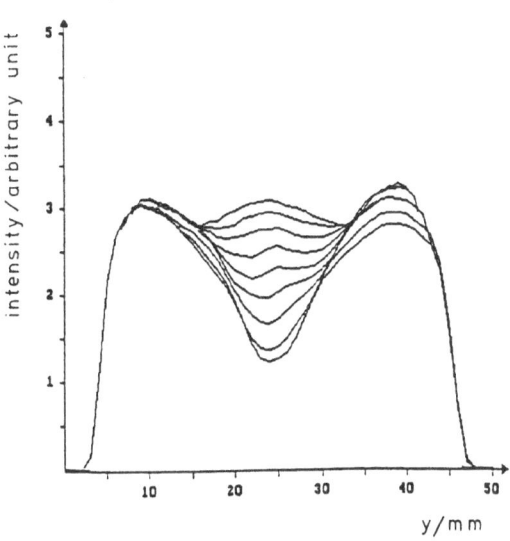

Fig.25 Cross sectional curves of the CPC — model corresponding
to Fig.24. 2mm distance between the curves

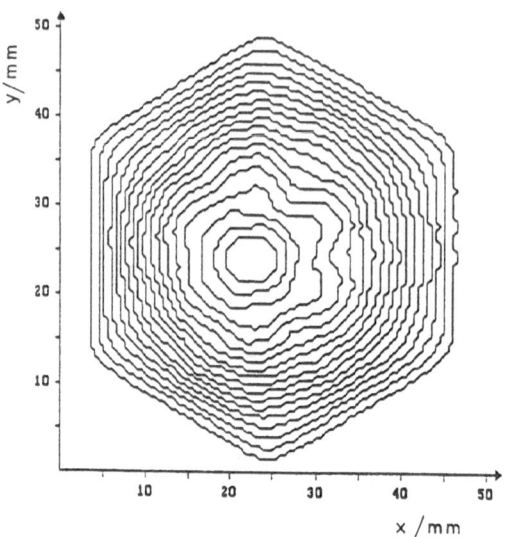

Fig.26   Level diagram of the CPC – model, measured 10mm
behind the exit aperture of the model

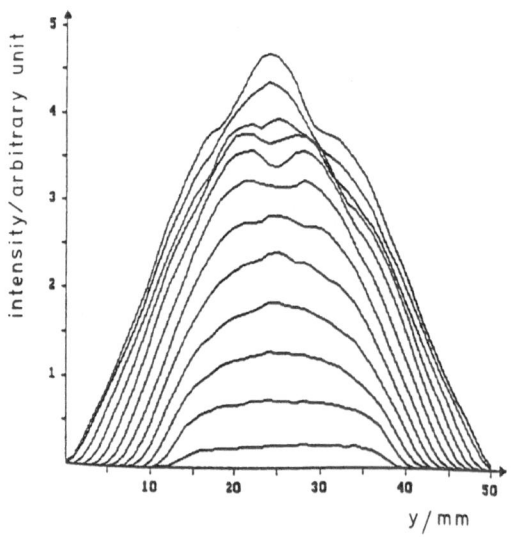

Fig.27   Cross sectional curves of the CPC – model corresponding
to Fig.26. 2mm distance between the curves

-186-

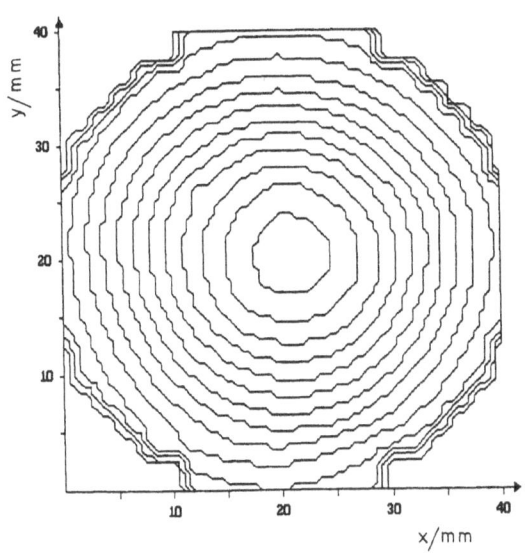

Fig.28 Level diagram of the reference with diameter 50mm
(Trumpet model)

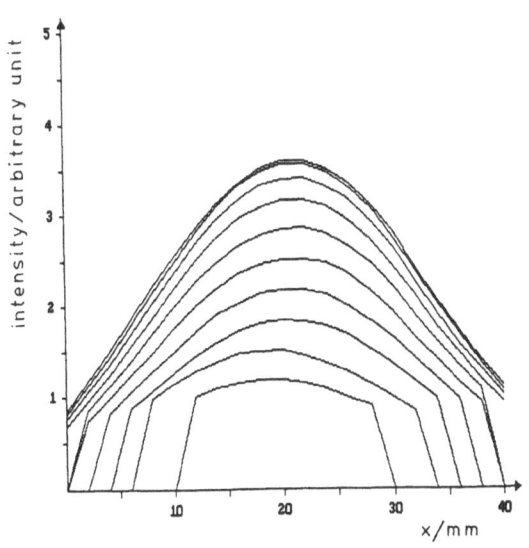

Fig.29 Cross sectional curves of the reference corresponding
to Fig.28. 2mm distance between the curves

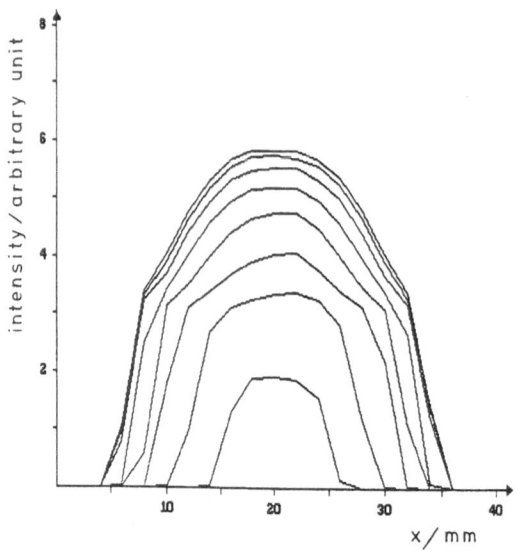

Fig.30   Cross sectional curves for the Trumpet model, measured
at the exit aperture of the model. 2mm distance

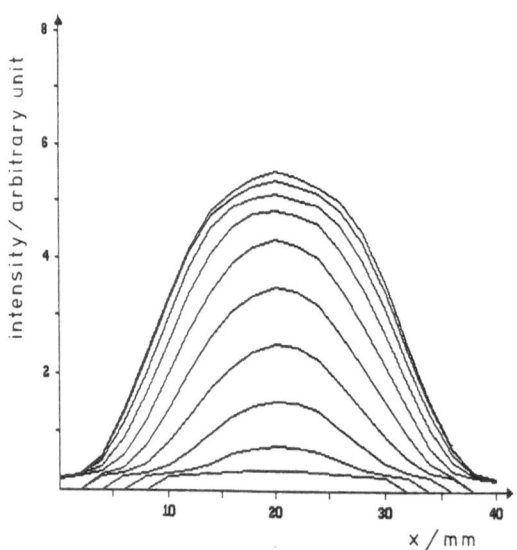

Fig.31   Cross sectional curves of the Trumpet model, measured
10 mm behind the exit aperture. 2mm distance

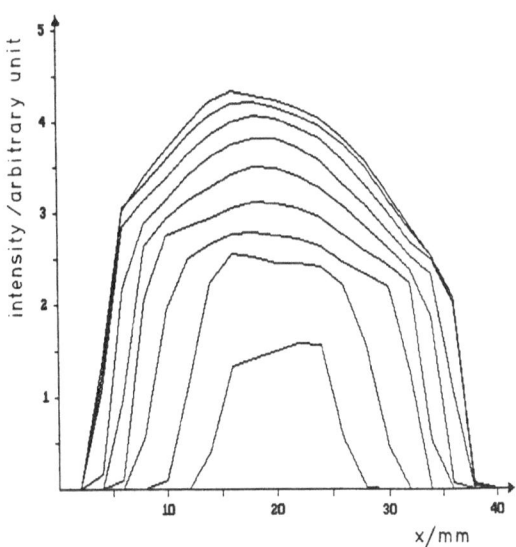

Fig.32 Cross sectional curves for the Trumpet model, measured
13.2 mm inside the model. 2mm distance

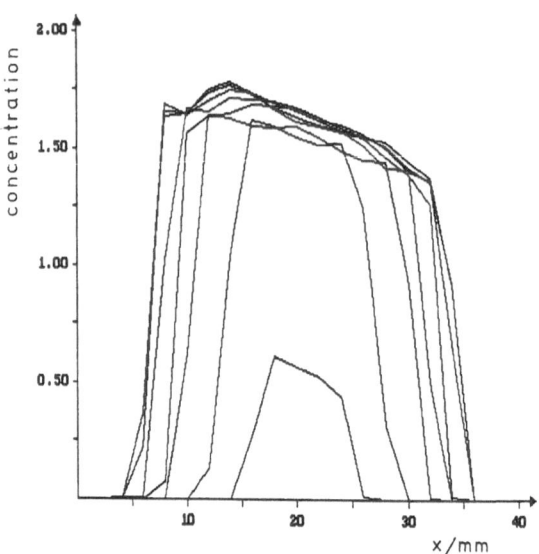

Fig.33   Curves of concentration for the Trumpet relatively
to the reference, measured at the exit aperture
2mm distance between the curves

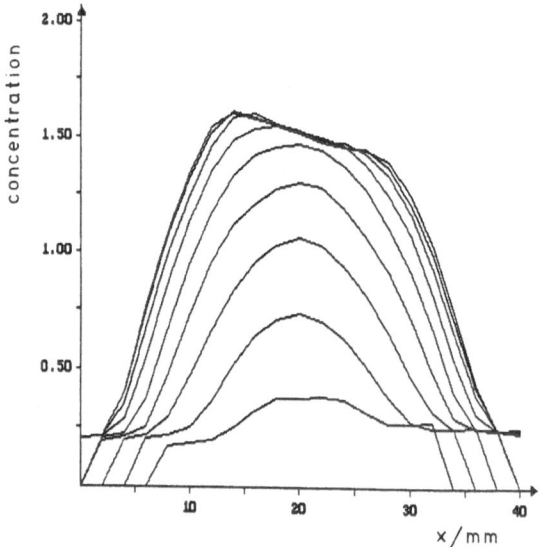

Fig.34   Curves of concentration for the Trumpet relatively
to the reference, 10mm behind the exit aperture
2mm distance between the curves

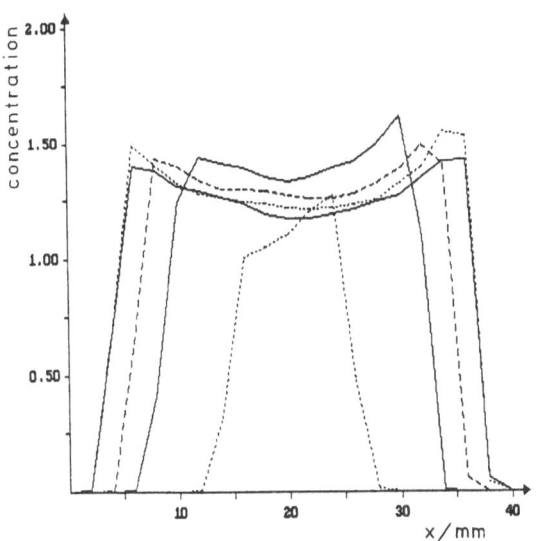

Fig.35   Curves of concentration for the Trumpet relatively to
         the reference, measured 13.2mm inside the model
         4mm distance between the curves

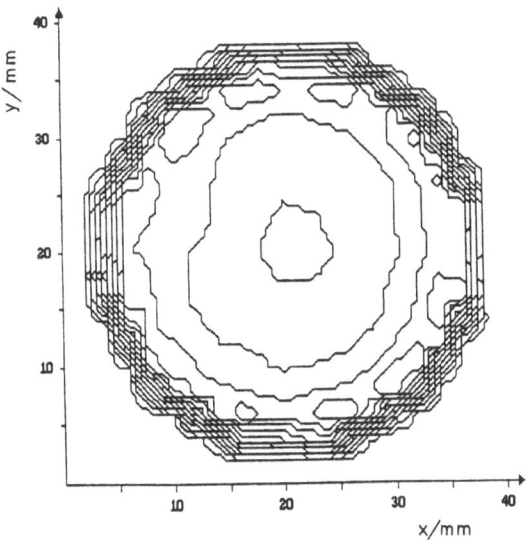

Fig.36   Level diagram of the concentration corresponding
         to Fig.35

plane	diameter mm	efficiency $\Phi_{meas.} / \Phi_{ref.}$
13.2	34.6	0.937
5.9	30.6	0.915
0.0	28.8	0.899
-5.0	-	0.895
-10.0	-	0.862

Fig.37   Table of efficiency for the measured planes of the
         Trumpet model

Corrosion Resistance of SiSiC Tube Material under the Specific Conditions of a Solar Heated Sulfuric Acid-Iodine Process Plant

W. Heider,

CESIWID Elektrowärme GmbH, Erlangen

R. Förthmann, A. Naoumidis,

Forschungszentrum Jülich GmbH

Abstract

As a part of the SOTA-Project AP 300 the corrosion behaviour of the SiSiC tubes as potential material for high performance heat exchangers has been investigated by isothermal exposure tests under the process conditions of a sulfuric acid hybrid process for water splitting.

A SiSiC standard grade (SK I) and a newly developed grade (SK II) with less free silicon were exposed to a SO_2/SO_3-containing atmosphere at 1050 °C as well as to concentrated sulfuric acid with and without addition of HJ at temperatures up to 350 °C by KFA Jülich.

Thermal exposure in SO_2/SO_3/air at 1050 °C results in an increase of bending strength. All of the ten test runs with the SK I standard grade carried out in boiling concentrated sulfuric acid for periods up to 120 days (= 2880 hours) show visible corrosion of the SiC phase, but no statistically demonstrable decrease in bending strength was observed. This also applies to the second grade SK II which has a higher strength than the standard grade partly influenced by the smaller fraction of free silicon.

1. Objectives

The scope of the R&D programme was to determine the corrosion resistance of a standard grade of Silicon-infiltrated Silicon Carbide (SiSiC) with about 20 wt% free silicon under isothermal test conditions corresponding to those for solar heated water splitting according to the sulfuric acid/iodine process. A further aim was to compare these results with the corrosion behaviour of a newly developed SiSiC grade with a maximum silicon content of 15 wt%, where a higher strength and a better corrosion resistance were expected [1, 2].

2. Test samples

SiSiC is very much a potential material for heat exchanger components due to its excellent thermal-mechanical properties (**Tab 1**) and extensive forming capability which allows the manufacture of large and complex articles as well in small tolerances at reasonable prices.

In principle for the manufacturing of such structural components different routes are used well-described in [3]. Especially for long heat exchanger tubes the extruding technique is preferred which has several advantages in comparsion with all other shaping techniques. For this reason the test specimens have been made by cutting off small pieces from SiSiC tubes 0 40/30 x 2000 mm manufactured as described in **Fig 1**. All samples were grinded and polished in such a way, that the surface roughness has no effect on bending strength and corrosion behaviour.

Properties		Symbol	Unit		Measuring temperature °C	°F	Material constant	
General properties	Bulk density	ϱ	g/cm³	g/cc			2.98	2.98
	Porosity, open		% by vol.	% by vol.			0	0
	Gas permeability		cm²/s	Darcy			10^{-7}	2×10^{-9}
	Si content (free)		% by wt.	% by wt.			~19	appr.19
	Vickers hardness Si		N/mm²	psi			12000	1.74×10^{6}
	Vickers hardness SiC						26000	3.77×10^{6}
Mechanical properties	Flexural strength (3 point)	σ_B	N/mm²	psi	20	68	300	43500
					1200	2200	300	43500
	Weibull's modulus	m			20	68	>10	>10
					1200	2200	>15	>15
	Young's modulus (dynamic measurement)	$E_{dyn.}$	kN/mm²	psi	20	68	340	49.3×10^{6}
					1000	1830	300	43.5×10^{6}
	Compressive strength	σ_D	N/mm²	psi	20	68	850	123000
	Shear modulus	$G_{dyn.}$	kN/mm²	psi	20	68	143	$20.7 \cdot 10^{5}$
	Poisson's ratio	ν					0.19	0.19
Thermal properties	Coefficient of thermal expansion	α	10^{-6}/K	10^{-6}/°C	20-1000	68-1830	4.3	4.3
	Thermal conductivity	λ	W/mK	BTU/ft-hr-°F	20	68	160	92
					1200	2200	40	23
	Specific heat	C_P	J/gK	BTU/lb-°F	200	400	0.880	0.21
					1000	1830	1.200	0.29
	Oxidation (100 h)		mg/cm²	mg/in²	1300	2370	0.7	4.5

Tab 1.: Typical properties of a commercial grade of SiSiC

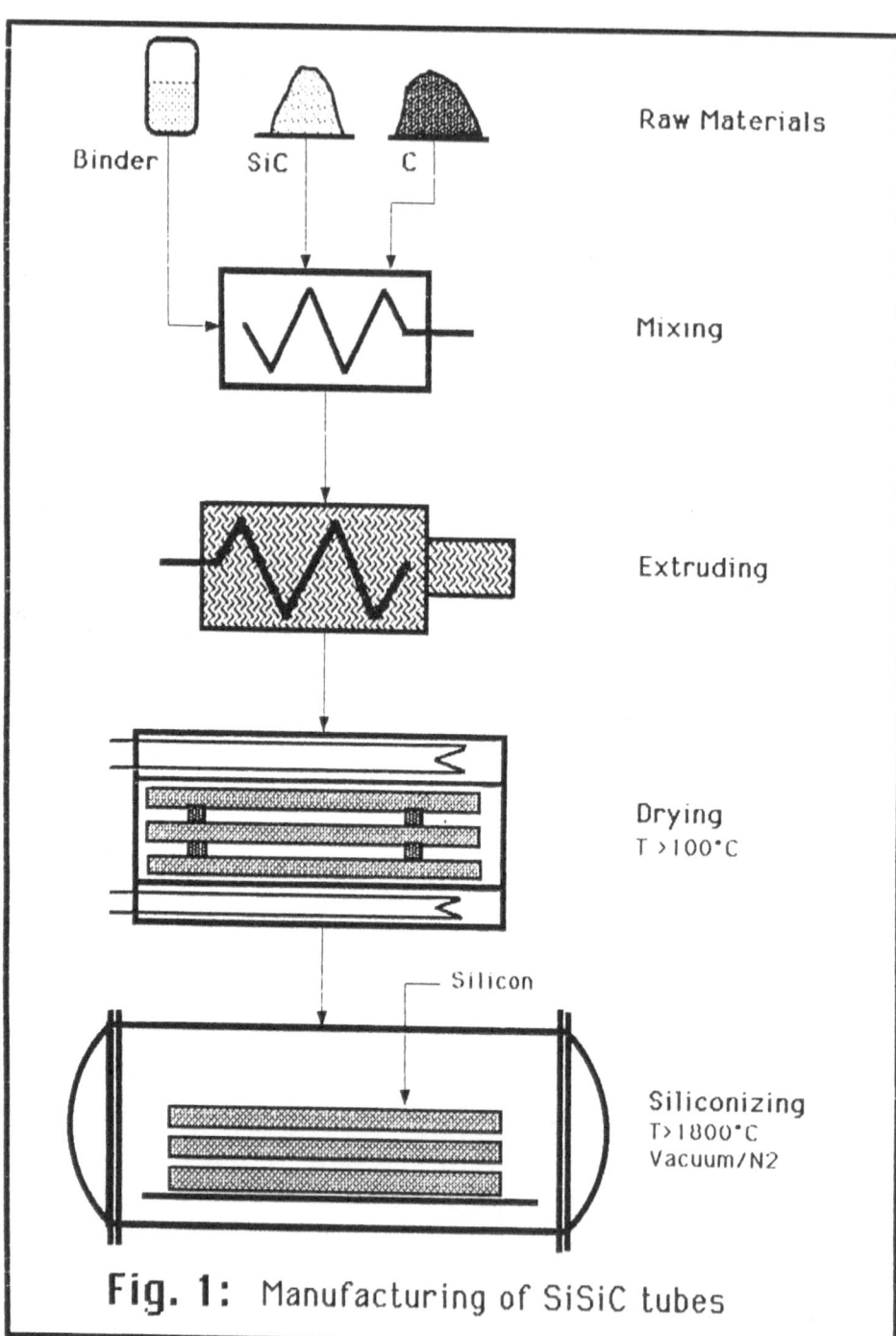

Raw Materials

Binder   SiC   C

Mixing

Extruding

Drying
T > 100°C

Silicon

Siliconizing
T > 1800°C
Vacuum/N2

Fig. 1: Manufacturing of SiSiC tubes

Fig 2 shows the dimensions and tolerances of the machined test
bars before polishing carried out by KFA.

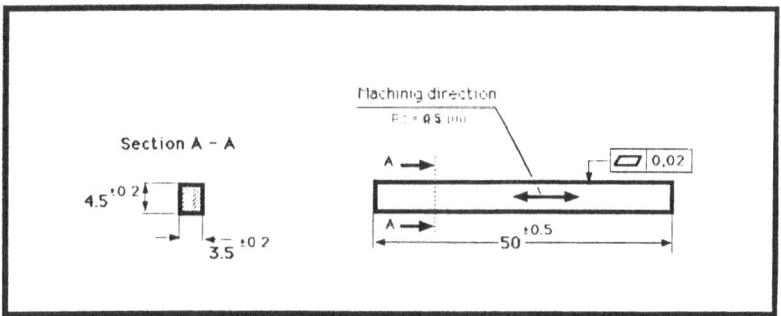

Fig 2: Configuration and tolerances of SiSiC samples for
measuring the 4-point bending strength before and
after exposure

Based on this manufacturing technique a large quantity of test
samples (appr. 340 pcs) of the SiSiC standard grade were prepared and marked as shown in **Fig 3** allowing the identification
of any position of the samples, where they originate. The microstructure, Si-content, and bulk density (BD) of this material
have been determined and the results are shown in **Tab 2** and
Fig 4.

Properties	Unit	SK I	SK-Test-material	SK II	Aim
Bulk density	g/cm³	2.93	3.0	3.05	≥ 3.0
Porosity	Vol%	3.9	2.0	1.5	≤ 0.5
Silicon content	Wt%	20	17	13	≤ 20
Bending strength	MPa	235	-	300	≥ 250

Tab 2: Properties of various grades of SiSiC used
for the exprosure tests

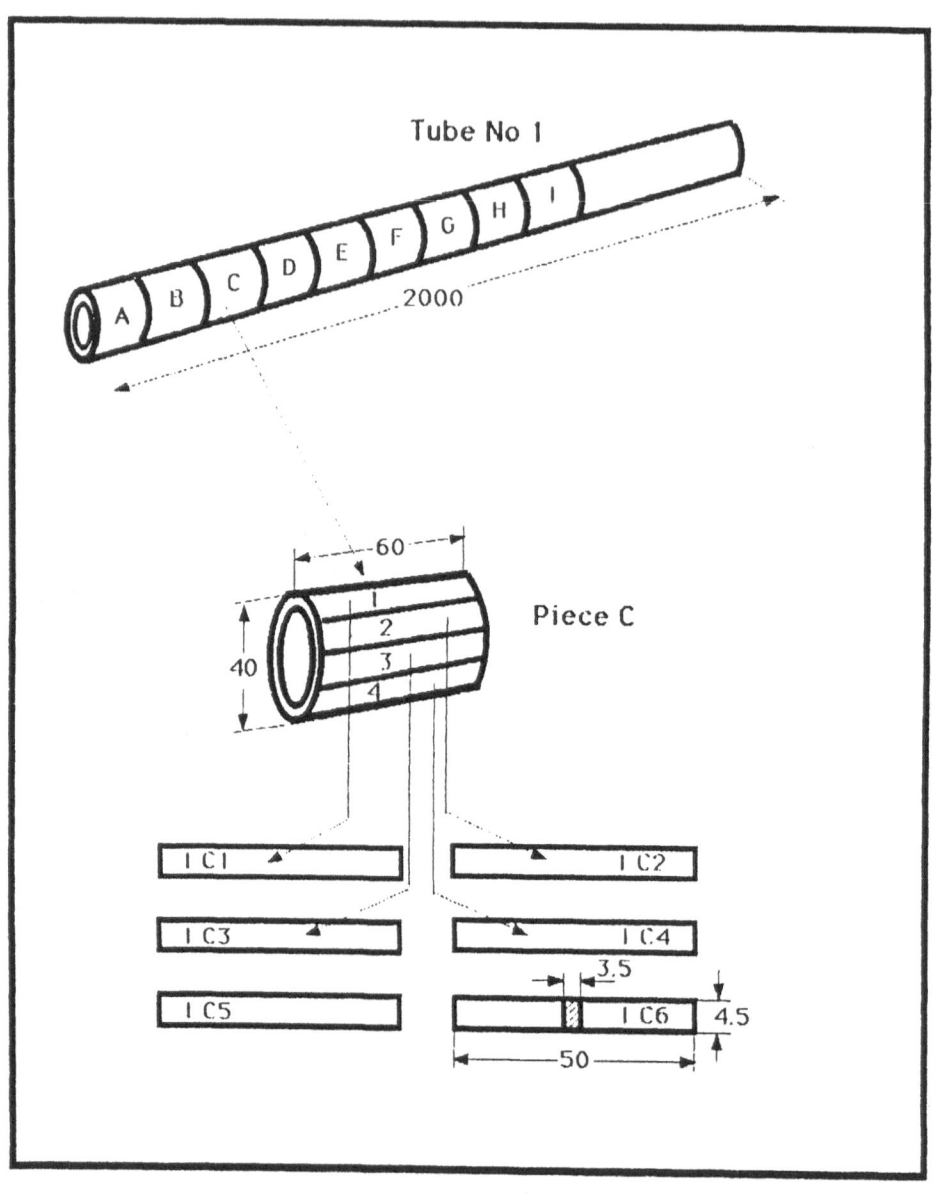

Fig. 3: Sample selection plan

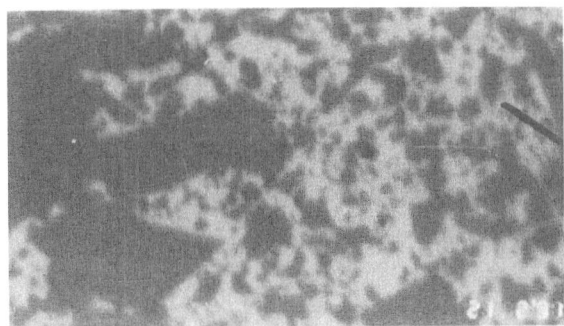

Si

SiC

SK I
with 20 Wt% Si

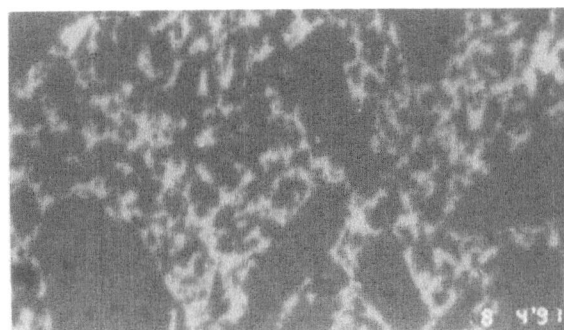

SK 5208
with 17 Wt% Si

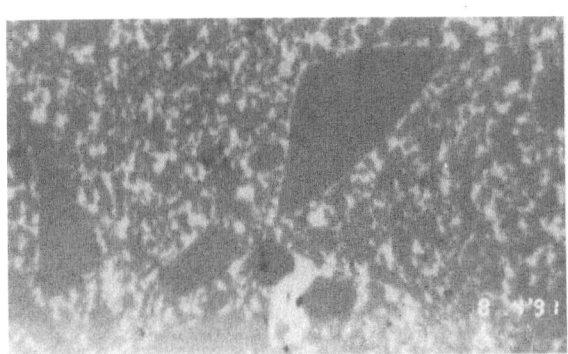

SK II
with 13 Wt% Si

Fig 4: Microstructure of SiSiC with various
fractions of exess silicon

In a second programme stage an improved extruded material with
a silicon content of max. 15 Wt% had to be developed by varying
the packing density of the SiC grains as well as the addition of
carbon which forms a second SiC phase by chemical reaction with
molten silicon during firing. The analysis of the microstructure, Si-content and bulk density of each material were also in-
investigated and results show a reduction of the amount of excess
silicon from 20 to 13 Wt% at least, while the bulk density was
increased from 2.93 g/^3 of the standard grade to 3.05 g/cm^3
referring to the material with the lowest Si-content also demonstrated in **Tab 2** and **Fig 4**. After finishing this work some SiSiC
tubes have been manufactured in the same way as described, but
unfortunately the microstructural analysis resulted in bad inhomogeneities (see **Fig 5**) due to large structural pores with
diameters up to 0.5 mm mostly filled with excess silicon.

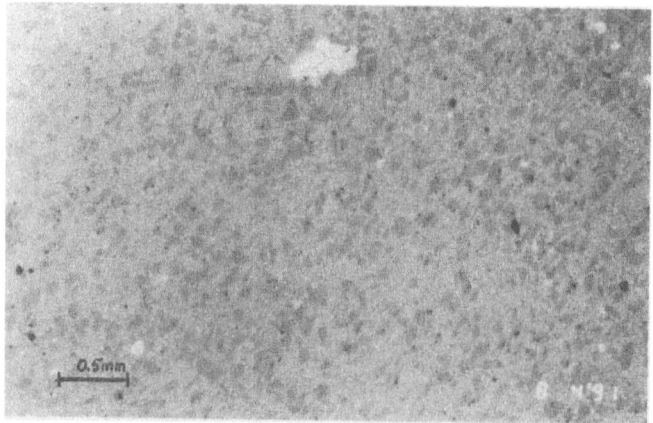

Fig 5: Microstructure of SK II showing a large
pore filled with exess silicon

Extensivly studies have shown that the reasons for these rela-
tivly large pores were mainly a defect at the used extruder and
partly the mixing and shaping conditions as well. Assuming that
these large inhomogeneities have a negative effect on properties, different trails were carried out to avoid this failure by
optimising the mixing and shaping technique. As result of these
trails the number and sizes of the silicon nucleates were reduced, but some smaller areas of free silicon were still existing and could not be removed completely. Because of the very
short available time further 300 test samples were prepared by
using such non defect-free material.

3. Corrosion conditions

The conditions for all corrosion tests were based on solar water
splitting process, which consists of the following four reactions:

(I) H_2SO_4 --------- $H_2O + SO_3$ [350 - 450 °C]

(II) SO_3 --------- $SO_2 + 1/2\ O_2$ [700 - 1100 °C]

(III) $2\ H_2O + SO_2 + I_2$---- $H_2SO_4 + 2\ HJ$ [25 - 50 °C]

(IV) $2\ HJ$ --------- $J_2 + H_2$ [300 °C]

The most critical steps are the SO_3 splitting at 700 - 1100 °C
and sulfuric acid concentration after hydrogen iodide seperation
following the reaction (III).

Based on these conditions the experiments were defined as follows:

a) temperature: 1050 °C

atmosphere: 10 Vol% SO3/SO2 and 90 Vol% air

b) temperature: up to 350 °C

corrosive medium: concentrated H2SO4

c) temperature: up to 350 °C

corrosive medium: boiling H2SO4 with HJ

4. Experiments

All corrosion tests were carried out by the Forschungszentrum Jülich GmbH under following conditions:

a. Isothermal corrosion tests in air with 10 % sulfur tri-oxide and sulfur dioxide at constant temperature of 1050 °C for periods of 30 to 120 days.

b. Isothermal corrosion tests in concentrated sulfuric acid with and without an addition of hydrogen iodide[*] at temperatures up to 350°C and as function of time between 7 and 120 days.

––––––

* In each case 5 ml of 50 % hydroiodic acid were added to 1.5 liter of concentrated sulfuric acid.

The test specimens made from both SiSiC grades were almost examined before and after corrosion as follows:

- measuring of 4-point bending strength
- characterization of the corrosion layer formed
- ceramographic studies of structural changes of the base material.

For these experiments two sorts of apparatus have been used described in Appendix A.

5. Results

5.1. Corrosion tests on SiSiC grade SK I

The corrosion bahaviour of SK I in SO_3/SO_2 containing air were carried out at constant temperature of 1050 °C by varying the exposure time from 30 to 60 to 120 days. Fig 6 shows the results of the strength measurements, where an increase of the mean value of the 4-point bending strength for the untreated samples from 236 MPa to 260 MPa after 30 days of corrosion, to 260 MPa after 60 days and to 284 MPa after 120 days can be observed. This means that corrosive damage of the base material with regard to strength due to the high SO_2/SO_3 fractions of the gas does not occur. On the other hand it is assumed based on former tests [4], that the strength increase is attributed to the nitrogen content in the air used as carrier gas.

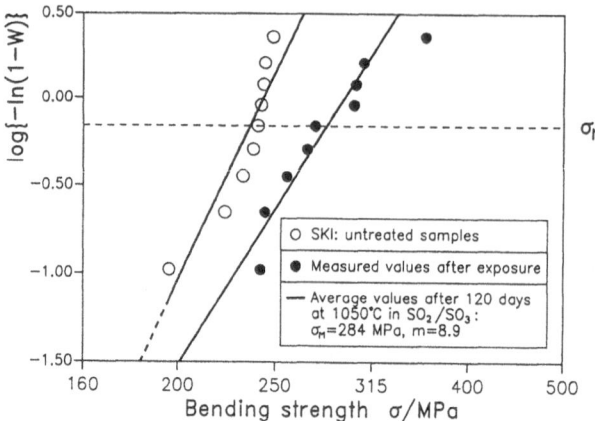

Fig 6: Weibull plot of measured bending strengths prior and after corrosion in 10 % SO2/SO3 - 90 % air
(SK I/CESIWID, 20 % free silicon)

Ceramographic studies of the exposed samples show a crack-free SiO2 layer with a thickness of 5 to 10 μm on the surface after 120 days. On **Fig 7** can be clearly seen that the Si phase is oxidized more rapidly than the SiC phase.

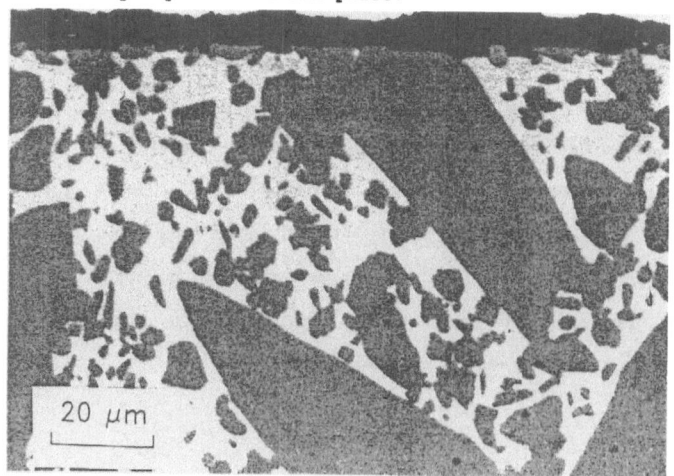

Fig 7: Micrograph of SK I after 2880 hours of corrosion in 10 % SO3/SO2 and 90 % air at 1050 °C

For examining the corrosion behaviour of SK I in boiling concentrated sulfuric acid strength and microstructural investigations were carried out after test over periods of 7, 14, 23, 30, 60, 90 and 120 days. The results of the strength measurements are plotted in **Fig 8**, where each value fit perfectly into the Weibull statistics of the untreated samples, whereby a corrosion-induced change in strength is not statistically demonstrable.

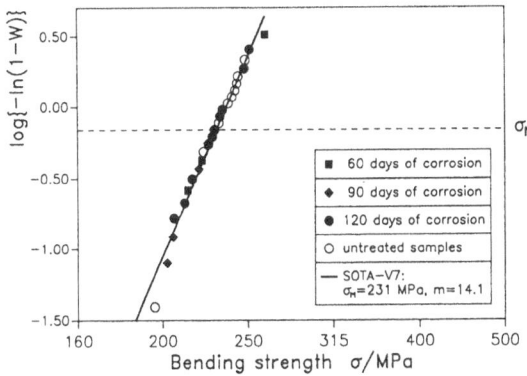

Fig 8: Weibull plot of measured bending strengths prior and after corrosion of SK I in boiling sulfuric acid

Scanning electron microscopy and ceramographic studies presented in **Fig 9** show that the surfaces are oxidized without forming coherent layers of uniform thickness. Particularly remarkable is that the SiC phase of the testing material is obviously more attacked by the acid than the Si phase i. e. opposite to the gas corrosion as discussed before.

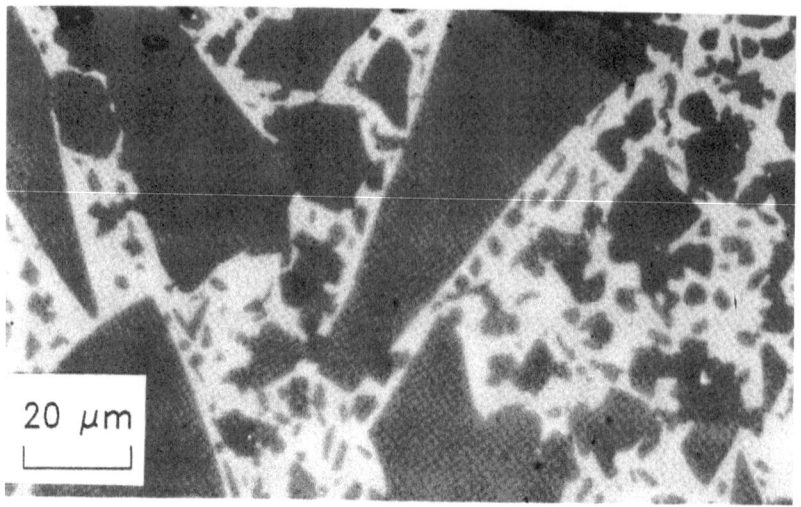

Fig 9: Micrograph of SK I after 2880 hours of corrosion in boiling concentrated acid.
Preferential oxidizing attack of the SiC phase (grey) can be seen in comparison with the Si phase (white).

In comparison with these results the corrosion behaviour of SK I was also investigated in boiling H2SO4 with addition of HJ for periods of 30, 60 and 120 days. Reffering to these tests it is important to note, that the added HJ was immediately oxidized to elementary iodine (J2), which evaporated completely within a few hours. As shown in **Fig 10** the yielded values of the measurements of the 4-point bending strength fit also perfectly into the Weibull plot of the untreated specimens. Thus even under the more critical conditions of this test no corrosion-induced change in strength could be found.

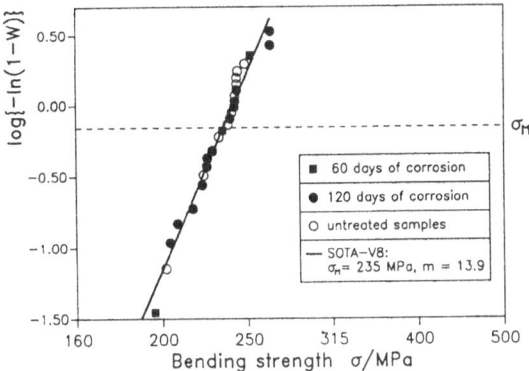

Fig 10: Weibull plot of measured bending strengths prior and after corrosion of SK I in boiling sulfuric acid with iodine addition

5.2. Corrosion tests on SiSiC grade SK II

The second SiSiC grade from type SK II has a smaller Si-phase than the standard grade amounting to 13 wt% free silicon, so that a better strength and corrosion behaviour were expected. In fact the 4-point bending strength of this material was 306 MPa (median value) and thus clearly higher than that of SK I with a mean value of 236 MPa. On the other hand the Weibull parameter of SK II was 7.8 only (**Fig 11**) and thus obviously lower than that of the standard material (m = 11.9). But even then **Fig 12** indicates a rather small standard deviation of the bending strength measured on samples which are taken from several tubes.

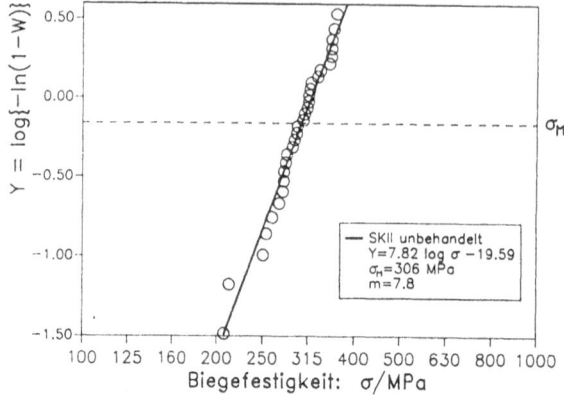

Fig 11: Weibull plot of measured bending
strengths of untreated SiSiC samples
(SK II/CESIWID, 13 % free silicon)

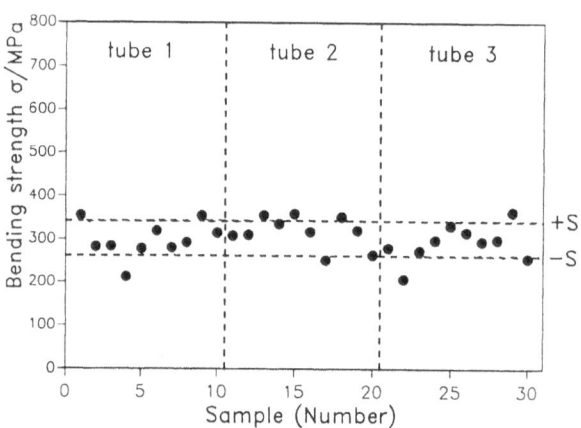

Fig 12: Bending strength of SiSiC grade SK II
(CESIWID) as a function of the posi-
tion in the three tubes

On this material three corrosion tests were carried out as follows:

- 61 days in 10 Vol% SO3/SO2 containing atmosphere at 1050 °C
- 61 days in boiling concentrated acid without HJ
- 61 days in boiling concentrated acid with HJ.

Fig 13 shows the Weilbull plot of the high temperature corrosion test in SO2/SO3 containing air of SK II resulted in an increase of the mean bending strength from 306 MPa to 433 MPa, which is much more pronounced compared with the observed strength increase of SK I under the same conditions.

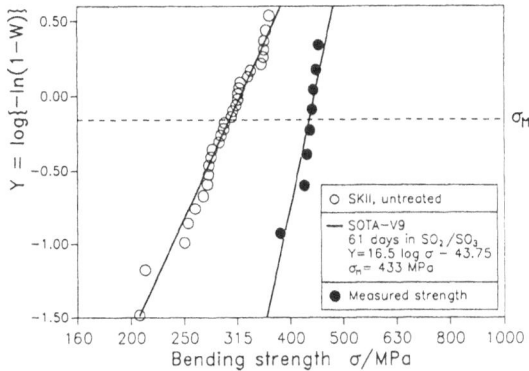

Fig 13: Weilbull plot of measured bending strength prior and after corrosion in 10 % SO2/SO3 - 90 % air at 1050 °C (SK II/CESIWID, Test time: 61 d)

The two corrosion tests in boiling sulfuric acid with and without addition of HJ on SK II resulted in some visible attacks on the surfaces similar to that of SK I, but without any effect on bending strength as shown in **Fig 14,** where all values measured after corrosion in H2SO4 fit perfectly into the Weibull plot of the untreated samples. **Fig 15** shows a tendency to higher bending strength indicated by an increase of the mean value from 301 MPa to 306 MPa after 61 days corrosion in HJ added H2SO4.

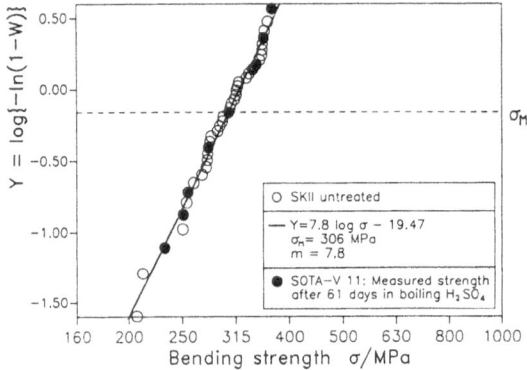

Fig 14: Weibull plot of measured bending strengths prior and after corrosion in boiling H2SO4 (SK II/CESIWID, 13 % free silicon)

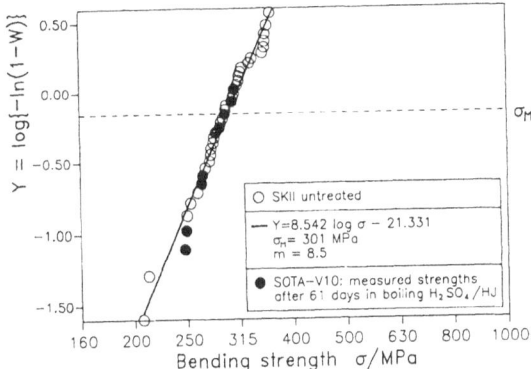

Fig 15: Weibull plot of measured bending strengths prior and after corrosion in boiling H2SO4/HJ (SK II/CESIWID, 13 % free silicon)

6. Summary and Conclusion

Based on the present experiments the following conclusions and observations are presented for further consideration.

- The corrosion behaviour of two qualities of SiSiC have been studied under the conditions of a water splitting equipment corresponding to the sulfuric acid/iodine hybrid process with the purpose of comparing the properties of both materials and to find out the suitability and possible limitations of SiSiC for this special application.

- In addition to a standard SiSiC grade with a fraction of 20 Wt% free silicon a second material with a reduced silicon content (13 Wt%) was developed based on a shaping and firing technique which is preferably used for the manufacturing of heat exchanger tubes.

- In the case of both SiSiC grades the corrosion tests in
  SO2/SO3 containing air at a constant temperature of 1050°C
  resulted in an increase of bending strength with time,
  whereas the strength increase of the improved material was
  much higher than that of the standard grade.

- All tests runs with the SiSiC standard grade SK I in
  boiling sulfuric acid with and without addition of hydrogen iodide showed visible corrosion on the SiC phase,
  which had no effect on the bending strength. This also
  applies to the second SiSiC grade after an exposure time
  of 61 days.

- Both materials indicate a high chemical resistance under
  the isothermal conditions as considered, and the test results have also shown that the chemical behaviour is not
  influenced by the amount of exess silicon.

- For a complete characterisation of the thermo-mechanical
  and chemical properties of SiSiC under consideration of
  the more practical and even more critical thermo-cycling
  conditions further investigations are necessary.

Testing equipments

The following apparatus were used to carry out the corrosion tests as described above.

1. Apparatus for gas corrosion tests:

Fig a shows the apparatus to carry out the gas corrosion tests in SO2/SO3/air. The furnace is heated with a Silit tube element and has an inner working tube made from quartz, where the samples are deposited. SO2 together with air as carrier gas flows continuously through the working tube. In front of the test specimens a Pt asbestos was installed for catalitic oxidizing of SO2 to SO3, which were absorbed again behind the furnace by using Na2SO3.

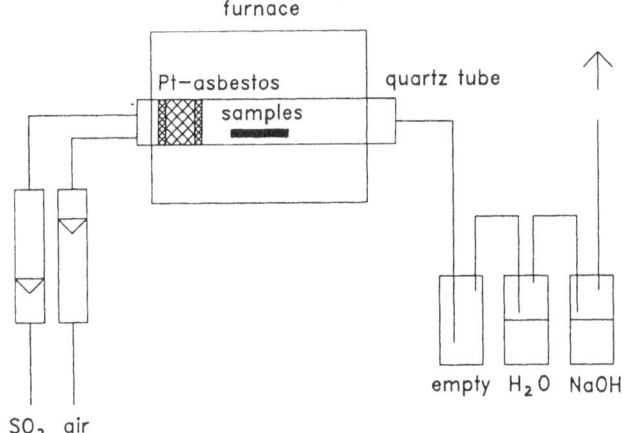

Fig a: Apparatus for the corrosion of SiSiC samples in a gas mixture of SO2, SO3 and air at 1000 °C. SO3 is produced at the platinum contact in the furnace inlet and absorbed as H2SO4 in water behind the furnace. SO2 is absorbed as Na2SO3 in NaOH.

2. Apparatus for corrosion tests in boiling H2SO4

Fig b shows a glass apparatus designed for boiling sulfuric acid, where 20 specimens in maximum can be inserted and exposed to temperatures up to 600 K.

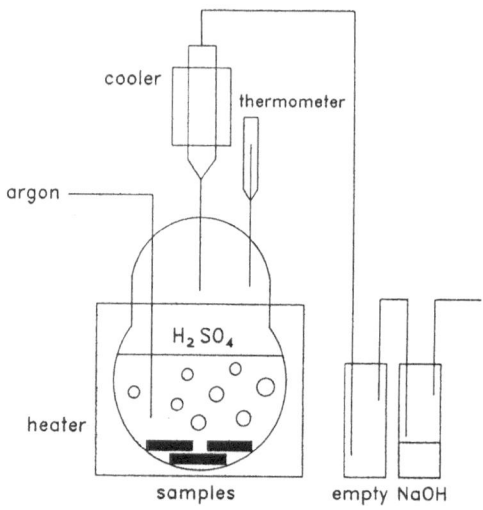

Fig b: Apparatus for the corrosion of SiSiC samples in boiling sulfuric acid

Schrifttum

[1] Thümmler, F.   Die keramischen Hochtemperaturwerk-

                          stoffe Si3N4 und SiC.

                          In: Keramische Komponenten für Fahr-

                          zeug-Gasturbinen.

                          Berlin/Heidelberg/New York:

                          Springer-Verlag, 1978.

[2] Knoch, H.   Eigenschaften und Anwendungsbeispiele
     Hunold, K.   nichtoxidischer Sonderkeramiken.

                          In: Technische Keramik.

                          Essen: Vulkanverlag, 1988.

[3] Willmann, G.   Herstellung von Komponenten aus Sili-
     Heider, W.   ciumcarbid für das Sonnenturmkraft-

                          werk GAST.

                          Sprechsaal 114 (1981) H. 10,

                          S. 758-765.

[4] Förthmann, R.   Hot Gas Corrosion of Silicon Carbide
     Naoumidis, A.   Materials at Temperatres between

                          1200 and 1400 °C.

                          Proceedings of the 2nd International

                          Symposium on High Temperature Corro-

                          sion of Advanced Material and Coa-

                          tings, Les Embiez (Frankreich),

                          22.-26. Mai 1989.

Fundamentals of Combined Thermal
and Photochemical Decomposition
of Sulfur Trioxide at High Temperatures

K.-F. Knoche, M. Dzubiella, M. Roth, D. Brüggemann,

RWTH Aachen,
Lehrstuhl für Technische Thermodynamik

Abstract

The influence of light illumination on the decomposition of sulfur trioxide has been investigated at temperatures up to 900 °C. The recorded spectra show that SO_3 and SO_2 have similar absorption regions in the ultraviolet part of the spectrum and are difficult to distinguish.

Comparative studies show that light illumination induces SO_3 decomposition even at ambient conditions and accelerates the thermal SO_2 production at higher temperatures.

Although the results are encouraging with a view to technical applications on a large scale, further experiments are necessary to quantify the data obtained until now.

Contents

1 Introduction

Thermochemical water splitting is accomplished by a chemical process which decomposes water into hydrogen and oxygen. In these processes high temperature heat is the major energy input; water electrolysis requires a large amount of electricity. Thermochemical processes are meant to achieve high process thermal efficiencies by avoiding the heat to electricity conversion. The endothermic high temperature reactions operate below 1000 °C due to material constraints. These reactions include the decomposition of sulfur trioxide.

In future, solar energy will be an interesting heat source for production of fuels and chemicals. Coupling of solar energy into chemical processes will strongly influence both, cost and efficiency of the overall system.

For a great number of well investigated cycles sulfuric acid decomposition has been selected as the high temperature step. One of them, the sulfur-iodine process, has the most considerable potential for thermochemical hydrogen production. General Atomic Inc. has investigated these cycles (fig. 1) for many years, demonstrated their feasibility and evaluated first chemical engineering flowsheets. The thermochemical water splitting process shows a quite interesting potential to optimize the adaptation of solar energy (fig. 2).

The dissociation of sulfuric acid occurs at temperatures higher than 800 °C. Although the efficiency is increasing with temperature the equilibrium mixture still contains a significant fraction of unconverted SO_3 even at 1000 °C. Using direct absorbers it is possible to achieve a better SO_3 conversion at higher temperatures. In this case carrier gases like air are not suitable for energy transfer because due to the very large value of reaction enthalpy large air mass flows would be required.

As a consequence, there is considerable interest in alternative methods supporting the thermal dissociation of SO_3. The studies described here are based on the idea that the reaction rate may be increased by exposure to solar radiation.

The utility of photochemical reactions has already been demonstrated in a variety of fields but not in the special case of SO_3. Moreover, even a reliable estimation by theoretical considerations is excluded until relevant spectral properties of SO_3 are not sufficiently known.

Therefore, before judging the possibility of solar photochemical application to SO_3 decomposition it has been necessary to determine position, shape and strength of SO_3 light absorption spectra and their temperature dependence.

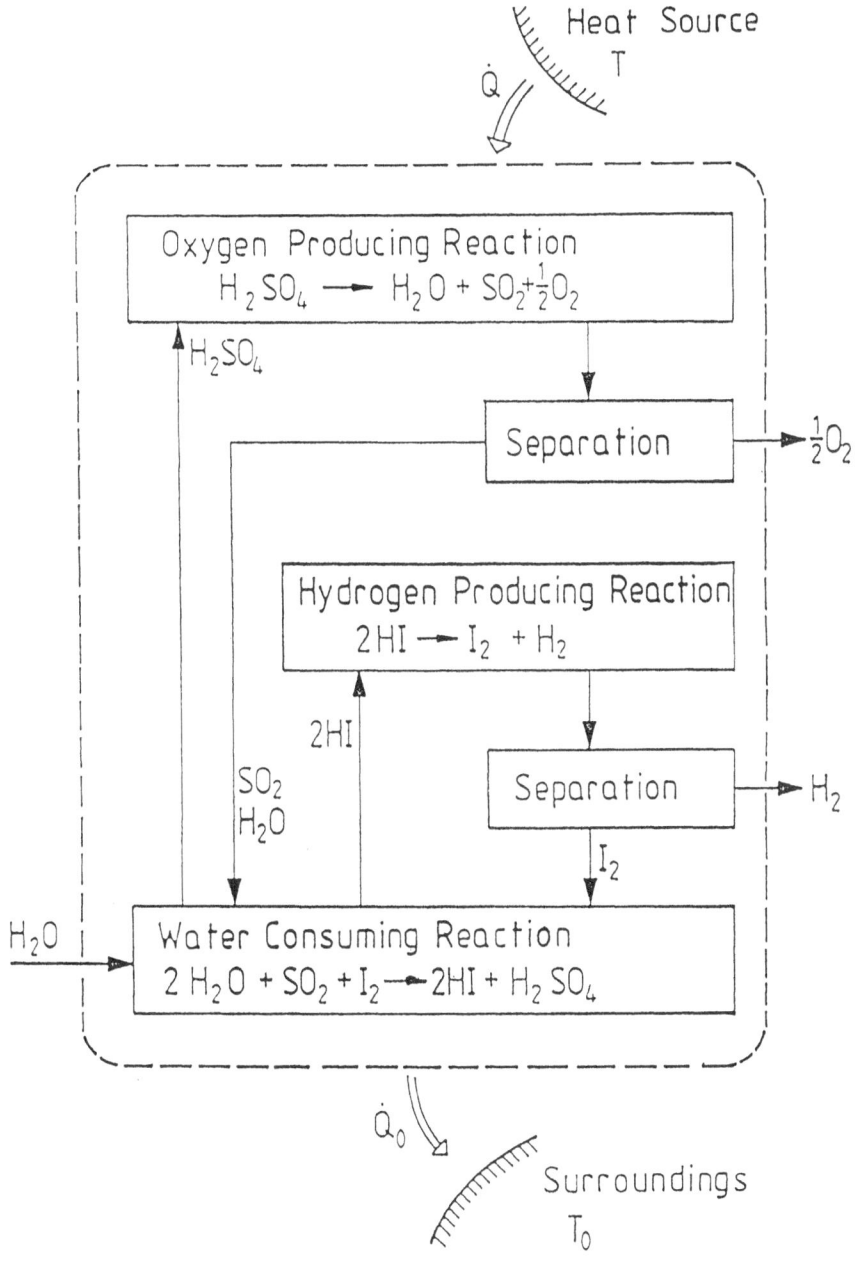

Figure 1: Reactions of thermochemical cycles

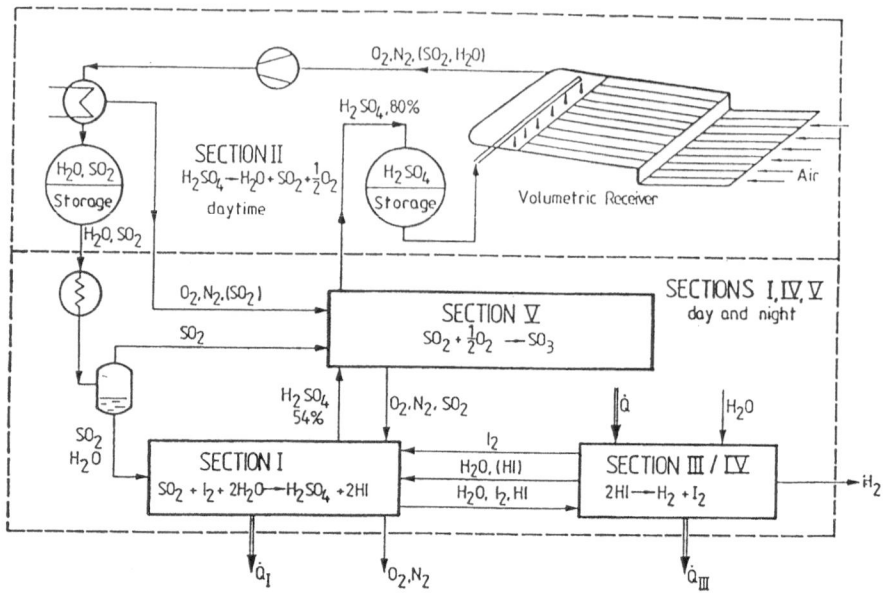

Figure 2: Solar driven GENERAL ATOMIC process with volumetric receiver

2 Physical and chemical properties of SO_3

Due to its high reactivity with water, sulfur trioxide is a substance not easy to produce, store and handle.

At room temperature sulfur trioxide is a colourless liquid which freezes at 16.8 °C and boils at 44.8 °C. For the gas phase in equilibrium it has been shown [1] that SO_3 also forms trimers which vanish with increasing temperature. Above 100 °C, S_3O_9 has not been observed [2]. Similar results were found for the liquid phase [3,4].

Solid sulfur trioxide forms three modifications, called α-, β- and γ-SO_3 [5]. The first two of them are results of a polymerisation in presence of small quantities of water. In practice, nearly pure γ-SO_3 is achieved by adding stabilisators like dimethylsulphate.

Due to this behaviour the experiments of this study were carried out at temperatures above 110 °C.

Very little information about electronical transitions of the SO_3 molecule is found in literature. One reason for the lack of data is the general difficulty in using SO_3 for experiments. In addition, it turns out that even small amounts of SO_2 conceal the ultraviolet spectrum of SO_3.

Until recently, only one very early study of UV absorption had been published [9]. In a new experimental work [10] it has now been shown that SO_3 may be excited at very short wavelengths (145–160 nm) to emit fluorescence. This vacuum-UV region of the spectrum is only observable by exclusion of air and therefore cannot be used for application of solar radiation on earth.

The vibrational transitions of SO_3 have been determined by infrared and Raman spectroscopy [1,2,11,12,13,14]. In future studies these data will be of help to identify and quantify the reactions in SO_3/SO_2 gas mixtures.

3 Preparation of SO_3 samples

The SO_3 used in the experiments described here has been delivered with a purity of 99.6 vol.% with 0.2 vol.% of the stabilisator dimethylsulfate, 0.15 vol.% sulfuric acid and a 0.05 vol.% impurity of iron (Fa. JANNSEN CHIMICA).

Bottling SO_3 samples with defined quantities into ampoules has to be done in closed vessels to avoid impurities and, most important of all, any reaction with room moisture.

To fulfill these demands a bottling line has been developed which is shown in fig.3 and described only briefly here. SO_3 is destillated from the storage vessel to the measuring vial in two steps.

The preparation starts with melting of solid SO_3 which is stored at 4 °C. During several days the reservoir stays in a thermostat of 35 °C. Gaseous pollutions are removed by sucking off the gaseous phase for several times. SO_3 will react to H_2SO_4 at the ice layer on the inner walls of the cooling traps. After this pretreatment SO_3 is destillated into the receiver under vacuum condition.

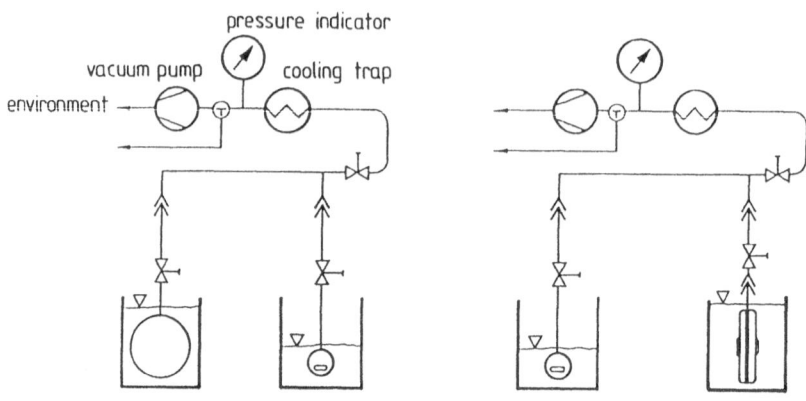

I. SO_3-storage, 35°C SO_3-receiver 10°C   II. SO_3-receiver 35°C measuring ampoule, 1°C

Figure 3: Preparation of SO_3 samples

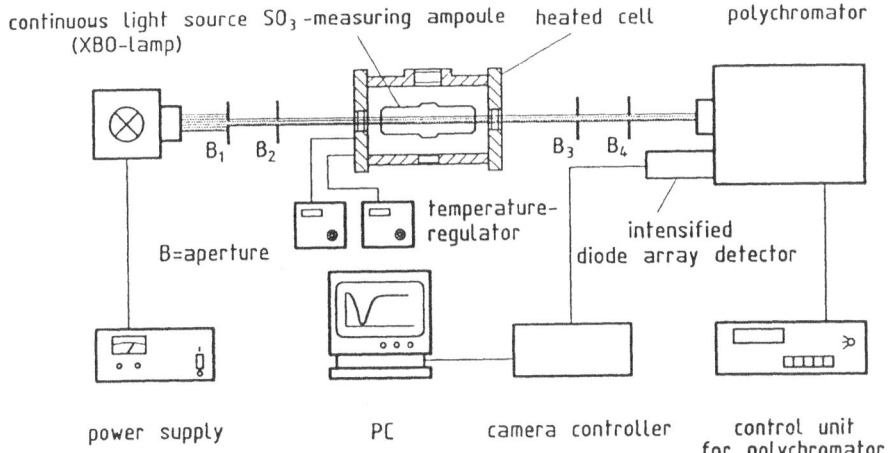

continuous light source SO_3-measuring ampoule   heated cell   polychromator
(XBO-lamp)

B_1 B_2 temperature-   intensified
B=aperture   regulator   diode array detector

B_3 B_4

power supply   PC   camera controller   control unit
                                                              for polychromator

Figure 4: Experimental facility for absorption measurements

4   Experimental configuration

In the experiments described in the following we used a configuration which allows light absorption measurements and, as a view into future investigations, may be extended to apply Raman spectroscopy.

An absorption spectrum represents the extent of light extinction as a function of wavelength. The percentage of absorbed light depends on several molecular and experimental parameters. In most of all cases, it is described by the Lambert-Beer's law, saying that the light is extinguished exponentially with respect to the pathlength of the light crossing the absorbing medium and its concentration given e.g. in units of mol/cm^3.

Instead of an absorption spectrum the recorded transmission spectrum will be displayed in this work. Both are, of course, complementary because extinction plus transmission is always unity.

The experimental configuration is shown in fig. 4. It may be divided into three units: the light source, the measuring vial and the detection system.

A Xenon high pressure light source (Osram XBO 150/4) bedded in a housing with Suprasil glass windows was chosen to allow measurement not only in the visible but also in the ultraviolet spectral regions. The total spectrum may be approximated by a black body radiation corresponding to a temperature of about 5500 to 6500 Kelvin. The light is directed by several optical lenses, mirrors and pinholes into the ampoule containing the gas sample.

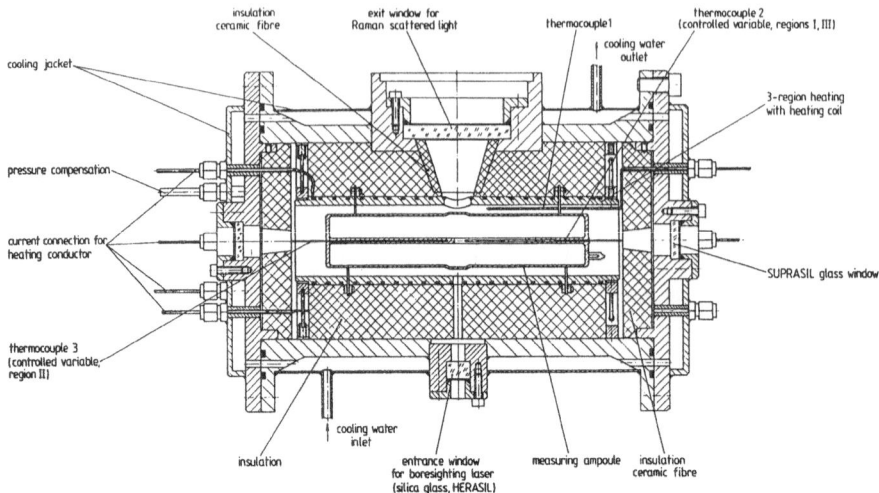

Figure 5: Arrangement of the heated cell

A heatable glass cell system had to be constructed considering several requirements due to the chemical and optical properties of SO_3. To mention only few examples, this implies that the cell is gas-tight at temperatures up to 1000 °C, optically transparent at ultraviolet wavelengths below 300 nm and of a length which is sufficient for absorption spectroscopy.

The constructed cell, shown in fig. 5 fulfills these requirements and is also suitable for applying further measurement techniques like Raman spectroscopy. The ampoule is provided with special fused silica (Suprasil) windows through which the light is directed and partially absorbed on its 15 cm path. Stress and heat transfer calculations were carefully performed regarding specifications given in literature [6,7]. The interspace of the double-walled vessel and its cap is cooled by flowing water to ensure an outer wall temperature of less than 40 °C.

Three thermocouples are used for measuring temperatures at different locations and for controlling the heating system. Temperature gradients along the absorption path are avoided by a variable three-zone heating system. Radial and axial heat losses by conduction or radiative transfer are reduced by using special fiber materials with low heat conductivity even at high temperatures for careful insulation. The pressure can be kept constant by a piping.

The transmitted light is spectrally dispersed by a polychromator. The resolution can be varied using gratings with different numbers of lines per mm. The spectrum is detected by an intensified photodiode array from which it is read out and digitally stored with the help of a computer control unit. A number of experiments, not documented here, have been performed to ensure the correct assignment of camera channels to wavelengths.

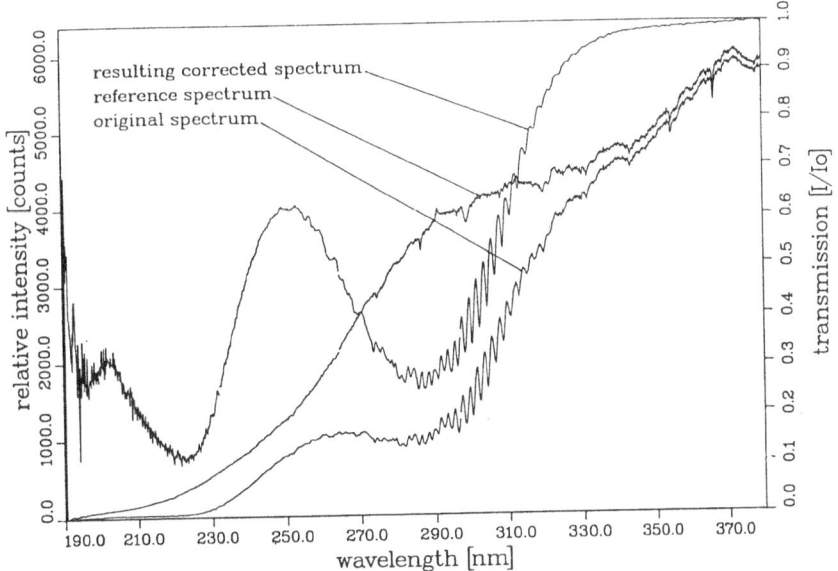

Figure 6: Original, reference and the resulting corrected spectrum

For a more general interpretation of the spectra it is suitable to correct the raw data by subtracting a background and dividing by a reference spectra to account for the spectral intensity distribution of the light source. As an example, fig. 6 shows an original, a reference and the resulting corrected spectrum. Only the corrected transmission spectra are displayed in the following.

5 SO₃ absorption spectroscopy

SO$_3$ absorption spectroscopy is complicated by two difficulties:

- the handling of SO$_3$ due to its high reactivity with air moisture;

- the strong interference of UV absorption regions between SO$_2$ and SO$_3$ bands.

While the careful preparation of SO$_3$ samples has already been described, the second obstacle has to be discussed in more detail.

It is known from literature [9,10] that even very small impurities of SO$_2$ in SO$_3$ cause absorption which conceals any spectral features of SO$_3$. Fajans mentioned an effective ratio of SO$_2$ and SO$_3$ absorption coefficients between 10^1 and 10^3 Suto failed in determining SO$_3$ reaction rates due to this complication.

In our experiments there are several origins of SO$_2$ in SO$_3$ samples:

- production during the closure of the filled glass cuvette by melting;

- thermal dissociation of SO$_3$ at higher temperatures;

- photodissociation of SO$_3$ by light.

Investigating the dissociative behaviour of SO$_3$ also means production of SO$_2$ and, as a consequence, attempts have to be made to distinguish between the absorption of both components.

For increasing the chance to observe a nearly pure SO$_3$ absorption spectrum measurements were tried at low temperature (50 °C) with only few light for detection and without melting the cuvette. The result is shown in fig. 7. Light is strongly absorbed in spectral regions below about 250 nm with a minimum of transmission between 210 and 220 nm. Further experiments support the theoretical prediction that only the ultraviolet region is of interest. Note, that at wavelengths below 200 nm light is absorbed by passage through air, the lines can be attributed to oxygen.

The conclusion that the absorption observed in fig. 7 is really that of SO$_3$ is supported by a comparison with fig. 8, showing spectra of pure SO$_2$ at different temperatures. Band systems occur around 290 nm and 200 nm. According to the SO$_2$ energy level diagram which is based on molecular data of [8] and shown in fig. 9, the bands belong to transitions labeled as $\tilde{A}$-$\tilde{X}$ and $\tilde{D}$-$\tilde{X}$/$\tilde{D}$-$\tilde{X}$.

In fig. 10 a spectrum of a probe containing SO$_3$ and impurities of SO$_2$ is compared to a pure SO$_2$ spectrum at the same temperature of 150 °C. The most important feature is the steep decay of transmission below about 280 nm. Obviously, the absorption in the region around 250 nm is only slightly interfered by SO$_2$ and is an indicator for SO$_3$. On the other hand the absorption above 300 nm is nearly completely caused by SO$_2$ and can serve as a measure for decomposition.

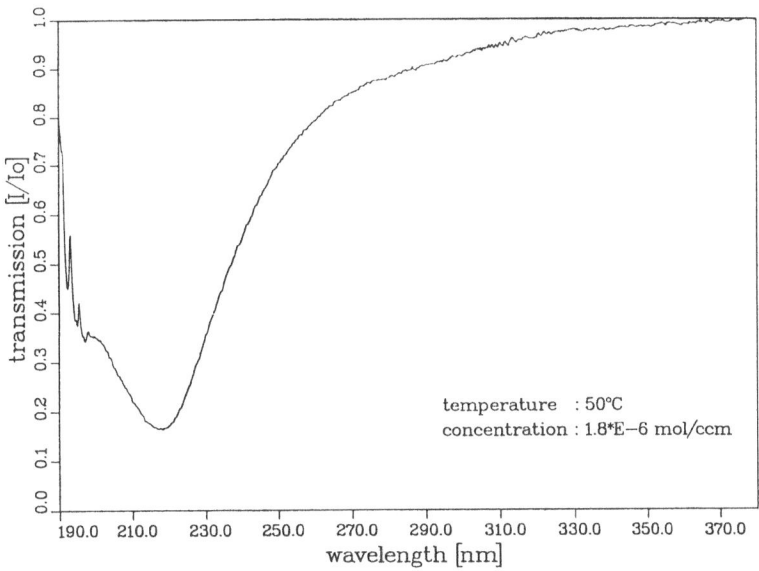

Figure 7: Pure SO_3 transmission spectrum

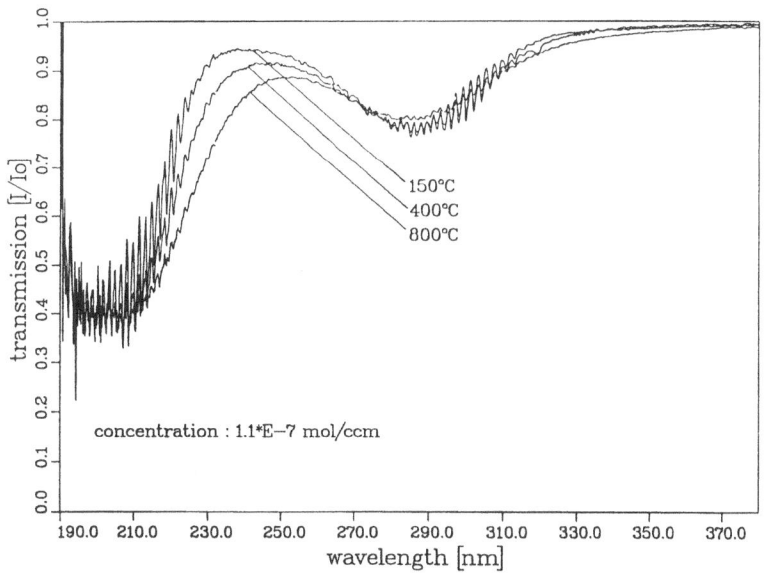

Figure 8: Pure SO_2 transmission spectra at different temperatures

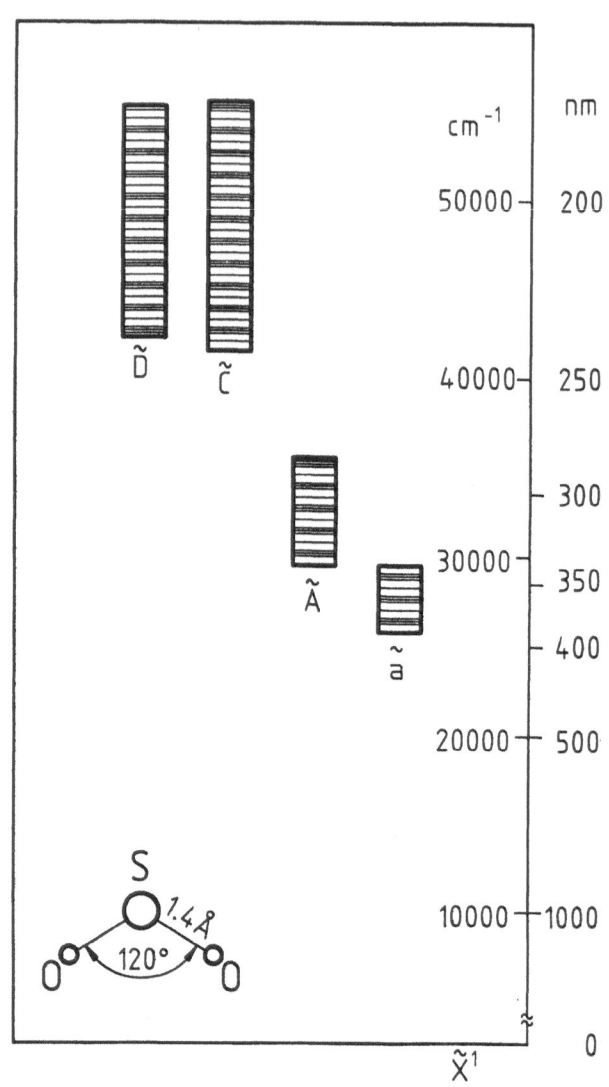

Figure 9: Energy term scheme of SO_2

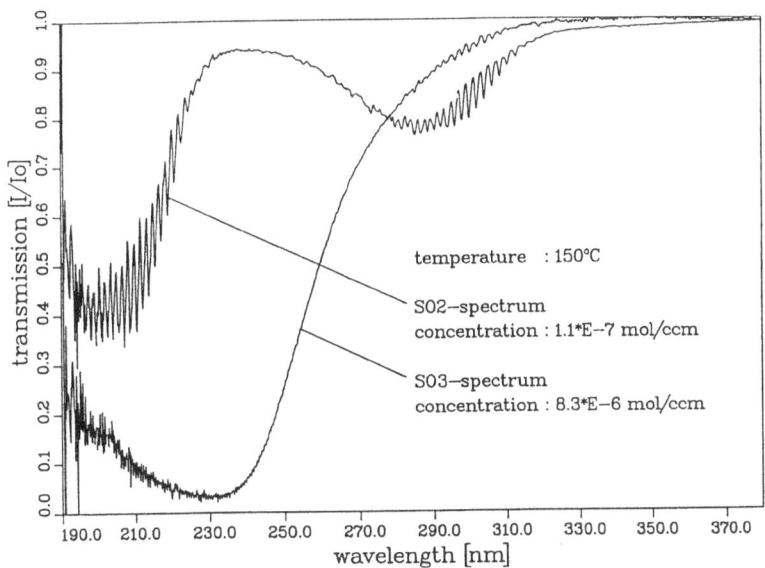

Figure 10: Spectrum of a probe containing SO_3 and impurities of SO_2 compared to a pure SO_2 spectrum

5.1 Thermal decomposition

As a next step, the time devolopment of thermal decomposition has been investigated. The results for two temperatures, 700 °C and 800 °C, are displayed in fig. 11 and fig. 12. In both cases the transmission spectra were recorded in time steps of 30 minutes. The increase of SO_2 becomes obviously by the decreasing light transmitted above about 260 nm. The curves reflect that thermal decomposition of SO_3 is very slow at these temperatures. Even after two hours no equilibrium state has been reached.

A comparison of the two pictures indicate a remarkable change of the absorption spectra with temperature. This has been further studied by increasing the temperature from 150 °C to 900 °C in several steps. The results are shown in fig. 13. Note, that in addition the sample has been kept at the constant temperature of 700 °C for 2 hours. This measurement helps to distinguish between the real temperature dependence of the spectrum and the changes due to increased (thermal) SO_2 production. It demonstrates that the content of SO_2 mainly influences spectral regions above the "branching point" at 250 nm.

As an important result inferred from these measurements it may be stated that SO_3 absorption is shifted towards higher wavelengths with increasing temperature. This behaviour is positively influencing the photochemical decomposition by solar radiation.

5.2 Influence of illumination by a light source

By the following experiments we could confirm that it is possible to support the decomposition of SO_3 photochemically. We used the broadband light source described with the experimental set-up in a previous section.

Fig. 14 shows that it is possible to decompose SO_3 even at temperatures as low as 150 °C, where no thermal decomposition occurred. The curves show the increase of SO_2 absorption just before and after 60, 150 and 240 minutes of irradiation. Another spectrum which has been taken after 120 more minutes without irradiation is nearly identical to the last one. This proves that SO_3 dissociation is stopped after blocking the illumination.

It is difficult to quantify the SO_3 decomposition for reasons like spectral overlapping of SO_2/SO_3 absorption bands and only rough knowledge of the initial SO_3 content in the absorption cell. Using only the longer wavelength parts we obtained a temporal production of SO_2 which is shown in fig. 15. It indicates that with irradiation it is already possible to achieve SO_2 concentrations at 400 °C similar to those after pure thermal decomposition at 800 °C.

5.3 First experiments with exposure to sun light

Encouraged by the results showing decomposition of SO_3 after illumination with a technical light source, first attempts using sunlight have recently been made.

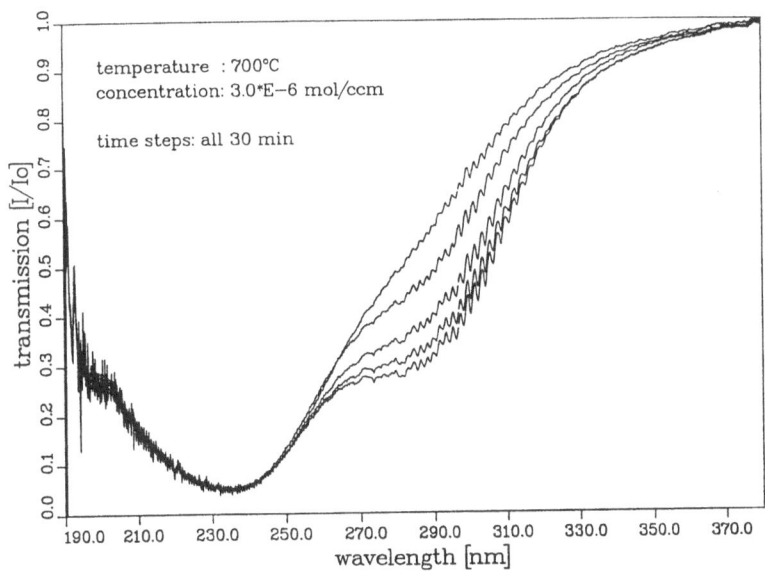

Figure 11: Thermal decomposition of SO_3 at 700 °C

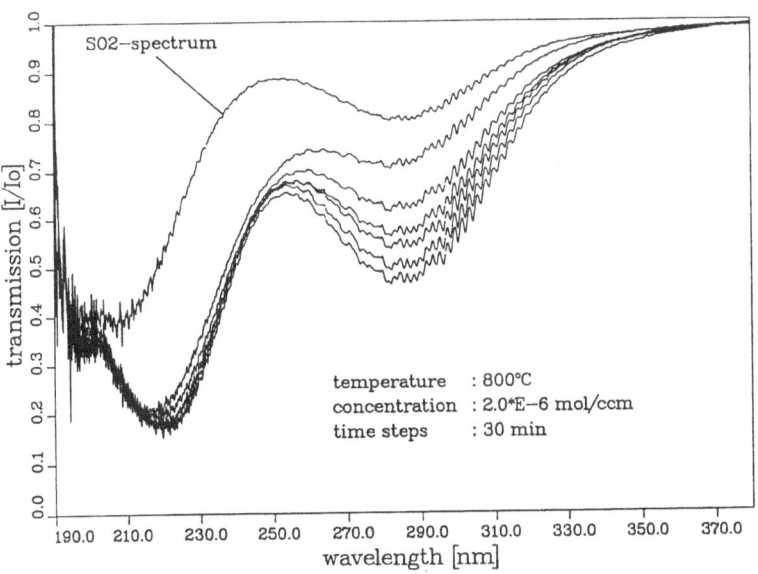

Figure 12: Thermal decomposition of SO_3 at 800 °C

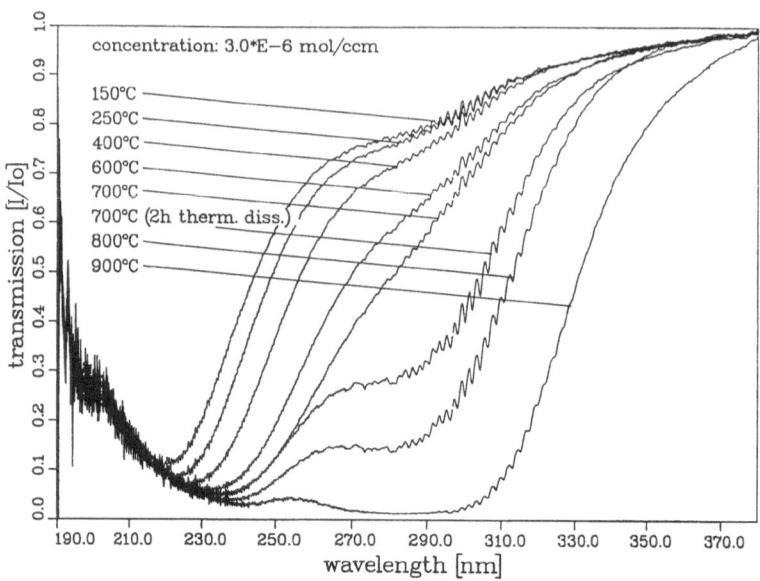

Figure 13: Temperature dependence of the transmission spectra

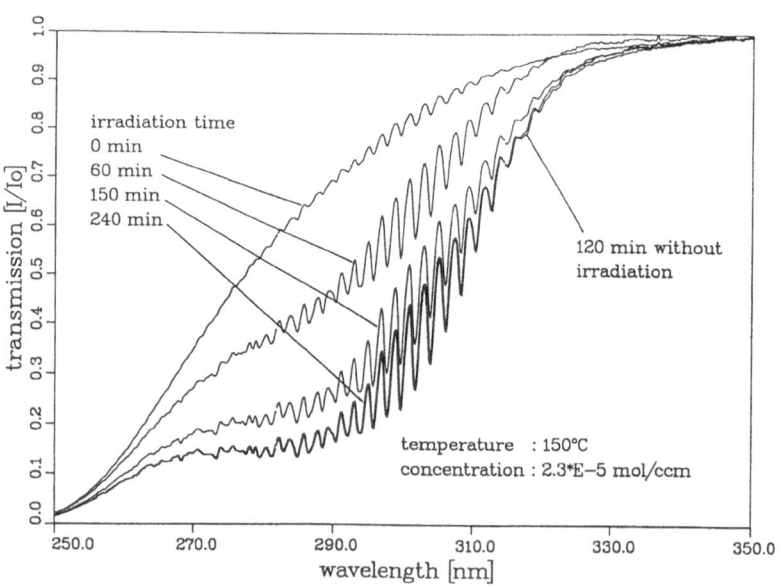

Figure 14: Photodissociation of SO_3 at 150 °C

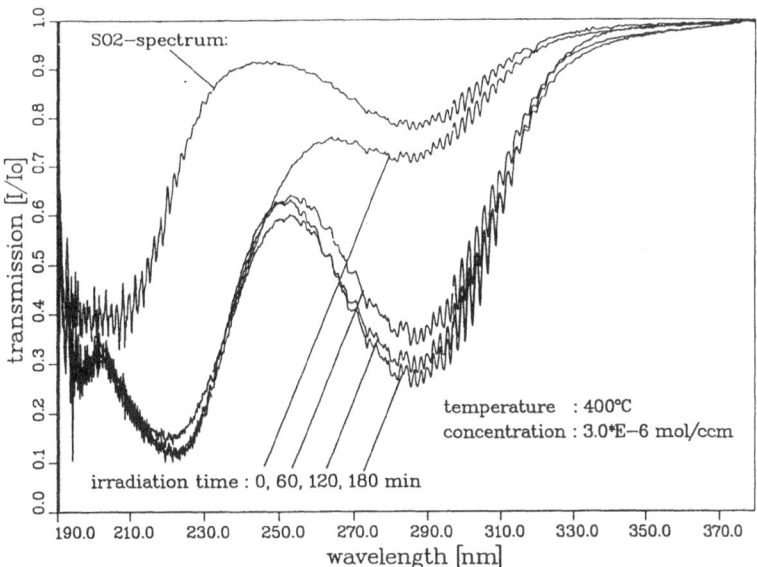

Figure 15: Photodissociation of SO_3 at 400 °C

A simple solar receiver, shown in fig. 16, has been built using a reflector with an aluminum oxide surface. The resulting transmission spectra after several time steps of sunlight illumination is shown in fig. 17. A decomposition of SO_3 is clearly observable!

At this stage it can not be excluded that SO_2 is formed by a catalytical reaction of the gas with the silica fused glass walls during light illumination. But even if the decomposition occurs on this way technical applications would be conceivable.

Although the decomposition seems to occur only after hours three points should be kept in mind:

- a very simple solar receiver has been used,
- a faster decomposition may be expected at higher temperatures,
- the sunlight illumination at Aachen is typically poor.

As a consequence, further laboratory and large scale experiments are necessary to give a final judgement about the technical applicability of sunlight for decomposition of sulfur trioxide.

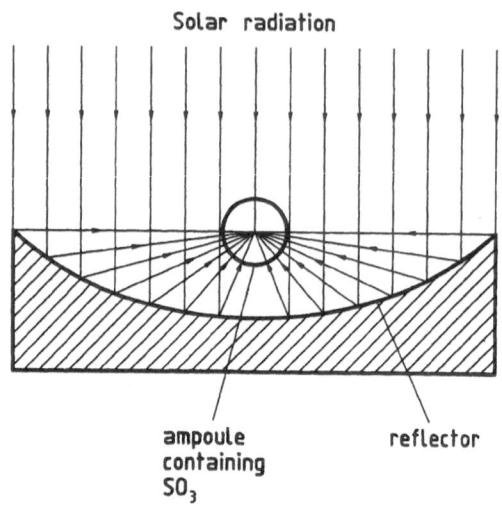

Figure 16: Simple solar receiver

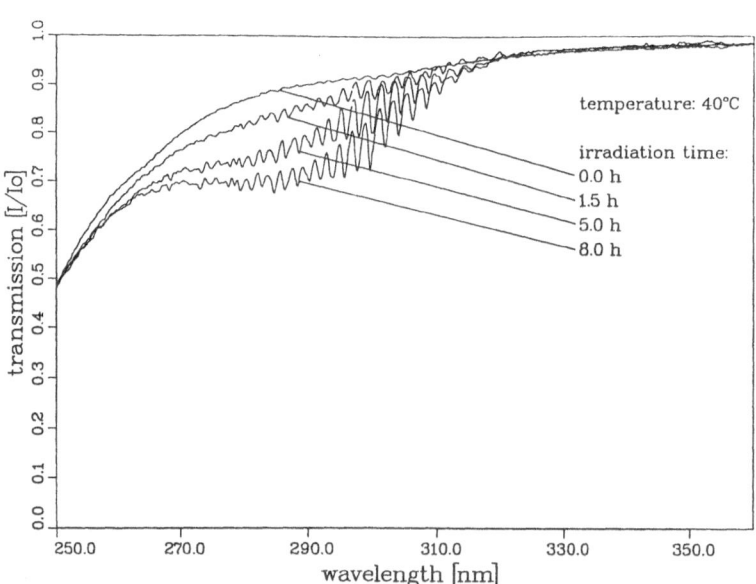

Figure 17: SO_2 production by solar irradiation

6 Conclusions and view on work in progress

The results of our experiments may be summarized as follows:

- It is possible to decompose SO_3 by illumination.

- The SO_2 production is increased at higher temperatures.

- The ultraviolet part of the spectrum is responsible for the absorption of light by SO_3.

- Because of the similarity of absorbing wavelengths it is difficult to separate and quantify SO_3 absorption even if only small fractions of SO_2 are present.

- Although absorption takes place predominantly at wavelengths below about 250 nm, the total range seems to reach more than 300 nm.

- Their seems to be a temperature shift towards higher wavelengths favouring photochemical reactions by solar light absorption.

- A first test by exposing a SO_3 sample to direct sun light indicates that it is possible to decompose SO_3 even at low temperatures.

The results obtained until now are encouraging enough to continue our research. Several points have to be clarified:

- For reasons already mentioned it has not been possible to quantify SO_2 or SO_3 concentrations by absorption spectroscopy. By Raman spectrocopy it will be possible to determine reaction rates as a function of temperature, illumination and other parameters.

- The way of decomposition and reformation of SO_3 at various conditions will also be investigated by this additional technique.

- Further experiments by solar light illumination will help to judge the possibility of SO_3 decomposition for large scale industrial applications.

7 References

[1] Lovejoy, R.W.; Collwell, J.H.; Eggers, D.F.; Halsey, G.D.: *Infrared spectrum and thermodynamic properties of gaseous sulfur trioxide*, J. Chem. Phys., **36** (1962) 612/7

[2] Stopperka, K.; Filz, F.: *Die Zusammensetzung der Gasphase über dem flüssigen System H_2SO_4-SO_3 in Abhängigkeit von der Temperatur*, Z. anorg. allgem. Chemie, **370** (1969) 59/66

[3] Walrafen, G.E.: *Raman spectral studies of oleums*, J. Chem. Phys., **40** (1964) 2326/41

[4] Amelin, A.G.; Illarinov, V.V.; Borodastova, Z.B.: Zh. Fiz. Khim., **25** (1951) 524/6

[5] *Gmelins Handbuch der anorg. Chemie*, 8. ed., Schwefel Erg. Bd. **3**, Springer, Berlin (1980)

[6] Schwaigerer, S.: *Festigkeitsberechnung im Dampfkessel-, Behälter- und Rohrleitungsbau*, 1983

[7] VDI-Wärmeatlas,2. Auflage, 1974

[8] Herzberg, G.: *Molecular Spectra and Molecular Structure*, vol. **2**, p. 178, p. 302, van Nostrand, New Jersey (1964)

[9] Fajans, E.; Maki, A.G.: *The Absorption Spectrum of Sulphur Trioxide*, Trans. Faraday Soc., **32**, 511, (1936)

[10] Suto, M.; Ye, C.; Ram, R.S.; Lee, L.C.: *SO_2 Fluorescence from Vacuum Ultraviolet Dissociative Excitation of SO_3*, The Journal of Physical Chemistry, Vol. **91**, No. 12, 1987

[11] Stopperka, K.: Z. Chem. (Leipzig), **6**, 153 (1966)

[12] Kaldor, A.; Maki, A.G.: *The Assignment of ν_2 and ν_4 of SO_3*, J. Mol. Spectrosc., **45**, 247–252 (1973)

[13] Dorney, A.J.; Hoy, A.R.; Mills, I.M.: J. Mol. Spectrosc., **45**, 253 (1973)

[14] Bodybey, V.E.; English, J.H.: J. Mol. Spectrosc., **109**, 221 (1985)

The Use of Thermal Solar Energy
to Treat Waste Materials Part 2

H. Effelsberg, E. Kirchner, B. Barbknecht,

Krupp Industrietechnik GmbH,
Systemtechnik,
Essen

Contents

1 Introduction

Ecology-minded measures to dispose of problematic waste are getting more and more difficult. By careful high temperature treatment, toxically and biologically relevant compounds have to be converted into compounds that are neutral towards the environment. The first part of the study had the purpose to make a statement about the technological applicability of thermal solar energy for the treatment of certain substances. For said purpose, different groups of waste (filter dusts coming from garbage incinerating plants, exhaust gas cleaning residues from the production of special steel, paint sludge from the automobile industry and sewage sludges) were investigated with regard to their thermodynamic features. The treatment of paint sludges turned out to be interesting. These substances permit the efficient utilization of the energy offered by a solar plant. The solvents can be separated in the low temperature range. In the medium temperature range, the organic components can be treated. Filter dusts can be treated in the high temperature range. After the adequate amount of energy has been supplied the harmful substances are stripped through evaporation and the resulting sludge can be disposed of in a manner that is not harmful to the environment.

The most difficult point in utilising solar thermal energy for the detoxification of harmful waste is to transfer the energy to the substances. There are a lot of receiver concepts to utilize solar energy, but only a few ones that worked on solid substances.

The objective of this study is the search for a receiver/reactor concept to treat waste materials in a solar-thermal plant.

In Section 2 the process-technological conditions for receivers applied for waste treatment are examined.

In Section 3 receiver found in literature for the thermal treatment of solids are under discussion.

Section 4 shows proposals and layout of receiver/reactor configurations for the treatment of waste. In Section 4.1 a heat exchanger reactor for the treatment of paint sludges is described. Section 4.2 analyses adaptions to field sector and reciever aperture. Section 4.3 developes configuration and lay-out of a melting reactor-receiver for the detoxification of filter dusts from waste incinerating plants.

In Section 5 operating parameters are discussed.

2 Process-Technological Conditions for Receivers Applied for Waste Treatment

In part 1 of this study some thermodynamic-caloric material values were compiled for the solar thermical treatment of industrial dusts based on oxidic compounds. The compositions of paint sludges, UHP-dusts and filter dusts coming from garbage incineration are shown in the following tables (1,2,3).

Table 1: Composition of a selected paint sludge sample from the motor industry [MOHL Stuttgart] (calculated as oxides)

Substance	Composition (in $\frac{kg}{100kg}$)
Al_2O_3	3.58
BaO	13.80
CaO	5.29
$CHCl_3$	1.20
Cr_2O_3	0.37
Fe_2O_3	4.80
H_2O	17.40
K_2O	0.81
MgO	1.63
P_2O_5	4.10
PbO	0.07
SiO_2	8.22
SO_2	5.91
SrO	0.32
TiO_2	28.00
WO_3	0.30
ZnO	1.21
ZrO_2	0.04
Miscellan.	1.47

Table 2: Composition of two selected waste gas cleaning residues from stainless steel production (UHP-dust) [KRUPP-Stahl AG] (calculated as oxides)

Substance	Composition (in $\frac{kg}{100kg}$)	
	Sample 1	Sample 2
Al_2O_3	0.40	1.10
CaO	7.50	18.60
Cl^-	0.50	–
Cr_2O_3	18.10	0.65
CuO	0.10	–
Fe_2O_3	48.70	19.90
K_2O	1.73	2.00
MgO	2.80	5.80
Mn_3O_4	3.60	22.90
MoO_3	0.10	–
Na_2O	1.21	1.33
NiO	1.40	0.03
PbO	0.39	0.89
SiO_2	7.90	0.29
SO_2	2.00	–
TiO_2	0.20	–
V_2O_5	0.10	–
ZnO	8.50	16.6

Table 3: Composition of a selected filter dust sample from refuse incinerators [MVA-Bielefeld] (calculated as oxides)

Substance	Composition (in $\frac{kg}{100kg}$)
Al_2O_3	11.204
BaO	0.277
$Ca(NO_3)_2$	0.173
$Ca_3(PO_4)_2$	3.128
$CaCl_2$	10.867
$CaCO_3$	2.740
CaO	2.384
$CaSO_4$	10.348
Fe_2O_3	2.806
K_2O	6.127
MgO	1.900
Na_2O	5.336
PbO_2	1.011
SiO_2	32.263
SnO_2	0.299
TiO_2	1.584
ZnO	3.668

The thermal treatment of such dusts depends upon the desired final product and thus upon the required reaction balances. These are a function of temperature, pressure and gas- or molten mass composition respectively and reaction time, i.e. the reaction kinetics. The following treatment aggregates are possible:

- mass reducing pyrolysis furnaces
- oxidically working melting furnaces
- mass reducing slag furnaces

In the following, the required reaction conditions and possible reactor- and plant configurations will be outlined on the basis of the solar thermical energy supply.

The following furnaces seem to be eligible:

- Mass reducing pyrolysis furnaces with indirect heat supply through heat exchangers with temperatures not exceeding < 800 °C.

- oxidising working melting furnaces with direct solar insolation for temperatures of ~ 1200 °C.

2.1 The Medium Temperature Range

The medium temperature range is limited by temperature gradient of the heat transmission through the solar-receiver and the possible heat exchanger. A temperature of 800 °C, reduced by the heat exchanger gradient, results as upper temperature limit for the GAST-alloy receiver.

2.1.1 Reaction Levels of Paint Sludges

Due to their sufficient quantity and finest distribution of reduction gas generating polymer resins or hydrocarbons paint and sewage sludges can be subjected to direct pyrolysis.

Pyrolysis tests with paint sludges have been described by Hedden and Vesper [19]. The aim of such tests was to convert the heavy metal compositions of the pigments in residues neutral to environment and thus qualified for deposition on dumping grounds.

The observed reactions with increasing temperatures are as follows:

- decomposition of polymer   >300   < 600   °C
- carbonate separation   >200   < 800   °C
- chromate reduction   >300   < 600   °C
- sulphate reduction   >600   < 900   °C
- oxide reduction   > 900   °C
- metal vaporization

The tests were carried out in a constant circulation gas atmosphere with 20 % CO, N_2 with a heating-up rate of 5 °C/min at the least and a dwell time of 30 min. under final temperature.

Figure 1 shows the measured reactions of pure pigments having applied N_2 for carbonate separation and N_2 plus 20 % CO for reduction reactions. Furthermore, ZnO was heated-up in a CH_4- atmosphere and the generation of CO was observed; the reduction of ZnO took place at 900 °C.

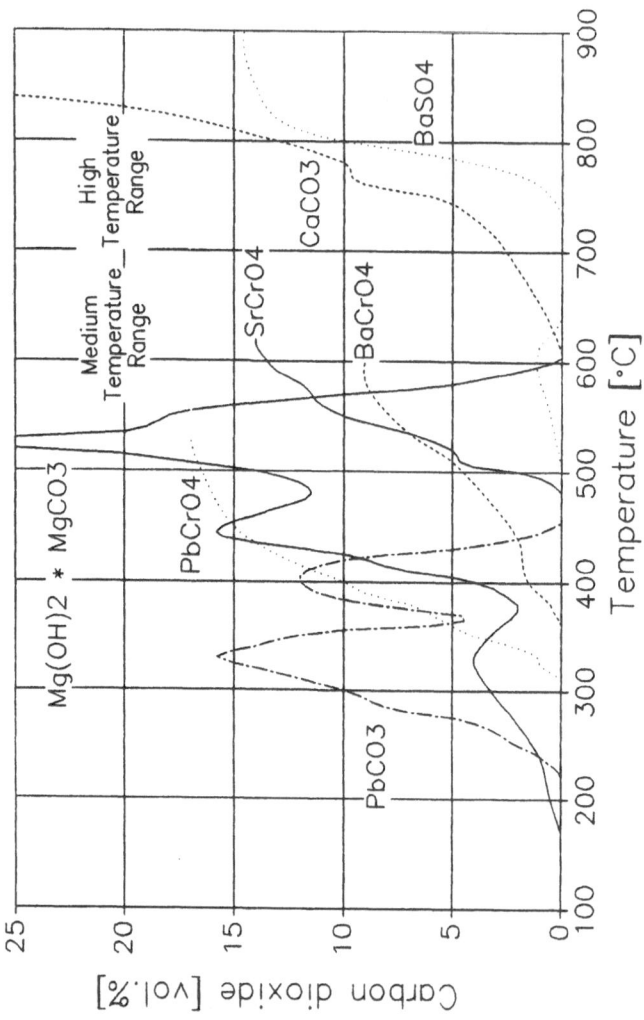

Figure 1: Reactions of Paint Sludge Content Matters as Function of Temperature

The carbonate decompositions are pure temperature functions and only depend upon the CO_2 partial pressure.

$$MCO_3 \longrightarrow MO + CO_2$$

M stands for various metals. Except for ZnO, heavy metals are as a rule difficult to dissolve.

The chromate reductions require CO or reactive hydrocarbons as reduction gas.

$$2\ MCrO_4 + 3\ CO \longrightarrow (2\ MO)\ Cr_2O_3 + 3\ CO_2$$

The generated chrome(II)-oxides are difficult to oxidize, that means only at very low pH-values, and are thus soluble again. M stands for various heavy metals of chromate colour pigments.

In fact, oxide reduction is only essential for zinc because being an amphoteric hydroxide it returns to solution. The reduction to volatile Zn, however, is very incomplete for CO at temperatures not exceeding 1000 °C. The tests show that an increased reduction takes place, probably due to the hydrocarbon radicals.

$$ZnO + (CH_3)\cdot \longrightarrow Zn + CO + \tfrac{3}{2}\ H_2$$

K_2O and Na_2O turn to KO and NaO-gases. Fe_2O_3 and heavy metal oxides are already reduced before ZnO.

The sulphate reduction results in black ash being elutable and thus requiring reoxidation.

$$BaSO_4 + 4\ CO \longrightarrow BaS + 4\ CO_2$$

The described reaction balances take place in the diffusion and Boudouard balance within the finely ground grain. In the course of the drying process and during further heating by gassing (CO_2 separation, cracking) the grain becomes porous but can also pass composition depending fusion points.

- polymer softening < 300°C

- slag softening > 800°C

For reaction control it is of advantage to arrange for the CO_2 separation prior to the polymer decomposition.

- polymer decomposition 300 − 440°C

- cracked gases $(CH_3)\cdot > 550°C$

The following treatment possibilites for paint sludges result from the reaction analysis:

- reduction treatment up to $\sim 600°C$ without ZnO and $BaSO_4$ reduction. ZnO is protected against elution by the carbon network.

- reducing/oxydizing treatment up to above 1000 °C with Zn and Pb/PbO vaporization and BaS oxidation. This treatment makes the most heavy metal oxides turn to sintered metal/metal carbide compositions.

2.1.2 Thermal Output

The specific heat consumption of the reactor depends upon:

- the heat capacity of the reactor material

- the heat tone of the reactants

- the heat losses of the reactor

$$\Delta i_R = \sum \left[c_{p_n}(T)\Delta T \mid_u^0 + h_n(T) \right] + \sum r_n(T) + \frac{\dot{Q}_V}{\dot{m}} \tag{1}$$

The heat consumption related to the reactor material depends upon the composition n and generally increases linear to ΔT. The melting, vaporization and phase transformation enthalpies are as a rule small and distributed across the temperature field ΔT.

The carbonate decompositions are endothermic and the chromate or sulphate oxidations are exothermic. Depending on the composition, the caloric effects can offset against each other. The heat loss $\frac{\dot{Q}_V}{\dot{m}}$ of the reactor depends upon:

- temperature of the reactor

- insulation factor

- reactor surface

-250-

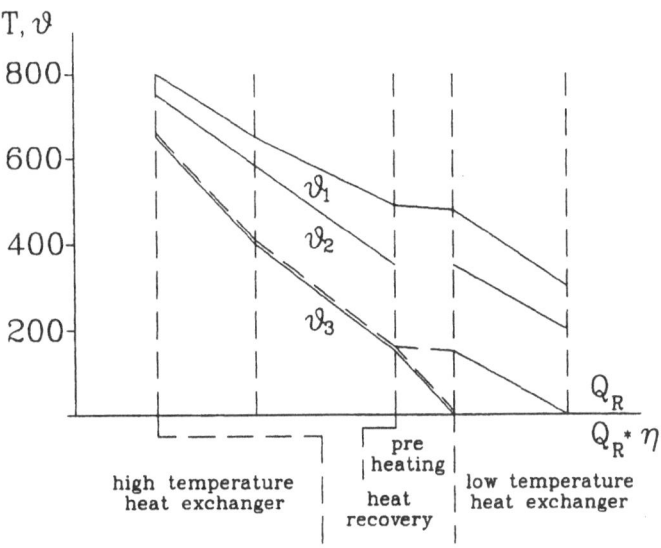

Figure 2: Required heat consumption for the system paint sludge as a function of temperature

The heat loss does not depend directly upon the process flow. It is assumed to be an integral heat efficiency grade η_H.

To optimize process control, the following conditions should be aimed at:

- quick heating-up to reach the reaction temperature

- keeping the reaction temperature on its level

- keeping heat penetration as uniform as possible without excess or insufficient temperatures

- long contact periods of reaction gases in the grain

Considering the granulation condition and the reaction control a direct gas/packed bed reactor seems to be the optimum aggregate.

The heat to be supplied by the solar plant via the heat exchanger depends upon:

- the supply temperature T_e, depending upon receiver and transfer temperatur ϑ

- the upper reactor temperatur T_o exceeding the solar temperature level and to be reached by combustion

- the proportional heat loss.

2.1.3 Temperature Gradients

With the following caloric data, the GAST-receiver can serve as energy supplier (table 4):

Table 4: GAST-Receiver Data

Receiver		Alloy	Ceramics
Temperature In	°C	300	300
Temperature Out	°C	800	1000
Pressure	bar	9,5	9,3
Pressure loss	bar	0,5	0,23
Air per module	$\frac{kg}{s}$	2,45	0,48
Air per tube	$\frac{kg}{s}$	0,136	0,048
Number of tubes		18	10
Heat flow per module	MW	1,31	0,36
Heat flow per m²	kW	72,4	78,3
Surface per module	m²	18,1	4,6

It is the higher temperature level of the ceramic receiver that represents the most important difference between the two receiver types. The required equipment, however, especially the internal insulation and the gas flow controls from the tower are much more sophisticated and expensive.

Therefore, the alloy receiver with an operating temperature of 800 °C should form the basis of the plant's arrangement.

The transmission temperature of the equipment is determined by:

$$\dot{Q} = F \cdot \frac{\lambda}{s} \cdot \vartheta = \dot{m} \cdot c_p \cdot \Delta T \tag{2}$$

$$
\begin{aligned}
\dot{m} \cdot c_p \cdot \Delta T &= \text{heat flow of the medium} \\
F \cdot \frac{\lambda}{s} \cdot \vartheta &= \text{heat flow through the boundary} \cdot \text{s} \\
\vartheta &= \text{temperature gradient above the boundary layer}
\end{aligned}
$$

The boundary layer conductance $\frac{\lambda}{s} = \alpha$ represents a material value for firm heat exchanger surfaces whereas it depends upon flow conditions in case of flowing media.

$$Nu = f(Re,Pr) \qquad (3)$$

ϑ is reduced by:

- enlargement of the heat transmission surface F

- increase of the boundary layer conductance $\frac{\lambda}{s}$ or α respectively

- reduction of the heat flow or the mass stream respectively.

An assessment of the system immediately reveals the following:

- A pyrolysis drum will require a very high ϑ because both the drum's surface F and the boundary layer conductance $\frac{\lambda}{s}$, especially from the drum's interior wall to the granular material gets very small.

- An additional circulation medium between the solar receiver circulation and the reactor circulation will only result in additional temperature trans mission losses. And this will even be the case when applying a high conduction or high capacity medium like sodium or over-critical gases because the required heat exchanger surfaces and the excess temperatures will be determined by the resistances existing on the receiver or reduction gas interface. Apart from that, over-critical gases only find an efficient application within a small temperature range lying far below the desired final temperature and are difficult to operate due to their high operating pressures.

- Solid bed and fluidized bed reactors show very low temperature losses in case of finely granulated grain. In this case, however, a reducing gas has to be supplied to circulate via a heat exchanger.

- An appropriate conditioning of the reducing gas mainly consists of N_2, CO and lighter hydrocarbons enables a direct heat exchange in the solar receiver.

In section 4.1 construction and operation parameters of a heat exchanger and a pyrolysis reactor plant as well as a heat flow balance are developed.

-254-

2.2 The High Temperature Range

Within the high temperature range, the solar radiation energy is directly transmitted to the reaction material. This entails constructional and material-technological adaptations of the solar insolation geometry and the material flow control. Furthermore, the open reactor atmosphere limits the chemical reaction control.

2.2.1 Reaction Levels of UHP-Dust

UHP-dusts exist of oxides stemming from exhaust gas residues of the austenitic steel production. Iron and chrome are predominant with the latter being an essential precious metal. To recirculate UHP-dusts, that means to return said dusts to the steel production process, the following treatment is necessary:

- removal of zinc, lead as well as of sodium and potassium

- reduction of precious metals, especially of chrome

To deposit UHP-dust and at the same time protect it against elution the following treatment is necessary:

- removal of zinc

The reduction of ZnO requires minimum temperatures of 1000 °C. Processing it performed with carbon to pellets in a dry distillation, e.g. in a multiple-hearth, seems possible always provided that a softening of $Fe_2O_3 \cdot SiO_2$ or a formation of Fe_3C is being avoided by careful reaction control.

$$Fe_2O_3 \cdot SiO_2 + 3\ CO \xrightarrow{\sim 1500 K} 2\ Fe + SiO_2 + 3\ CO_2$$

$$Fe_2O_3 \cdot SiO_2 + C_n + 3\ CO \xrightarrow{< 1500 K} 2\ Fe + SiO_2 + C_{n-3} + 6\ CO_2$$

Even at a temperature of 1000 °C the vaporization of Zn and Pb or PbO turns out to be very slow and incomplete but it can be accelerated by negative pressure.

$$ZnO + CO \longrightarrow Zn_{(g)} + CO_2$$

$$2\ PbO + CO \longrightarrow Pb_{(g)} + PbO_{(g)} + CO_2$$

Disadvantages for solar treatment are as follows:

- The reaction temperature and thus the major heat consumption are above the solar energy of $< 800°C$ that can be supplied indirectly

- Due to long treatment periods at high temperatures energy consumption is very high

- A processing to pellets is very cost-intensive

- This treatment does not permit recycling

If the UHP-dust is to be processed in order to be reapplied for special steel production, the reaction temperature is determined by the reduction of chrome.

$$4\ Cr_2O_3 + 4\ C \longrightarrow Cr_4C + 6\ CO$$

This reaction only starts in the fused state determind by the slag composition, especially by the portion of $Fe_2O_3 \cdot SiO_2$ or similar compounds, and lies at temperatures $> 1300°C$. Following the reduction of FeO to Fe_3C a gravity-induceddepending separation of carbidic and metallic phase begins ($> 1400°C$). To prevent Cr_2O_3 and other metal oxides from getting lost in the slag higher temperatures are required. To lower the carbide contents which is undesired in steel, UHP-dust is being reduced in the C-balance by using an argon atmosphere with electric or plasma treatment during which Zn and Pb are stripped as vapour. A solar thermal treatment is opposed by major operational difficulties.

- The reducing atmosphere in the furnace can only be insufficiently generated in a solar melting furnace that is open by necessity

- As far as material technology and construction are concerned the transmission of the high temperatures to the molten mass is rather difficult

- As an alternative, nothing but a carbidic Fe-Cr can be won

2.2.2 Reaction Level of Filter Dust

Filter dusts originating from garbage incinerating plants contain manifold and often different oxides and salts.

The oxides, ZnO as well as Na_2O, K_2O and CaO are relevant for their hydroxide solubility. The alkaline chlorides, sulphates and nitrates, too, are soluble. Phosphates, however, are not. The elutable portion of these dusts amounts to approx. 40%. The elutability can be drastically reduced by melting.

- Thermal decomposition, especially of K_2O, nitrates and carbonates

- Evaporation of PbO, As_2O_3 and NaCl

- Reduction of solubility by slagging with non-soluble components

The oxidic slag phase is likely to form only above 1200 °C depending on the portions of NaCl, $CaCl_2$, $Na_2O \cdot SiO_2$ and $Fe_2O_3 \cdot SiO_2$, whereby the latter portions must not necessarily exist as compound in the dust or have a higher melting point as individual compounds. For the fused mass behaviour it is favourable if a compound already melting at low temperatures like $Na_2O \cdot SiO_2$ (870 °C) or NaCl (800 °C) initiates the fused phase through the formation of a solution. Though NaCl is water-soluble, it can be stripped out of the molten phase by vaporization. $Na_2O \cdot SiO_2$ is not water-soluble. The vaporized masses are small (table 5):

Table 5: Vaporized substances

K_2O	$\longrightarrow$ 2 KO	> 880°C	7,2 %
$Ca(NO_3)_2$	$\longrightarrow$ NO_2	> 600°C	0,1 %
$CaCO_3$	$\longrightarrow$ CO_2	> 600°C	1,2 %
PbO_2	$\longrightarrow$ PbO	> 885°C	< 1,0 %
NaCl	$\longrightarrow$ NaCl	> 800°C	N.A.
As	$\longrightarrow$ As_2O_3	> 400°C	< 0,1%

2.2.3 Thermal Output and Temperature Gradients

Due to the direct solar energy insolation the reactor is at the same time the receiver. Heat consumption is determined by the reaction and operation data as it is the case with the indirect reactor. The thermal output is a function of the field and reactor geometry. Both have to be matched by the operation mode of the reactor according to material-technological temperature and reaction limits set. In section 4.2, the field and reactor parameters for a radiation reactor are worked out.

The requirements set forth by the reaction control are as follows:

- Material supply and material flow control according to the aggregate condition and the flow properties of the material to be treated

- Transformation to the molten state

- Formation of thermal reaction surfaces or favourable reaction conditions respectively

- Avoidance of operatively unstable states of process

In section 4.3, the construction and operation parameters of a reactor as well as a heat flow balance are being worked out.

3 Receiver for the Thermal Treatment of Solids

In this chapter, we want to discuss some receiver concepts for the solar-thermal treatment of waste. Due to a graduated temperature- exposed treatment these receivers are classified as low, medium and high temperature receivers. In this part of the study, we want to consider concepts for the treatment of filter dusts coming from local garbage incinerating plants (high temperature range) and for the treatment of paint sludges generated in the automobile industry (low and medium temperature range). In the low temperature range, a removal of solvents and water is eligible. In the medium temperature range, the carbon contents could be carbonized and in the high temperature range, the material to be detoxified could be fused. The receiver concepts were broken up in indirectly and directly absorbing receivers. In the matter on hand, indirectly absorbing receivers are those aggregates that heat-up a heat transfer medium (gaseous or liquid) by concentrated solar radiation with a heat transfer medium transmitting the heat energy to the material to be treated (figure 3).

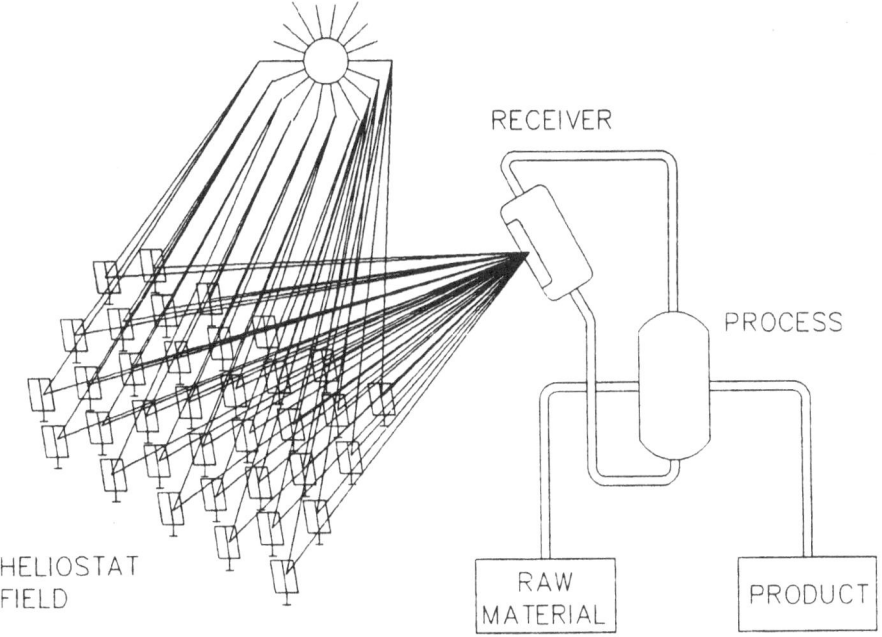

Figure 3: Indirectly solar energy input [10]

Directly absorbing receivers are aggregates in which the material to be treated absorbs the solar radiation directly and transforms it into heat (figure 4).

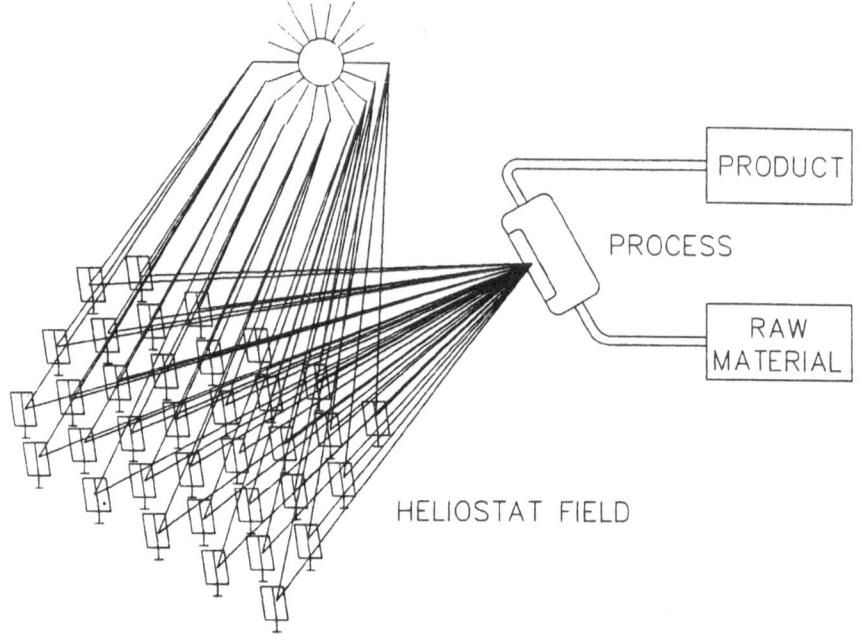

Figure 4: Directly solar energy input [10]

As the latter ones as a rule allow higher temperatures and make it possible to utilize the UV-portion of the solar radiation for detoxification these concepts will be considered separately.

3.1 Low Temperature Heat

The most suitable method for drying, i.e. removing the solvent and water portions at approx. 150 °C is the utilization of the lost heat from the medium or high temperature part respectively. A supply of hot gas via heat exchangers from parabolic through-collectors arranged on the bottom seems to be promising, too. Some concepts for the solar drying of different materials were elaborated. Due to the too low drying temperatures of approx. 70-90 °C, however, these are not eligible in this case. In California, parabolic through-collectors are today applied for electric power generation (SEGS-Solar Energy Generating System) [5,17]. These reflectors are long parabolic tubes having a tube in their focal line that accomodates flowing heat transfer oil. By means of the reflectors, sun radiation is focussed on the tube thus heating-up the heat transfer oil up to 350 °C. Through heat exchangers, this heat can now be transmitted to a carrier gas that in turn can be applied for drying sludges. The water and solvent content of paint sludges lies between 10 and 60%. This process will be described hereafter. In case of contact drying the required heat is directly transmitted from heated-up surfaces to the material which has either permanent firm contact to the heating surface or is again and again rearranged on said surface. Due to the risk of caking on the heating surfaces this approach is less suitable for drying paint sludge because caking blocks the dryer or at least reduces its drying capacity. In case of convectional drying - on the other hand - the heat necessary for vaporizing the liquid is transmitted to the material by means of a gaseous or vapour-like drying agent. The individual processes for convectional drying differ in their ways to bring the material to be dried in contact with the hot gas. The material to be dried is either overflowed, cross-flowed, whirled up, dragged or attrited. In this study, we want to describe two different processes for drying paint sludge using hot gas produced by an indirectly working solar receiver.

Fluidized bed dryers are characterized by their high drying capacities [32]. A fluidized bed dryer (figure 5) makes the material to be dried flow through the fluidized bed duct in horizontal direction whereas the drying medium gas passes the blower stream bottom in a vertical direction and thus causes a fluidization of the material. Conveyance is then automatically effected due to the fluidized material's proper motion and supported by the vibration movement of the fluidized bed duct in its lengthwise direction. The material's dwell time can be changed by modifying the vibration drive. As the design of such a dryer can also have a closed structure it is also suited for processes during which the solvent is to be recovered.

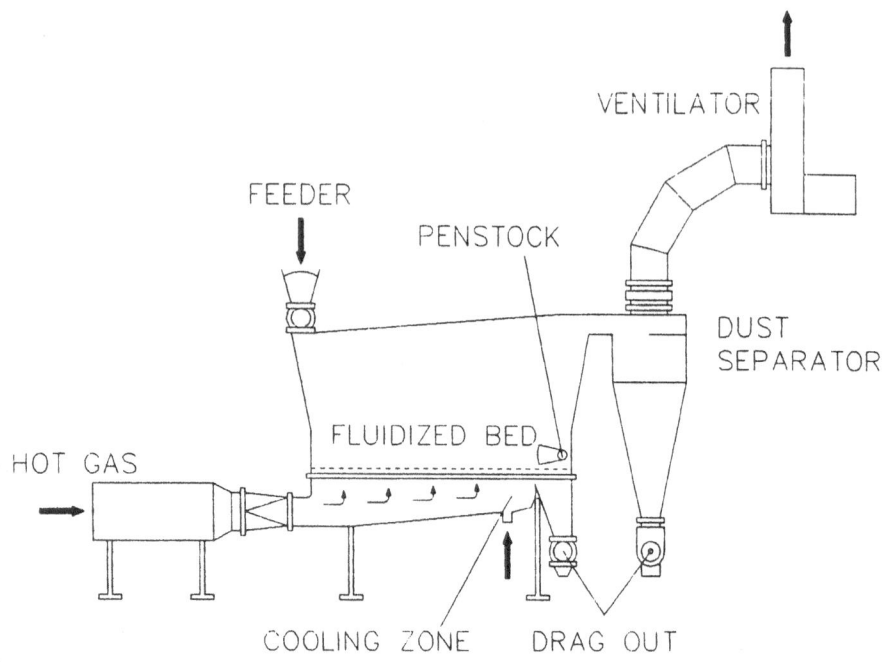

Figure 5: Fluidized Bed Dryer (Büttner-Schilde-Haas Design) [32]

Material permitting pneumatic conveyance can be dried during transport in a stream dryer. The most simple dryers for this procedure consist of a vertical ascending pipe in which granulated or broken substances are dried in a gas stream during suspension (figure 6).

Drying only takes some minutes so that only finely granulated substances permitting a fast heat and material exchange or coarse grained products to be dehumified from their retained water can be dried. For dry material whose interior moisture has to be diffused away the stream dryer is only conditionally qualified. Often, the time required for the diffusion processes is not sufficiently available. For the latter products, multistage steam dryers can be applied, if necessary. Any stream dryer employs direct-current. The supplied moist material contacts the hottest drying gases and the material and heat exchange conditions are favourable because the dry material particles are swept by the drying gas at any side. The material maintains its cooling limit temperature as long as surface evaporation takes place. Only after this has finished the heating of the dry material particles starts. Heating, however, finds its limits because at this point of time the gas stream has already been cooled down by the evaporation of water. For this reason and due to the material's low dwell time in the dryer even materials sensitive to high temperatures can be dried by means of high gas entry temperatures without suffering damage. Centrifuge- or filter-moist inorganic and organic compounds, synthetic powders and granulates, food and feeding stuff as well as other products like sawdust, sand, quartz etc. can be dried in the stream dryer.

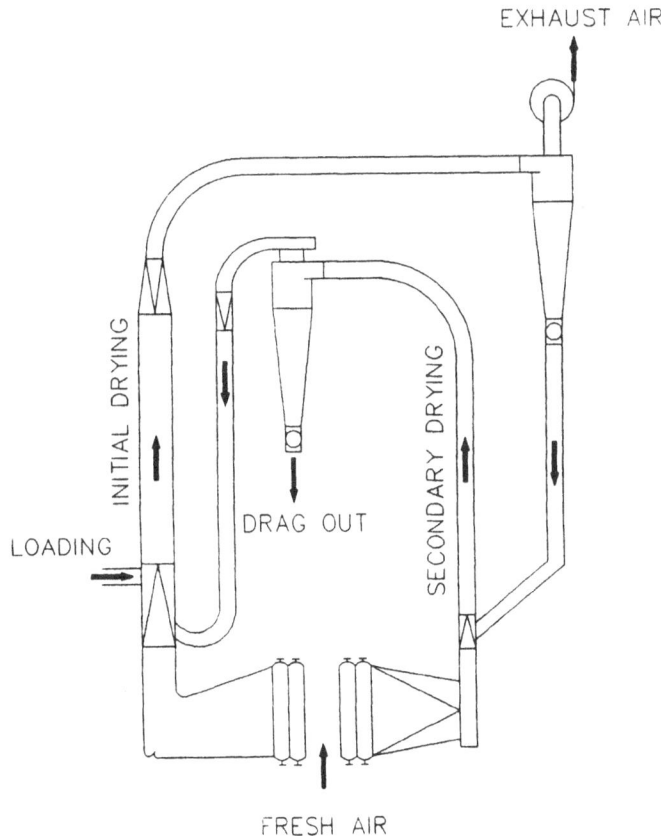

Figure 6: Two-stage Stream Dryer (H. Orth, Böhl/Pfalz Design) [32]

The water and solvent portions contained in the carrier gas are removed by condensation. Depending upon the pureness of the condensate, this can be processed further or has to be disposed of. After the heating-up step the carrier gas can be reapplied for drying.

3.2   Receiver for Medium Temperature Range

In the medium temperature range (up to approx. 800 °C) indirectly operating receivers can be applied. Here, the temperature is related to the material to be treated. Losses caused by transmitting heat to the gas have to be added so that receiver temperatures of approx. 900 through 1000 °C are to be preferred.

3.2.1   Receiver with Radiation-Absorbing Tube Walls

Tube receivers (figure 7) consist of heat exchanger tubes planely arranged in a cavity. The opening has an oval shape and is downwardly inclined towards the heliostats. Compressed air enters the receiver and is led through the heat exchanger tubes. The solar radiation absorbed by the outer sides of the tubes heat up the air that is led through them.

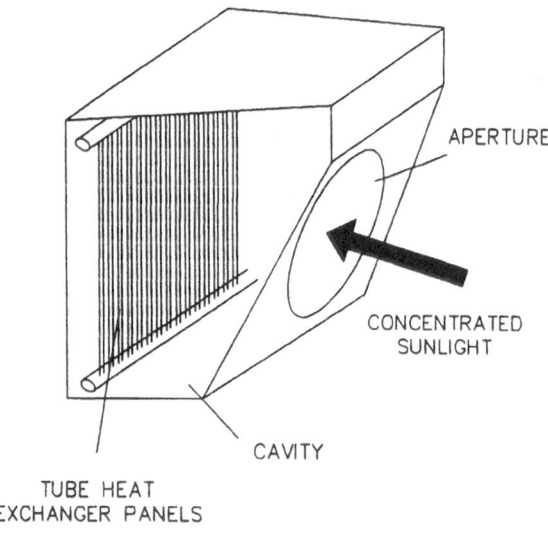

Figure 7: Schematic Diagram of a receiver with radiation absorbing tube walls

3.2.2 Metal Tube Receivers

If the tube consists of metal the receiver can only be applied for temperatures below 1000°C. With the Boeing Bench Receiver [12] gas having a temperature of 540 °C, 9 bar, was heated-up to 820 °C with surface temperatures lying below 870 °C. The metal type of the tube receiver within the GAST- project (gas-cooled solar tower) was designed for the heating of a gas mass stream of 2,45 kg/s, 9,3 bar from 625 °C up to 800 °C [10].

3.2.3 Ceramic Tube Receivers

Ceramic tube receivers consist of silicium carbid U-tube heat exchangers and can thus also be applied for temperature slightly above 1000 °C [12]. The employment of ceramic absorber tubes was tested with the BMSR (Bench Model Solar Receiver) showing the following design data: gas entry temperature 482 °C, gas outlet temperature 1065 °C, mass stream 0,93 kg/s and pressure 9 bar. The ceramic type of the GAST-tube receiver [10] was designed for heating-up an air stream of 0.48 kg/s and 9.3 bar up to 1000 °C.

3.2.4 Volumetric Receivers

During the past years, development focussed more and more on volumetric receivers [11,16,24,29]. The decisive difference between these receivers and tube receivers lies in the three-dimensional absorption of the concentrated solar radiation. This takes place in a ceramic matrix transmitting the heat to a heat transfer medium (figure 8). On the one hand, this can be the surrounding air taken up by the matrix. On the other hand, also systems with helium closed by a window were designed as heat transfer medium. A closed cycle makes smaller heat transfer surfaces possible. This concept, however, has the disadvantage that a window absorbs a part of the solar radiation and has thus to be cooled.

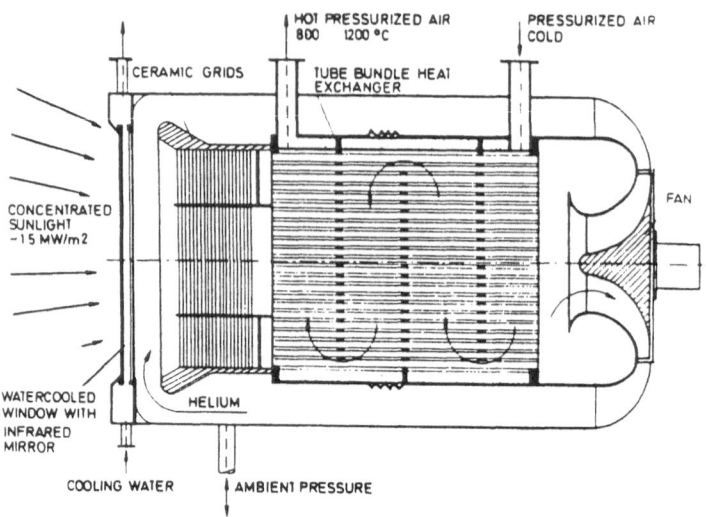

Figure 8: Schematic Diagram of a Volumetric Receiver

3.3   Receivers for the High Temperature Range

3.3.1   Direct Absorption Receivers

The direct absorption of the highly concentrated solar radiation at the medium to be
heated up, that takes place in the DAR (Direct Absorption Receiver), permits a more
efficient and less expensive construction of receivers [9,13,20,27].

3.3.2   Fluidized bed Solar Receivers

In case of the described test arrangement [4,15] (figure 9) the fluidized bed serves to
improve the heat transfer from the hot absorbing surface to a heat transfer gas rather
than serving as absorber medium. The fluidized bed is 2 cm thick, 90 cm high and 78 cm
wide. The fluidized bed material consists of 14 kg of aluminium oxide particles with an
average particle size of 850 μm. The fluidized bed garantees a good heat transfer from the
absorber surface to the carrier gas [1,6]. The walls of the fluidized bed chamber consist

of steel with a very high temperature stability. The front surface serves as absorber area. This receiver can be applied for temperatures of up to 900 °C with a temperature of the absorption area being less than 1050 °C.

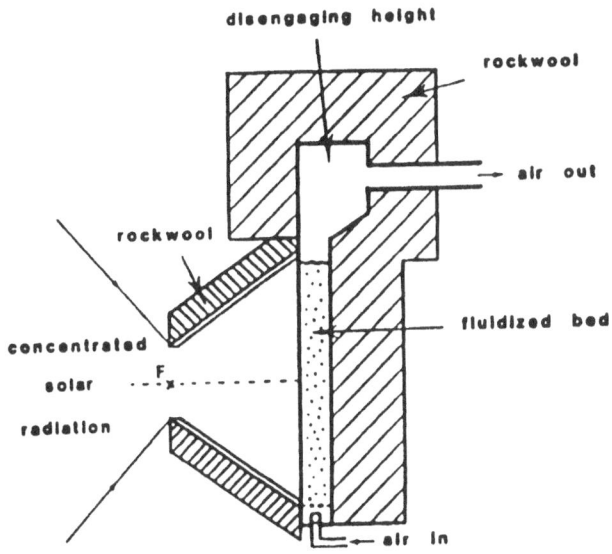

Figure 9: Fluidized bed solar receiver

3.3.3 Small Particle Heat Exchange Receiver - Mark I (SPHER)

This receiver design [14,21,23] makes use of small carbon particles serving as absorber medium. Under operation conditions, these particles oxygenize in the gas stream, and particle-free gas leaves the receiver. The required volume of carbon amounts to 0.1 % in weight of the employed air volume. The receiver is separated from the surroundings by a window.

3.3.4 Solar Thermally Activated Radiant Reactor (STARR)

Due to the SPHER concept the possibilities to carry out chemical reactions by directly irradiated particle suspensions were investigated [22].

Particle as gas heater: The particles absorb the concentrated solar radiation and heat up the surrounding gas

Particle as feedstock: The particles are the reactands. The carrier gas only serves as transport medium for the solid

Particle-gas reaction: The particles react with the surrounding gas

Particle as reaction catalyst: The particles absorb the concentrated solar radiation and heat up the surrounding gas; this, however, serves at the same time as photo-catalyst.

3.3.5   Solid Particle Receiver

The Sandia National Laboratories developed a concept for a receiver consisting of a curtain of free falling particles (figure 10). Here, silicium carbide particles of a particle size of approx. 500 μm were applied as absorber medium. Solar radiation could heat up these particles to a temperature of 1000 °C. As far as higher temperatures are concerned the particle size becomes a question of critical importance. On the one hand, the particles must show a very good absorption capacity. On the other hand, they have to maintain their pourability in spite of said temperatures, that means that they must not tend to agglomerate.

3.3.6   Molten Salt Receiver

In the framework of the DAR concept, a receiver applying the absorber medium molten salt was developed. The so-called molten salt receiver (figure 11) [7,18]. This molten salt consists of a eutectic mixture of lithium, sodium and potassium carbonates. Being a relatively thin film of 1-2 mm the molten salt partly absorbs the concentrated solar radiation. Anotherpart is absorbed by the absorber surface.

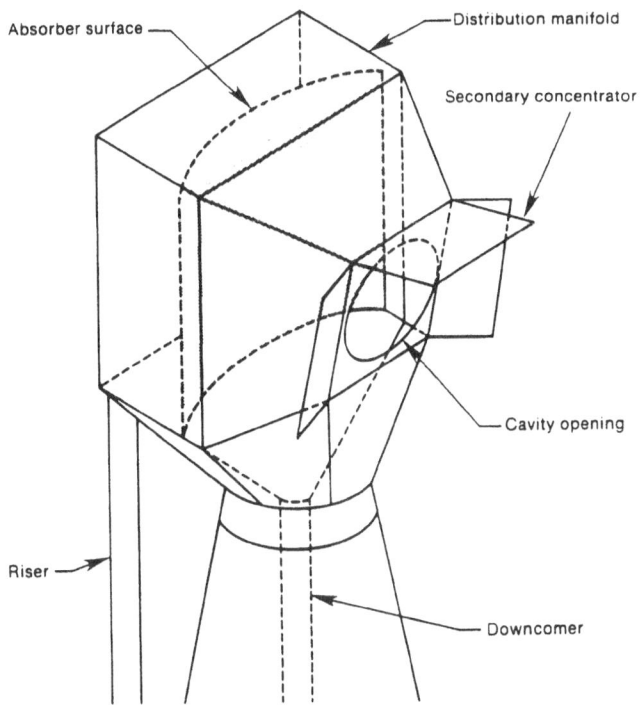

Figure 10: Solid Particle Receiver

3.4 Application for the Solar Treatment of Waste

A solar thermal treatment in a fluidized bed receiver seems to be eligible [13] because the substances to be treated exist in form of powder or arise as dust-like product due to the increasing drying. But as the absorption features of these, particles are not optimal and the receiver equipment should not become too big the material would have to be supplied into the radiation process repeatedly. The necessary melting temperature has to be reached and maintained for a corresponding time period because the vaporization of the harmful substances takes a certain period of time. It is conceivable to convert the particle stream to a kind of fluidizing roller by guiding the flow correspondingly. It would be absolutely necessary, however, to separate the reaction chamber from the surroundings by a window because otherwise it could not be guaranteed that vaporized harmful substances do not enter the atmosphere.

This receiver concept does not qualify for the thermal treatment of the waste considered here because filter dusts coming form garbage incinerating plants show a relatively high salt content resulting in low softening temperatures of approx. 850°C, with melting tem-

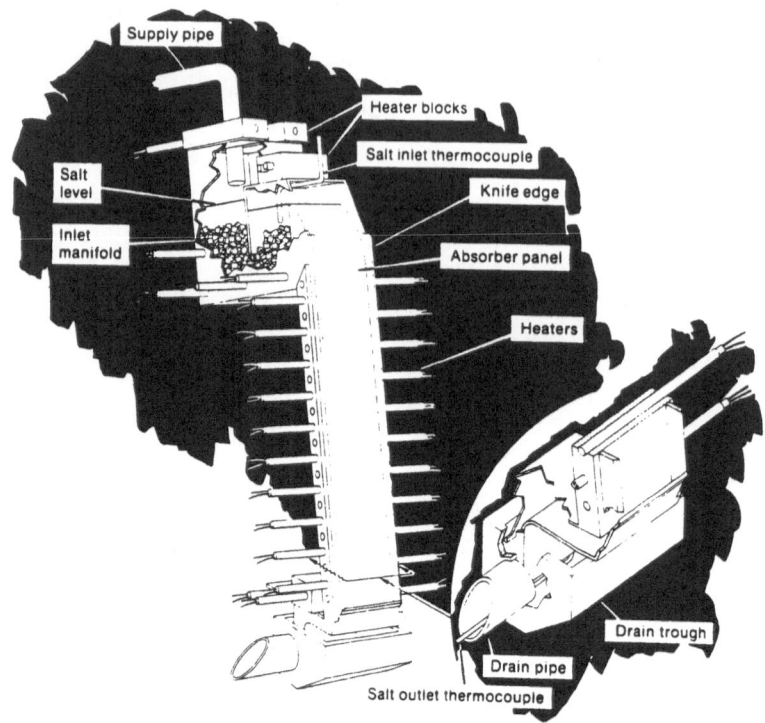

Figure 11: Molten Salt Receiver

peratures of more than 1000°C. Apart from that, it is difficult to keep the shape of such a fluidizing roller and thus maintain a defined way of the particle clouds in the radiation path. The window represents another critical issue. It has to be cooled in order to dissipate the heat resulting from self-absorption. Furthermore, it has to be protected by an air film from dirt accumulation by the particles or from soiling by the condensation of the vaporized components.

3.4.1 Rotating Receiver with Direct Heating

Figure 12 shows a rotating receiver with direct heating [10]. The concentrated solar radiation enters to heat the medium that due to the rotation sticks partially to the walls. This receiver/reactor was tested at the Odeillo furnace. Sand was used as heat carrier which was fed into the reactor at a mass flow of almost 1000 kg/h. The sand was heated up to almost 1000 °C. After leaving the receiver, it gave its heat to the air in a fluidized bed heat exchanger.

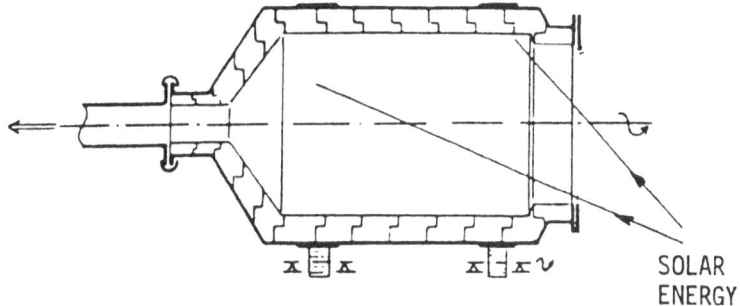

Figure 12: Rotating Receiver with Direct Heating [10]

The concept of a melting reactor seems to be the most promising one for a direct absorption receiver/reactor to treat waste in a thermal solar plant. Such a reactor will be investigated more precisely in the following.

As certain waste like paint sludges coming from the automobile industry require defined reaction atmospheres and as the necessary reaction temperatures amount to approx. 800 °C indirectly working receivers can be applied here. Heat energy will be supplied by means of a carrier gas transmitting the heat to the material.

4 Proposals and Layout of Receiver/Reactor Configurations for the Treatment of Waste

Waste materials have to be processed and worked up by using the following treatment phases.

- Drying, if possible, already dried when delivered

- Size reduction according to process requirements

- Thermal treatment according to reaction requirements

- Material flow control and conditioning during the necessary treatment phases

The necessary treatment phases require the supply of energy as:

- electric energy

- low temperature heat < 200°C

- medium temperature heat < 700°C

- high temperature heat > 1200°C

Low temperature is mainly required for drying the charging material; to a limited degree, it can also be made available directly from the medium temperature phases in form of cascaded heat. In the medium temperature range, carbonizing at low temperatures and in the high temperature range, reactions in the molten mass stream are possible.

Any energy demands inclusive of eletric energy should be satisfied by the solar plant. This requires different receiver systems.

The optimized GAST field with 170 m tower height was chosen for reference sizing of the rector configurations.

Heat Exchanger - Reactor

The existing GAST receiver supplies $2 * 28$ MW$_{th}$ with temperatures amounting to:

- 800 °C for the alloy receiver

- 1000 °C for the ceramic receiver (optional)

By adding temperature losses during the indirect heat removal treatment temperatures in the medium temperature range can be made available with freely selectable reaction conditions. Sections 2.1 contains the energetic process requirements. The dimensions of the plant can be selected freely and have no effect on the solar operation because any portion of the receiver performance can be tapped for the thermal treatment plant.

Heat exchanger and reactor configurations are described in Section 4.1. They process suitably pretreated paint sludges according to reaction levels of the medium temperature range described in section 2.1. The pre-treatment include drying and sizing.

Radiation Reactor

The heliostat field configuration is not qualified for a full supply of radiation to a radiation reactor.

- The insolation performance of 28 MW_{th} per half field is too great corresponding to a reactor output of approx. 60 $\frac{t}{h}$ of slag production

- The reactor's insolation opening relative to the fixed field focus is too great for a single material flow controlling treatment reactor

- The timely change of the insolation energy is too great for a constant process control

The possibilities to limit insolation will be treated in Section 4.2, the energetic process requirements in section 2.2 and the proposed possibilities of the reactor configurations are described in Section 4.3.

Radiation reactors permit a high temperature treatment > 1200°C. As open furnaces they can only be operated in an oxidizing phase. With direct feed they require material pretreatment of drying, sizing and, possibly, preheating.

4.1 Heat Exchanger-Reactor

A heat exchanger reactor is a plant configuration with the following phases:

- Solar tube receiver; available as GAST - receiver

- Heat exchanger to transmit convective energy to an intermediate carrier medium

- Reactor plant for the pyrolytic treatment of industrial dusts inclusive of conditioning plants

Of essential importance for the design of a heat exchanger reactor is the reaction temperature of the material to be treated which as a waste product will not show a uniform composition.

4.1.1 Heat Transfer

Figure 13 shows the heat consumption of a reduction reactor. In case of high temperatures, the transmission temperature ϑ to the solar receiver heat flow is determined by the pre-set final temperature and is always high in case of low temperatures.

The transmission temperature gradient is composed of:

- ϑ_1 = Transmission from solar gas

- ϑ_2 = Transmission to reduction gas

- ϑ_3 = Transmission to reduction product

For the convective heat transfer in a packed bed an analysis by Ranz in [33] between single sphere and packed bed heat transfer leads to the following approximate relation:

$$\vartheta_3 \sim \frac{1}{0,165} \cdot Re^{0,5} \cdot Pr^{0,7} \cdot \frac{A}{F} \cdot \Delta T \sim 0,44 \cdot Re^{0,5} \cdot \Delta T \qquad (4)$$

The quotient $\frac{A}{F}$ = flow cross section/bed surface is a constant for equal grain packed bed and low flow velocities, in [33] assumed to be 0,093. The heat transfer in the packed bed is very good in case of small grain, that means ϑ_3 is small in comparison to $\vartheta_1 + \vartheta_2$. To

-274-

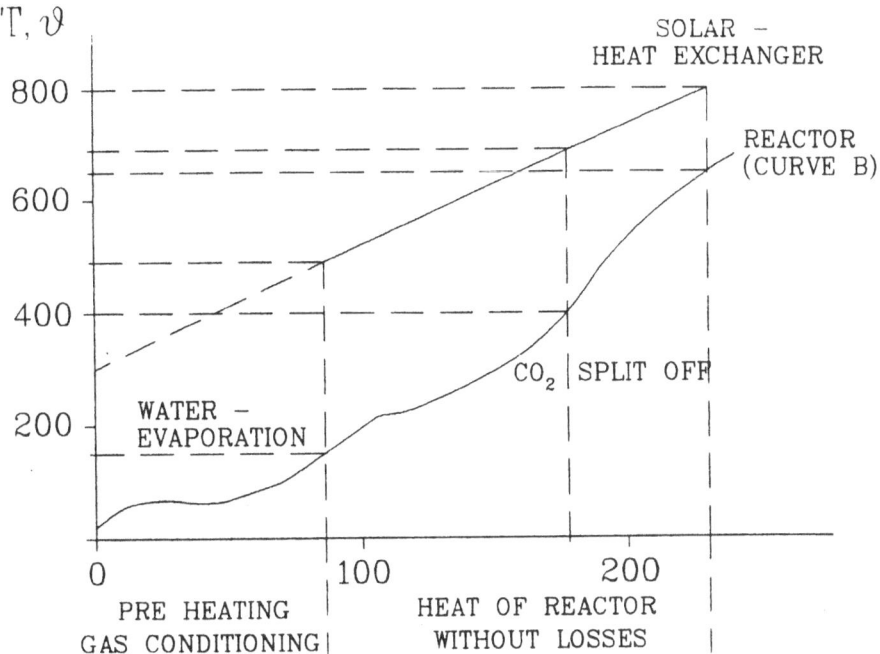

Figure 13: Heat Consumption of the Reduction Process

guarantee a slow heating-up it is necessary to subdivide the packed bed into partial beds and to supply them in staggered form with heat flows of different temperatures.

The heat transfer temperatures in the heat exchanger are (with Pr = 0,7):

- through the tubes: $\vartheta_1 \sim 37 \cdot Re_1^{0,2} \cdot \frac{A_1}{F_1} \cdot \Delta T_1$

- around the tubes: $\vartheta_2 \sim 14 \cdot Re_2^{0,2} \cdot \frac{A_2}{F_2} \cdot \Delta T_2$

On both sides, similar material values, but different flow conditions apply. The flow cross section A will as a rule be determined by the flow density and viscosity, that means the system pressure influencing the circulation work to a high degree.

- Receiver-circulation ~ 10bar

- Reduction gas-circulation ~ 2 bar

The reduction gas should be led around the tubes because the flow cross section around the tubes is greater and an additional surface requirement F_2, can be compensated for by ribbed tubes.

Due to the higher temperature drap in the reduction circle ΔT_2 becomes approx. twice as great as ΔT_1.

The proportion of the transmission temperatures leads to:

$$\frac{\vartheta_1}{\vartheta_2} = 2,6 \cdot \left[\frac{F_2 \cdot A_1 \cdot \Delta T_1^{0.8}}{F_1 \cdot A_2 \cdot \Delta T_2^{0,8}}\right] \qquad (5)$$

In case of $\frac{A_2}{A_1} = 5$, $\frac{\Delta T_2}{\Delta T_1} = 2$ and $\frac{F_2}{F_1} = 1,1$ in case of smooth tubes $\frac{\vartheta_1}{\vartheta_2} = 0,5$. If the circulation pump energy is to be lowered, $\frac{A_2}{A_1}$ has to become greater, $\frac{\vartheta_1}{\vartheta_2}$ becomes smaller and the heat exchanger thus becomes greater unless $\frac{F_2}{F_1}$ is enlarged by the ribs.

With $\dot{Q}$ and ϑ being preset, F is determined to be:

$$F = \frac{\dot{Q}}{\alpha \cdot \vartheta_{log}} \qquad (6)$$

$$\vartheta_{log} = \frac{\vartheta_A - \vartheta_E}{\ln \frac{\vartheta_E}{\vartheta_A}} \qquad (7)$$

ϑ_E and ϑ_A are the corresponding one-sided transmission temperatures at the entry and the outlet of the heat exchanger and consists of convection and conduction.

α as heat transfer coefficient is calculated on the basis of the functions of $Nu = f(Re, Pr)$.

The position-depending heat consumption of the pyrolysis reactor as shown in figure 14 offers a subdivision of the heat consumption.

- High temperature heat exchanger $> 500°C$

- Low temperature heat exchanger $< 500°C$

Since due to the distribution of the reaction heat a slightly lower heat consumption arises in the upper temperature range of the pyrolysis reactor a part of the reactor gas flow has

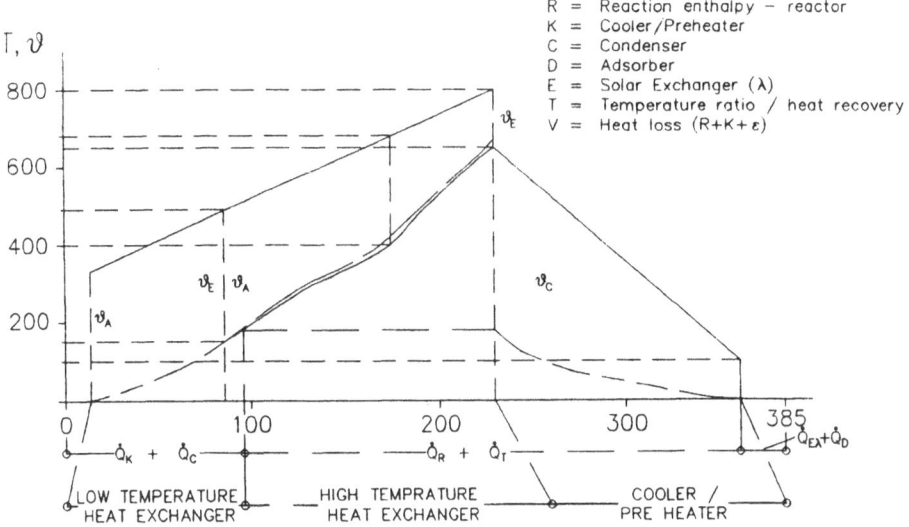

Figure 14: Aggregate-depending Heat Consumption of the Pyrolysis Reactor

to be subdrawn and supplied to corresponding positions of the reactor bed to control the heat consumption and temperature of the pyrolysis furnace.

The size of the heat exchanger is determined by the heat output and the available transmission temperature. With a solar temperature of 800 °C a very high $\vartheta_{1,2}$ is available in case of a pyrolysis furnace temoerature of 650 °C. As far as material technology is concerned it could be more favourable to lower the solar temperature by cooling and to put up with slightly greater heat exchanger surfaces instead.

4.1.2 Determination of Heat Exchanger Dimensions

In case of a treatment temperature of 650 °C the following heat output is required for a throughput of 1 $\frac{t}{h}$ paint sludge granulate:

$$\dot{Q}_P = \dot{q}_R(\vartheta) \cdot \frac{1000}{3600} \cdot \frac{1}{\eta_P} \tag{8}$$

$$\dot{Q}_P = 230\frac{kW}{kg} \cdot \left[\frac{1}{\eta_V} \cdot \frac{1}{\eta_C} + \frac{1}{\eta_D} - 1\right] \tag{9}$$

$$\eta_s = 0,60 \tag{10}$$

$$\dot{Q}_P = 385\frac{kW}{kg} \tag{11}$$

$$\begin{aligned} \eta_V &= 0,8 \quad \text{heat loss of reactor plus heat exchanger} \\ \eta_C &= 0,8 \quad \text{condensation loss } \frac{\vartheta_C}{\vartheta} \sim \frac{150}{650} \sim 0,25 \\ \eta_D &= 0,9 \quad CO_2 \text{ absorption/desorption with } T_u \end{aligned}$$

As the hot reactor material has to be cooled prior to coming into contact with the air an inert gas circulation permits a heat recovery as follows:

- Preheating of the charging material to < 150°C

- Reactor shell heating

- Preheating of the reduction gas

Applying a temperature ratio of $\frac{\vartheta_T}{\vartheta} < \frac{650 - (2 \cdot 100)}{650} = 0,7$ the maximum amount of hot reactor material to be exchanged is:

$$\dot{Q}_T = \dot{m}_t \cdot c_p \cdot \vartheta_t \tag{12}$$

$$\dot{Q}_T = 0,9 \cdot \frac{1000}{3600} \cdot 1,1 \cdot 650 \cdot 0,7 \tag{13}$$

$$\dot{Q}_T = 125\frac{kW}{kg} \tag{14}$$

The effective heat consumption in case of heat re-exchange is:

$$\dot{Q}_S = \dot{Q}_P - \dot{Q}_T = 260\frac{kW}{kg} \tag{15}$$

The caloric operating data of the heat exchangers are:

High Temperature Heat Exchangers A, B

T_{max}	800°C	
$\Delta T_{R_{A,B}}$	650 − 400 − 150°C	Reactor temperature
$\Delta T_{S_{A,B}}$	800 − 680 − 490°C	Receiver temperature
$\vartheta_{A,B}$	150 − 280 − 340°C	Transmission temperature
ϑ_{log}	197 − 194°C	Average transmission temperature
$\dot{Q}_{S_{A,B}}$	$65 + 100 kW$	Reactor

Low Temperature Heat Exchanger C

T_{max}	450°C	
ΔT_{R_C}	0 − 150°C	Reactor temperature
ΔT_{S_C}	300 − 490°C	Receiver temperature
ϑ_C	275 − 340°C	Transmission temperature
ϑ_{log}	300°C	Average transmission temperature
$\dot{Q}_{S_C}$	$70 + 25 kW$	Condensator/Desorber

The heat transmission coefficient α is calculated for different sections according to constructional flow conditions. The VDI-map [31] gives calculation supporting documents for:

- turbulent flow through tubes

- cross flow tube bundles

Tables 6, 7 show the calculation results whereby the same tube diameters were assumed for the ceramic and the alloy receiver and a tube grid for $\frac{A_2}{A_1} = 6,5$ was assigned. The transmission temperatures were calculated with $\frac{F_2}{F_1} = 0,9$.

$$\frac{\vartheta_1}{\vartheta_2} = 0,9 \cdot \frac{\alpha 2}{\alpha 1} \tag{16}$$

Inspite of the high transmission temperatures of 300 °C 17,5 m² tube (interior) corresponding to 146 m tube are required. With 1.8 kW per tube meter the heat tranfer is approx. half as high as in the GAST - receiver corresponding to the fact that the tranfer resistance here, too, lies completely at the air side.

Table 6: Calculation Results

High Temperature Heat Exchanger

		Alloy-Receiver			Ceramic- Receiver		
		solar		reactor	solar		reactor
Density ρ	$\frac{kg}{m^3}$	3,43		0,87	3,92		1,26
Caloric conductibility $\lambda *10^{-3}$	$\frac{W}{m \cdot K}$	69,2		60,0	63,0		45,2
Dynamic viscosity $\eta *10^{-6}$	$\frac{kg}{m \cdot s}$	42,2		36,9	28,5		
Pressure p	bar	10		2	10		2
T_{max}	°C		800			800	
Gas temperature t_m	°C	740		530	590	100	280
Heat volume $\dot{Q}$	kW		65				
Mass stream $\dot{m}$	$\frac{kg}{s}$	0,47		0,24	0,47		0,38
Tube diameter d	mm	38	(6)	42	38	(6)	42
Division ψ			1,3	0,40		(1,3)	0,40
Cross section	$10^{-3}m^3$	6,8		43	6,8		43
Flowing velocity $\dot{u}$	$\frac{m}{s}$	20		6,4	17,6		7,0
REYNOLDS–figure $*10^3$		62		7,3	69		13
PRANDTL–figure		0,69		0,66	0,67		0,67
NUSSELT–figure		120		100	130		145
Heat transfer figure α	$\frac{W}{m^2 \cdot K}$	218		91	216		99
$\frac{\vartheta_1}{\vartheta_2}$			0,46			0,59	
Surface F	m^2	4,8	1,1	5,3	4,7	1,1	5,2
l			(6)			(6)	

Table 7: Calculation Results

Low Temperature Heat Exchanger

	Unit	solar		reactor
Density ρ	$\frac{kg}{m^3}$	5,15		1,26
Caloric conductibility $\lambda *10^{-3}$	$\frac{W}{m \cdot K}$	52,6		45,2
Dynamic viscosity $\eta *10^{-6}$	$\frac{kg}{m \cdot s}$	32,6		28,5
Pressure p	bar	10		2
T_{max}	°C		500	
Gas temperature t_m	°C	400		280
Heat volume $\dot{Q}$	kW		95	
Mass stream $\dot{m}$	$\frac{kg}{s}$	0,47		0,38
Tube diameter d	mm	38 (19)		42
Division ψ			(1,3)	0,40
Cross section	$10^{-3}m^3$	6,8		43
Flowing velocity $\dot{u}$	$\frac{m}{s}$	13,4		7,0
REYNOLDS–figure $*10^3$		80		13
PRANDTL–figure		0,66		0,67
NUSSELT–figure		150		145
Heat transfer figure α	$\frac{W}{m^2 \cdot K}$	208		99
$\frac{\vartheta_1}{\vartheta_2}$			0,40	
Surface F	m^2	5,3	1,1	5,2
l			(6)	

Calculations were made for 6 parallel tubes. Compared with the GAST - receiver, the flow length is 3 times and the mass stream 0.6 times greater.

For straight tubes the heat exchanger dimension ($\frac{F_1}{F_2}$) is:

$$21 \text{ m smooth tubes (6)} * 0,15 \text{ m}^{\prime}$$
$$14 \text{ m ribbed tubes (6)} * 0,15 \text{ m}^{\prime}$$

The required performances for pumping will be very high in both cases.

For helical coil type tubes, the heat exchanger dimension is:

2 * 3 double spirals (16 spirals * 0,13 m $\simeq$ 126 m) 2,1 m long * 580 / 270 mm$^{\not}$

The flow path around the tubes of 2.1 m is considerably shorter, the heat exchanger construction is more compact and thus, pump over performance and heat losses are lower. Apart from that, high temperature stresses can be better compensated for by the helical coil type construction.

4.1.3 Pyrolysis Reactor

Preconditions for a pyrolytic treatment of waste are:

- a reduction furnace atmosphere, that means reactive reduction gases CO, H_2, C_nH_{2n}, CH–radicals, and possibly existing harmful gases like HCL etc., too

- a sufficiently high and properly graded reduction temperature depending on the material's composition

- a sufficiently high reduction time depending upon the material's consistency

- a closed treatment plant preventing oxygen from entering and able to be operated with pressure or low pressure respectively, if necessary.

Further features for the selection of the process are:

- an energetically favourable process control, if possible in closed energy cycles

- favourable heat transfer conditions to minimize the scope of devices

- operational effectivity and reliability, especially with regard product quality and plant emissions

- operational adaptability to different charging materials

4.1.4 Determination of Plant Dimensions and Operational Data

The following figure shows a proposal for a possible plant flow chart suited for the reducing treatment of organo-metallic waste granulates by means of thermal solar energy.

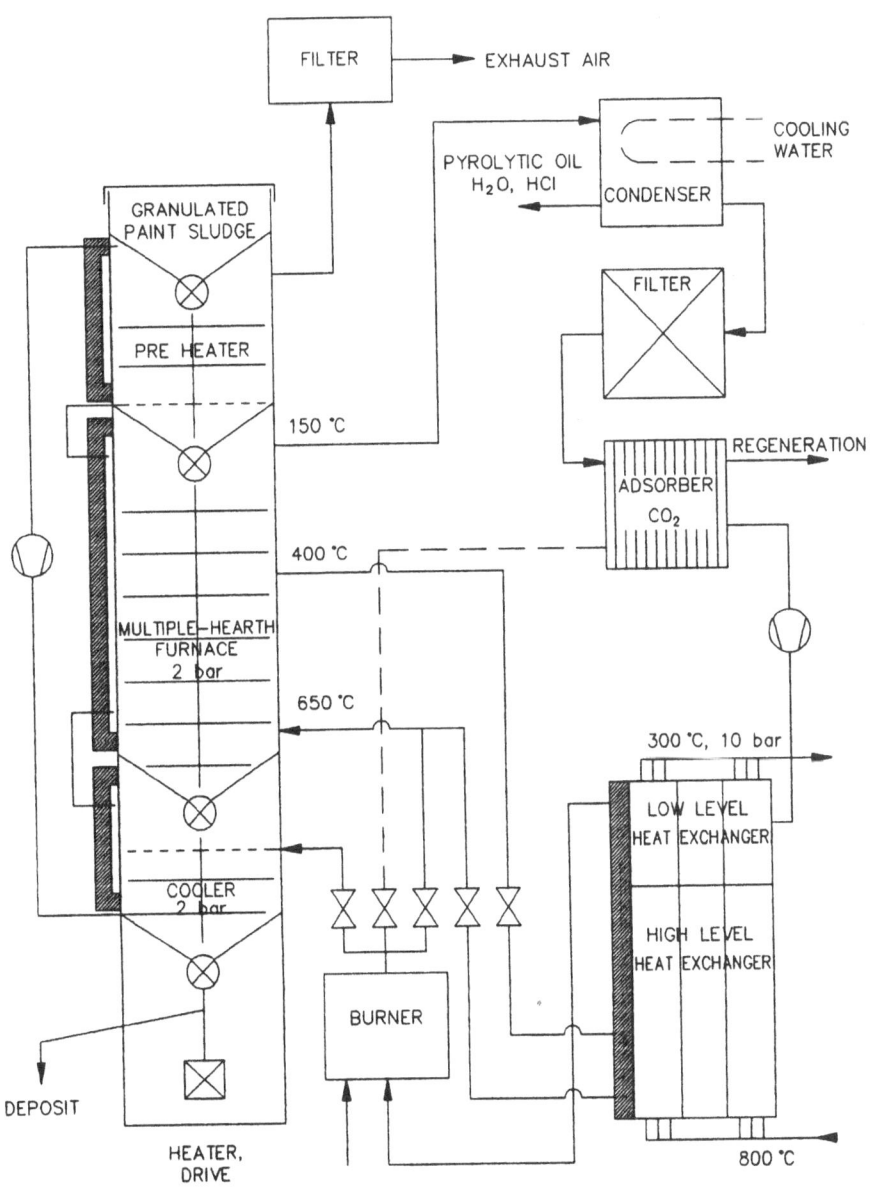

Figure 15: Reactor Configuration

1. Receiver Heat Exchanger:

Designed as serpentine tube heat exchanger for 260 kW.

$$\text{Receiver side} \quad 0{,}047 \, \tfrac{kg}{s} \quad 800 - 300°C$$
$$\text{Gas side} \quad 0{,}24 \, \tfrac{kg}{s} \quad 650°C \text{ - } T_u$$

with partial streams on the gas side for:

High temperature range	650 °C
Medium temperature range	400 °C
Low temperature range	$\sim 300°C$

Total heat exchanger surface	15 m² (Receiver)
Weight	0,6 t
Insulation (porous chamotte)	8,0 t

2. Multiple-Hearth Reactor:

With a throughput of 1 $\frac{t}{h}$ a dwell time of 90 min is to be achieved in the multiple-hearth reactor.

$$\text{Filling} = \frac{\dot{m} \cdot \tau}{\rho_e} \tag{17}$$

Granulate density (with $\varepsilon_a = 0{,}65$) input	$1{,}3 \, \frac{kg}{m^3}$
Granulate density of product	$1{,}0 \, \frac{kg}{m^3}$
Filling	1,13 m³
Reactor size	1 m $\ast$ 3,5 m = 2,8 m³
Storeys	10 @ 0,30 m = 3,0 m
Gas throughput (2 bar)	$> 400°C = 0{,}24 \, \frac{m^3}{s}$
Gas throughput	$< 400°C = 0{,}38 \, \frac{m^3}{s}$

The even heating-up of approx. 5.5 °C/min has to be controlled by a careful adjustment of the granulate stream and the matching of the gas throughput with the reaction heat consumption. This is especially true for the daily starting cycle. In order to keep the temperature in the bed on even levels a shell heating and a good insulation are necessary. A quasi-packed bed multiple-hearth reactor operates more favourable than a fluidized bed reactor as far as energetics is concerned because the latter generates high flow losses that even become excessive with long dwell times. More favourable are mechanic distribution and weir section permitting a great and even bed cross section with a correspondingly low flow resistance. In case of long dwell times, heat transfer is effected without problems. The mechanical stability of the grain (uniform grain bed), however, is critical because with increasing grain abrasion the flow resistance and grain segregation increases very strongly.

Bed control can only be designed if the granulate input is known. As the total height of the bed is very great due to the long dwell time a graduated heating in partial beds with intermediate material reservoirs through which the flow does not pass or only passes to a minor degree or which are heated by conductivity respectively may be possible.

The input and output of the granulate is effected by cellular wheel transfer tubes with the granulate being distributed on sieve plates by rotating conveying arms.

The following values result per storey:

Reactor cross section 1,0 m²	F =	0,78 m²
Central tube - reservoir 0,6 m²	F =	0,28 m²
	H =	0,22 m
	V =	0,06 m³
Fixed air-run distribution plates	F =	0,5 m²
1,2 / 0,7 m² + 1 m²	V =	0,03 / 0,05 m³
	H =	2 * 0,05 m
		2 * 0,03 m
	V =	0,06 m³

Distribution Mechanism

Rotating mixing/conveying arms

Pressure Loss

Gas velocity (free)	u =	0,48 / 0,75 $\frac{m}{s}$
Bed height	h =	0,1 m
Granulate	d =	2 mm
Free volume	ε =	0.35
HT-furnace per hearth		2,6 mbar
LT-furnace per hearth		5,6 mbar
Furnace with 10 hearths		41 mbar

$$\Delta p = 1,75 \cdot \frac{h \cdot \ddot{u}^2 \cdot \rho}{d} \cdot \frac{1-\varepsilon}{\varepsilon^3} \qquad (18)$$

3. Cooler and Preheater:

The reducing paint sludge treatment aims at the thermal setting of non-elutable compositions in an inorganic/organic matrix. Correspondingly, the reactor mixture has to be recooled prior to coming into contact with air. For this reason, a closed cycle between preheater and cooler is available. This closed cycle transmits the thermal energy at max. 150 °C to the charging material or operates the shell heating of the multiple-hearth furnace respectively. Thus, energy is at the same time optimized.

The constructions of cooler and preheater are similar to that of the multiple-storey furnace and thus can form a connected unit.

Filling of preheater and silo		=	1,2 m³
Size	1 m⁴* 2,0 m	=	1,6 m³
Hearths	1 * 0,40 m	=	0,40 m
Filling of cooler		=	0,6 m³
Size	1 m⁴* 2,0 m	=	1,6 m³
Hearths	3 * 0,40 m	=	1,2 m
Gas throughput (2 bar)	< 650°C	=	0,24 $\frac{m^3}{s}$

Preheater, Multiple-Storey Furnace, Cooler

Overall size	1 m⁴* 8,5 m	
Weight furnace		3,2 t
1-hearth drive		0,3 t
4 transfer tubes		0,6 t
Insulation		8 t
Filling		3,9 t
Total weight		**16,0 t**
Furnace distribution capacity		2 kW

4. Condensator, Filter, Adsorber:

The gas with a temperature of approx. 150 °C leaving the multiple-hearth furnace is circulated and processed prior to the heating-up by the receiver heat exchanger.

Condensator

Performance	70 kW
Temperature	150 °C $\longrightarrow T_u$
Condensate	Pyrolyseöl, H_2O, HCl
Weight	0,5 t
Water	0,3 t
Electrical energy (pump, cooler)	1,5 kW

Bag Filter

Cleaning precision	$> 3\mu$
Throughput	$0,38 \frac{m^3}{s}$
Temperature	T_u
Surface	$2 * 1 \ m^2$
Pressure loss	25 mbar
Weight	150 kg

CO_2 – Adsorber (activated carbon filling)

Throughput	$0,10 \frac{m^3}{s}$
Temperature	T_u
Regeneration	300 °C, air
Time of exposure	6 h
Filling	$2 * 0,25$ t
Pressure loss	5 mbar
Weight	700 kg

5. Pyrolysis Gas Circulation:

A catalytic burner oxygenizing the pyrolysis gas serves to control the pyrolysis gas circulation.

- Control of the furnace's entry temperature
- Control of the gas volume stream
- Throw-out of inerted excess gas
- Control of adsorber regeneration

The circulation requires a blower:

Throughput $0{,}38 \ \frac{m^3}{s}$

Performance   6,8 kW

Δp 150 mbar

6. Cooler and Preheater Circulation:

Der Kreislauf erfordert ein Gebläse:

Throughput $0{,}24 \ \frac{m^3}{s}$

Performance   2,3 kW

Δp 60 mbar

7. Energy Consumption:

Total energy consumption per ton of paint sludge (inclusive of accessory units) amounts to:

Thermal energy $260 \ kW_{th} \ / \ t$

Electric energy $11 \ kW_e \ / \ t$

8. Accessory Units:

The following accessory units have to be planned:

- Storage and pretreatment of delivered paint sludges
- Drying and, if necessary, breaking
- Pelletizing of dusts
- Conveyor to the horal silo
- Conveyor for the reactor product and the pyrolysis oil

4.2 Field Section and Receiver Aperture

By cutting a limited radiation field in the panel of the GAST - receiver, the radiation performance and insolation angle can be blended to a fixed path in accordance with the reactor's positionals needs. Solar-geometrical and constructional requirements for such a field selection and receiver cut out will be described in the following.

4.2.1 Field Sector

Figure 16 shows the field-effectivity data for the specified GAST - field of heliostats. These data show the energy density in direction to the tower aperture, that means the GAST - aperture. The energy density existing at the rear field border amounts to 50% of the surface-related sun insolation and increases to 80% in the area near the tower. The tower aperture is inclined in a way that the greatest possible field is covered which of course increases to the outside according to square-law.

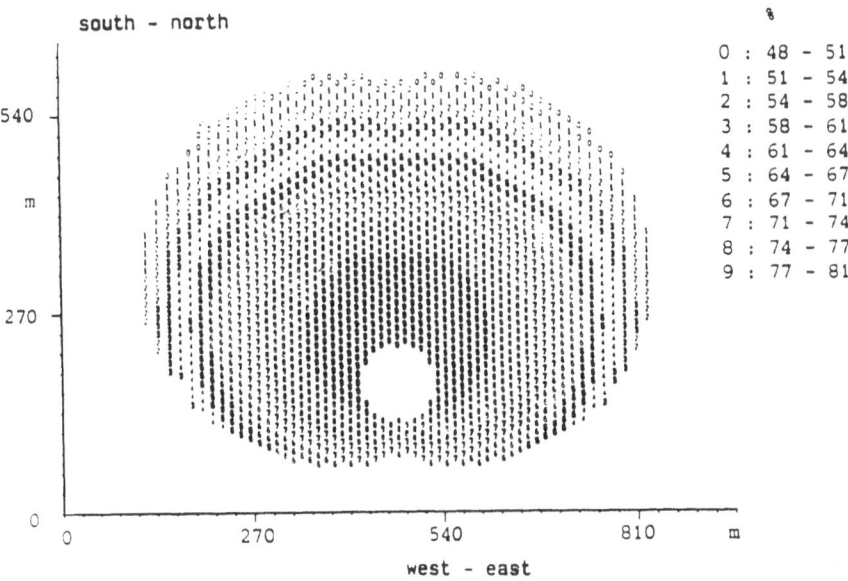

Figure 16: Heliostat Field Efficiency at Design Point [26]

The field sector selected for the receiver section has to meet the following conditions:

- Energy values should be as constant as possible as far as the annual rhythm is concerned

- The daily rhythm should be as long and constant as possible

- No blocking and shading through the tower

The heliostat surface needed varies for the design point ($700 \frac{W}{m^2}$ at 31 ° sun angle) and corresponding to the selected heliostat distance from the tower between 1,5 and 2,5 $\frac{m^2}{KW_{th}}$ (aperture), the average being 2,0 $\frac{m^2}{KW_{th}}$. The field area needed varies with the tower distance between 1,2 and 2,8 m², the average being 2,0 m² ground / m² heliostat or 4 $\frac{m^2}{KW_{th}}$.

Most suitable for this is the middle to rear section of the centre field.

For a heliostat position of approximately 29 ° is:

heliostat surface	2,1	$\frac{m^2}{KW_{th}}$
field area	4,5	$\frac{m^2}{KW_{th}}$

For a heat requirement of 450 KW a field area of 2000 m² is needed corresponding to an elipse of 62 m depth and 41 m width. The angle cone at the tower aperture has the following vertical and horizontal angle spread:

$$\Delta\varphi(26 \div 32°) \quad \simeq \quad 334 - 272m$$
$$\Delta\xi(\pm 3,8°) \quad \simeq \quad 41m$$

The heliostat field partially participating to the heat flux into the receiver aperture is approximately double the size.

4.2.2 Focus Length and Radiation Angle

For the GAST - configuration, all heliostats are aligned to the centre of the tower aperture. As all heliostats have a uniform focus length of 300 m the image of the aperture and on the tube receiver arranged behind changes with the position of the heliostats.

The focus points are lying on an evolute passing through the tower aperture. Figure 17 shows the dependence of image size and image position on the receiver as function of the vertical and horizontal field angle φ or ξ respectively.

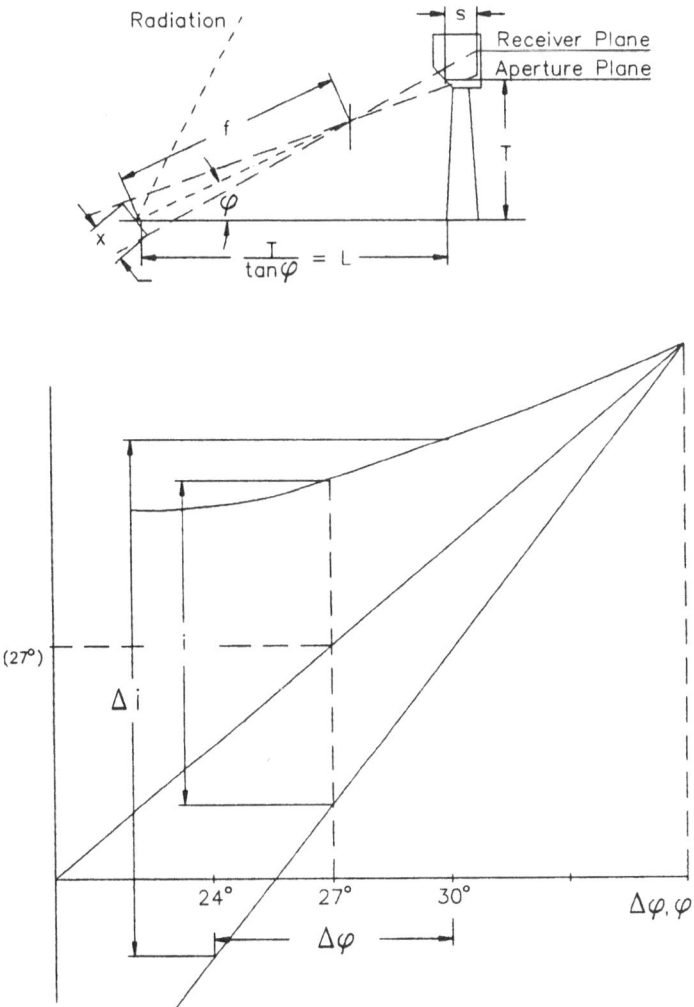

Figure 17: Dependence of Image Size (i) and Image Position (I) on the Receiver

$$T \simeq \text{height of tower aperture (170 m)}$$

T $\simeq$ height of tower aperture (170 m)

S $\simeq$ distance of receiver level (8 m)

f $\simeq$ focus of heliostats (300 m)

φ $\simeq$ vertical field angle ($> 23°$)

ξ $\simeq$ horizontal field angle

x $\simeq$ (projected) dimension of heliostats (< 7 m)

i $\simeq$ Image size

I $\simeq$ Image position

i and j are the vertical and horizontal image sizes of a heliostat whereby the apparent heliostat dimension x depends upon the actual angle of the sun position. I and J are the vertical and horizontal image positions of a heliostat, measured as projected distance from the centre of the receiver aperture.

$$i(\varphi, \xi) \approx \left[\frac{T}{\sin \varphi} + \frac{S}{\cos \varphi} - f \right] \frac{x}{f} \cdot \frac{1}{\cos \xi} \tag{19}$$

$$I(\varphi, \xi) = S \cdot \tan \varphi \cdot \frac{1}{\cos \xi} \tag{20}$$

$$j(\xi) = i(\varphi, \xi) \cdot \cos \xi \tag{21}$$

$$J(\xi) = S \cdot \tan \xi \tag{22}$$

The integral vertical field angle is determined by $\Delta\varphi$ and $\Delta\xi$ of the insolating field sector.

The vertical image position rises with φ, the image size decreases up to the angle φ_0 ($\sim 36°$), that places the focus point of the heliostats on the receiver plane and then decreases again.

In case of a cylindrical receiver plane, the horizontal image position is proportional to the horizontal field angle ξ; the image size is like the vertical expansion a function of φ. For a plane receiver the image changes with $\tan \xi$ and is a function of φ and ξ.

The combined insolation conditions of field, vertical and horizontal field angles und the reactor configuration result in an optimum in the middle field or in the central third of the receiver.

The field angle preset by the GAST - field can be modified by

- changing the focus length of the heliostats

- changing the insolation angle of the heliostats, that means deflecting the field angle of the heliostats from the central receiver aperture.

The purpose of such an adjustment is as follows:

- Modification of the image size to a geometrical and intensity proportion adapted to the reactor/receiver

- Modification of the (apparent) field angle to minimize the focal aperture

Position and size of the reactor/receiver can be varied and several reactors can be arranged for a greater field of heliostats by combining the two parameters.

4.2.3 Aperture Position and Radiation Duct

For the GAST - Receiver corresponding apertures have to be considered, which meet the solar geometrical conditions as well as the constructional reactor conditions:

- The aperture rim has to be cooled, or framed by the tube receiver or must lie outside the radiation field.

- The insolation opening at the reactor, that means the reactor seal at the end of the radiation duct, has to be circular and slightly inclined (inclination of the reactor).

- The radiation duct must not be too short in order to be able to control the air turbulences.

- Insolation angle and insolation intensity have to lie within the constructional reactor parameters.

The GAST tube receiver represents a good radiation aperture. In the vertical direction, the lower third of the receiver has to be rearanged. Either the tubes have to be diverted to form a collar or the tube module has to be subdivided by upper and lower manifolds to be assembled in a radiation-free area next the reactor aperture.

The radiation-reduced area can be reached by:

- lowering the insolation angle of the upper heliostat images within the GAST - receiver aperture (tower aperture)

- aligning the insolation angles of the heliostat images in an extended receiver aperture

- aligning the insolation angle of the heliostats with adapted (lengthened) focus within the GAST - receiver aperture

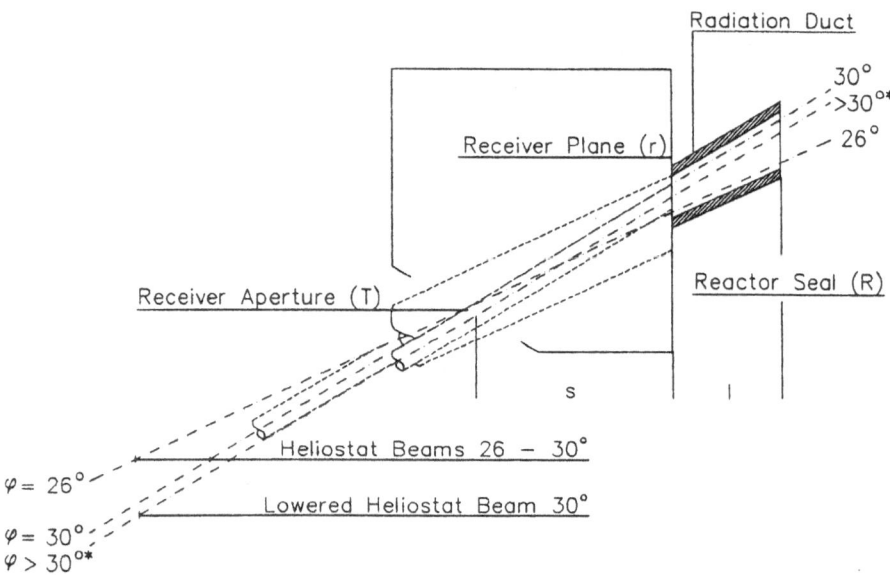

Figure 18: Duct Aligned Beam Path

The selected solution depends upon the aperture size and position and has to be calculated within a more precise programme. Figure 18 shows the beam path for a lower aperture position of $\varphi > 26°$ with a lowered insolation angle for $\varphi > 30°$, causing a radiation-free area at the upper radiation duct, though this variant is very limited.

By adaptation of the heliostat foci and of the insolation aperture, several reactor apertures could be arranged at the lower rim of the receiver.

4.2.4 Ground Reflector and Concentration

A direct insolation to a radiation reactor entails some operational, energetical and structural disadvantages that will be described in more detail in Section 4.3.

A reflection of the energy beam to ground level, if possible, may therefore have some advantages. For such a construction, the following conditions must be met:

- Different focus lengths for groups of heliostats to generate one or more concentrated beam bundles.

- Arrangement of the parabolic reflectors or reflector strips for beam diviation to ground level to a reactor.

- Modified guiding algorithm for the heliostats.

- A low assembly height for the reflectors to reduce energy losses and flow control precision.

The last point can of course result in an advantage because the heliostats arranged near the tower are critical for the receiver focussing. By means of a reduced focus they can easily be adjusted to match considerably lower mounted reflectors.

4.2.5 Energy Density

The direction-depending energy density is a function of the field effectivity, the image ratio and the image position [3]. The image ratio is determined by focus length, the image position and the aperture position.

The integral value of the energy density on the receiver plane depends upon the field sector:

$$e_r = q_H \cdot \bar{\eta} \cdot \frac{F_H}{F_r} \cdot \cos \bar{\varphi} \sim \frac{q_B}{\cos 31} \cdot \bar{\eta} \cdot \frac{F_H}{F_B} \cdot \frac{F_B}{F_r} \cdot \cos \bar{\varphi} \qquad (23)$$

$$\frac{F_B}{F_r} = \left(\frac{L}{s} \right)^2 \cdot \frac{1}{\tan \varphi} = \left(\frac{H}{s} \right)^2 \cdot \frac{1}{\tan^3 \varphi} \qquad (24)$$

$\eta(\varphi, \xi)$ is the heliostat field effectivity at the design point and means the energy released by radiation per field area F_H in direction of receiver F_r, with the receiver positioned at the angle $\bar{\varphi}$ in radiation direction. The area ratio can be written as the second power of the length ratio $\left(\frac{L}{s}\right)^2$ to the receiver aperture.

The heliostat ground area demand is according to section 4.2.1 $\frac{F_H}{F_B} = 0,47$. The area ratio $\frac{F_B}{F_r}$ can be written as the second power of the length ratio $\frac{L}{s}$ with the intersect at the tower aperture.

$$\frac{F_B}{F_r} = \left(\frac{H}{s}\right)^2 \cdot \frac{1}{\tan^3 \varphi} \sim 2095 \qquad (25)$$

The vertical receiver aperture F_r is approx. 1,15 m^2 (1,18 $^{\circ}$) at a distance of s of 9 m. The heliostat images for field angles φ of 26 $^\circ$ to 32 $^\circ$ are between 6 m^2 and 0,6 m^2 and their image centres are vertically (φ) 1,04 m and horizontally (ξ) 1,18 m apart. To the full radiation on F_r is also contributed by heliostats outside the field sector F_B due to the image size discussed in section 4.2.2. The resulting angles Θ and ζ of the radiation tube result from the condition that the radiation duct wall shall not be insolated directly. The circular reactor seal D_R is a function of duct geometry Θ, ζ, l and the furnace inclination α. The focal aperture of the receiver results from the image relation:

$$\tan \frac{\zeta}{2} = \frac{D_R - i_r}{2 * l} = \frac{i_r + i_T}{2 * s} = \frac{i_T + D_R}{2(s + l)} \qquad (26)$$

For Θ an analogeous relationship applies with correction of D'_R with $\cos(\alpha + \varphi)$ and variable $i_T(\zeta)$ (figure 17, 24).

Calculated for D_R with $i_R = 1,31$ and l = 2 m:

$$D_{R_\lambda} = i_r + (i_r + i_T) \frac{l \cdot \cos \varphi}{s} \qquad (27)$$

$$D_{R_\lambda} \sim i_r \cdot \left(1 + 1,88 \cdot \frac{l \cdot \cos \varphi}{s}\right) \qquad (28)$$

$$D_{R_\lambda} = 1,37 \cdot i_r = 1,62 m \qquad (29)$$

$$D_{R_v} = \left(i_r + (i_r + i_T) \frac{l \cdot \cos \varphi}{s} \right) \cdot \frac{\cos \varphi}{\cos (\alpha + \varphi)} \tag{30}$$

$$D_{R_v} = 1,37 \cdot i_r \cdot 1,12 = 1,80m \tag{31}$$

If the radiation tube is formed zylindrical/elipsoid, the collar of the receiver tubes could extend into the radiation duct and so receive the emission radiation from the furnace.

From the aforesaid the following dimensions are deduced.

Table 8: Calculations

φ	=	26 ÷ 32	°	Field angle
l	=	2	m	Radiant tube
D_R	=	1,6	m	Reactor seal
α	=	10	°	Reactor inclination
$\dot{q}_B$	=	700	$\frac{W}{m^2}$	Solar insolation
$\dot{q}_H$	=	810	$\frac{W}{m^2}$	Heliostat insolation

Results:

$\bar{\eta}$	=	0,60		according to [26,30] for $\bar{\varphi} = 26 \div 32°$
$\dot{Q}$	=	450	kW	Solar heat flux
$e_{r\perp}$	=	450	$\frac{kW}{m^2}$	Energy density at receiver aperture
e_r	=	390	$\frac{kW}{m^2}$	normal to receiver
Θ	=	3,5	°	
ζ	=	6,0	°	
d_r	=	1,18	m	Diameter of the receiver aperture
F_r	=	1,15	m^2	vertical area of receiver
F_R	=	2	m^2	normal area of receiver

The energy density at the upper end of the aperture is considerable higher than at the lower end. Density and distribution of energy can be intensified, modified and evened out by modifying the focus length.

4.2.6   Air Stream

Figure 19 shows an airstream model of a cavity receiver as it has been applied for the GAST - concept. It shows distinct air stream directed outwards in height of the aperture. The receiver aperture to the reactor is approximately at the same height.

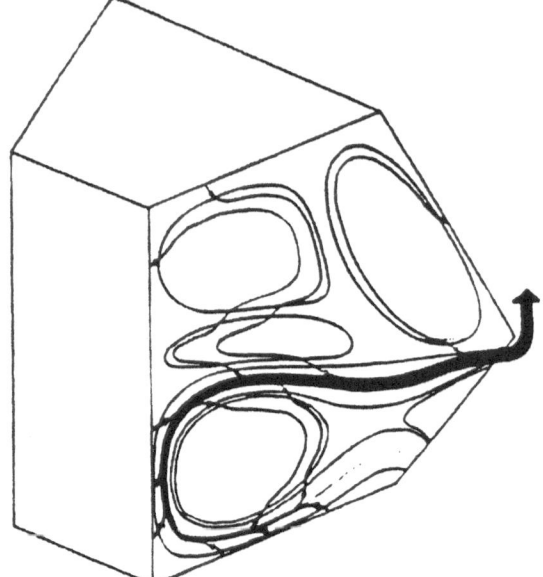

Figure 19: Air Stream Model in a Cavity [28]

It is to be avoided that turbulences or suction currents transport reactor gases into the receiver.

Constructional possibilities to avoid this are as follows:

- Deflection of the downward flow above the aperture through a screen in the radiation-free area of the receiver manifold arranged above

- Constant suction flow through the radiation duct to the receiver

- Longer radiation duct, at least up to the height of the lower reactor seal

- Airtight ceramic sliding seal

By corresponding heliostat-focus-grouping and correction of the insolation angle the radiation duct can be drawn as screens into the receiver chamber and thus reduce the circulation losses of the tube receiver.

4.3   Melting Reactor - Receiver

A melting reactor - receiver is a drum receiver utilizing the direct insolation to transfer heat to the reaction material. As both insolation geometry and reaction control have firm geometric, thermal and mechanic requirements for such an aggregate detailled constructional considerations have to be dedicated to the solar reactor configuration.

Section 4.2 analyzes the receiver section opening and the basical geometry of the heliostats. Section 4.3 discusses the receiver configuration and the reaction layout as well as accessory aggregates and operational requirements.

4.3.1   Reactor Configuration

Figure 20 shows the basic geometry of a direct radiation drum reactor.

The radiation beam passes the receiver aperture and the radiation duct through the circular reactor seal and then meets the upper wall of the drum. Radiation energy is transmitted to the wall or the material adhering to the wall or reflected to other reactor surfaces respectively and transferred to the ground material of the drum by drum rotation.

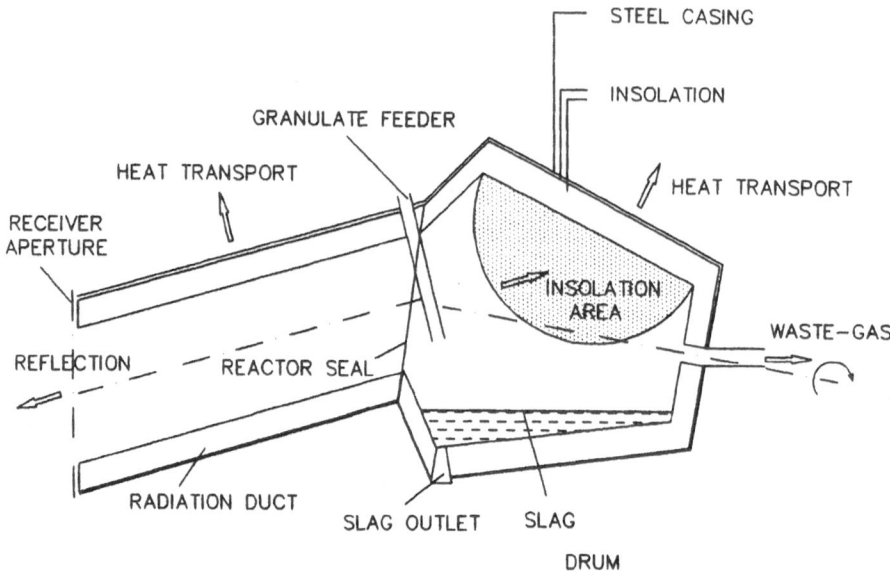

Figure 20: Drum Reactor

The material transport on the wall as well as the thermal conduction in this material layer is much higher if a fused mass is used. The thin molten layer on the ceiling wall is overheated during each drum rotation. By this, reactions like vaporization, thermal decomposition and - depending on the gas flow control - oxidation- or reduction reactions become possible.

Pure oxidic melting reactors as they are used for the melting of filter dust seem unproblematic when the melting point is set accordingly. At the start, the molten layer of the day before generates a catching effect as soon as the melting point is reached. Temperature control is done through the heliostats or the variation of the throughput respectively. Products released in form of gas are sucked off, thereby preventing their escape through the radiation duct. If the reactor sucks from the receiver aperture, hot air is drawn in. Thus, the reactor phase is not disturbed.

4.3.2 Determination of Reactor Dimensions

The following layout calculations serve to assess the reactor size.

Insolation Reactor Aperture

D_R	=	1,6	m	Reactor seal diameter
α	=	10	°	Aperture inclination
$F_{r\perp}$	=	1,0	m²	Radiation duct cross section Receiver aperture
$F_{R\perp}$	=	1,6	m²	Radiation duct cross section Reactor seal
$e_{R\perp}$	=	290	$\frac{kW}{m^2}$	Energy density at reactor seal
e_R	=	225	$\frac{kW}{m^2}$	Energy density normal to seal area

Reactor Mass Stream

$$\dot{m} = \frac{\dot{Q}_s \cdot \eta}{\bar{c}_p \cdot \Delta T}$$

$$\dot{m} = 0,28 \quad \frac{kg}{s}$$

$$\dot{m} \simeq 1 \quad \frac{t}{h}$$

with

c_p	=	1,2	$\frac{kJ}{kg \cdot K}$	Average heat capacity of the mixture
ΔT	=	1000	°C	Increase to melting temperature
η	=	0,75		Reactor heat efficiency
$\dot{Q}_s$	=	450	kW	

Heat demand in case of oxidic dusts consists to approx. 90% of capacitive heat and to 10% of vaporization heat. The increase of the melting temperature does not take into account a preheating of the reaction material. The reactor heat efficiency of 0,75 is assumed to be somewhat lower than the receiver efficiency.

Reactor Dimensions The geometric configuration of a solar rotary tubular reactor is shown in figure 21. The slightly inclined circular reactor seal serves as a sealing towards the stationary radiation duct by means of a ceramic sliding packing. By the inclination, the material circulation and the material charging from above or from the front is contained the reactor.

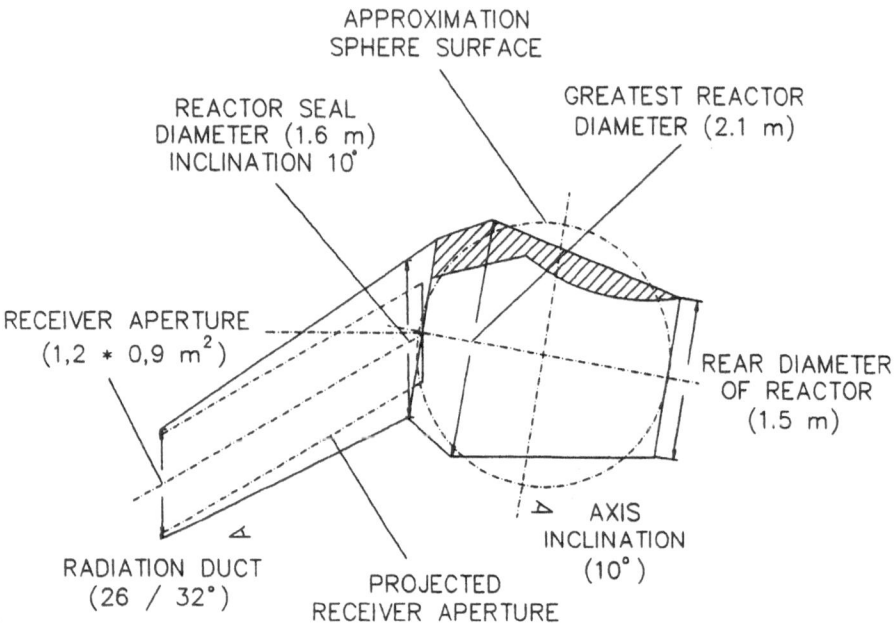

APPROXIMATION
SPHERE SURFACE

REACTOR SEAL
DIAMETER (1.6 m)
INCLINATION 10°

GREATEST REACTOR
DIAMETER (2.1 m)

RECEIVER APERTURE
(1,2 * 0,9 m²)

REAR DIAMETER
OF REACTOR
(1.5 m)

AXIS
INCLINATION
(10°)

RADIATION DUCT
(26 / 32°)

PROJECTED
RECEIVER APERTURE

Figure 21: Dimension Mastering Illustration Calculated for Reactor-, Isolation- and Position Data

In order to create a lower slag chamber and to adapt to the upper insolation angle the reactor room first has to expand and diminishes than with the inclination angle. The reactor length is a function of the seal diameter and insolation angle and has to consider the feeding and degassing equipment.

$$
\begin{array}{llll}
D_R & = & 1,6\ \text{m} & \text{Reactor seal diameter} \\
D_v & = & 2,1\ \text{m} & \text{Greatest reactor diameter} \\
& & & \text{for} \sim 20\ \text{cm bath height} \\
\alpha & = & 10\,° & \text{Axis inclination or approach angle} \\
& & & \text{respectively according } 2\alpha \text{ ceiling inclination} \\
L_v & \approx & 0,45\ \text{m} & \text{Reactor extension length corresponding} \\
& & & 30\,° \text{ insolation angle} \\
L_h & \approx & 1,7\ \text{m} & \text{Reactor insolation length corresponding} \\
& & & L_h = D_R \cdot \frac{\cos(\varphi+\alpha)}{\cos(90-\varphi-2\alpha)} \\
D_h & = & 1,5\ \text{m} & \text{Rear diameter of reactor} \\
I_B & \approx & 1,7\ \text{m} & \text{Widest image on the reactor wall} \\
& & & s = 1,05\ \text{m behind the aperture,} \\
& & & D' = 1,9\ \text{m reactor diameter} \\
F_W & = & 2,3\ \text{m}^2 & \text{Radiated reactor wall } (\beta = 100° \text{ radiated angle}) \\
& & & L_h \cdot I_w \cdot \frac{\pi}{4}
\end{array}
$$

Drum Filling

$$
\begin{aligned}
m & = \left(\frac{D}{2}\right)^2 \cdot \left[2\pi\frac{\alpha}{360} - \cos\alpha \cdot \sin\alpha\right] \cdot L_h \cdot \rho \\
& = 1,0\ \text{t},\ \alpha = 8° \text{ with } 0,2\ \text{m filling level} \\
& = 0,7\ \text{t},\ \alpha = 3° \text{ with } 0,15\ \text{m filling level}
\end{aligned}
$$

Dwell time in the reactor according to 0.7 or 1 hour respectively.

4.3.3   Heat Transfer and Heat Losses

Heat Transfer to the Reactor Wall

$$
e_w = \frac{\dot{Q}}{F_W} = 112\,\frac{kW}{m^2} \qquad \text{(average)} \tag{32}
$$

Due to the GAST - receiver energy distribution the insolation to the front port of the furnace is much higher than in the rear part. The differences can be corrected or controlled by aligning the heliostats.

Radiation Heat Transfer The absorbing and emitting surfaces of the furnace are assumed to be black emitters, which is an unfavourable assumption. Observed from equal distance the radiation density is independent of the observation angle, i.e. proportional to the surface of a (big) sphere, though not exact for a big radiator/sphere surface relation (figure 21).

Insolated reactor wall	F_W	=	2,3	m²	T_W	
Reactor shell surface	F_M	=	14	m²	T_M	$= T_W - \vartheta$
Reactor seal	F_R	=	2	m²		
Radiation duct surface	F_S	=	9	m²	T_s = 900 °C	
Receiver aperture	F_r	=	1,1	m²	T_r = 50 °C	

The radiating surface F_W emits against a sphere surface approximately equal to:

$$F_M + F_R - F_W = \sum F \tag{33}$$

F_R is lost into the duct, a section of $F_r \cdot \cos\varphi$ is radiating against the outside. Both have to be corrected by the relative distance of the receiving surfaces. E_λ is the conductive loss through the furnace wall. ΔT allows for E_λ and the heat transfer to the cold material by radiation as well as heat transfer into the bath by the superheated film material (see section 4.3.6).

The insolated energy is estimated as aforesaid:

$$\sigma \cdot F \cdot \Delta(T^4) = \dot{e}_x \cdot F$$

$$\sigma \cdot F_w \left[\frac{\sum F - F_R}{\sum F} \cdot \left(T_W^4 - T_M^4\right) + \frac{F_r}{\sum F} \cdot \left(T_W^4 - T_r^4\right) + \frac{F_R - F_r}{\sum F} \cdot \left(T_W^4 - T_s^4\right) \right]$$

$$= e_r \cdot F_r - \frac{F_W}{F_M} \cdot E_\lambda - \varepsilon \frac{\vartheta}{\Delta T} \cdot E_R \tag{34}$$

$$
\begin{aligned}
\vartheta &= 50 \quad ^\circ \quad (T_W - T_M) \\
\Delta T &= 1000 \quad ^\circ \quad (1350 \text{ K - } 350 \text{ K}) \\
\varepsilon &= 1 \qquad \text{convective heat transfer by film} \\
x &= 20 \qquad \text{superheating } \left(\frac{\Delta T}{\vartheta}\right) \cdot \varepsilon \\
e_r \cdot F_r &= 450 \quad \text{kW} \quad \eta = \frac{333 kW}{450 kW} = 0,75 \\
T_W &= 1400 \quad \text{K} \quad \simeq 1130^\circ
\end{aligned}
$$

In the equation listed above the first term is the radiation against the uninsulated reactor shell surface, the second the radiation loss through the receiver aperture, the third against the radiation duct wall. This is equal to the insulated energy reduced by the conductive loss through the insulated furnace surface and the proportionate heat transfer by the superheated film into the furnace bath.

Accordingly at a temperature of 1350 K the effective radiation plus convective surface loss amounts to 110 kW, which just covers the total heat loss, so that the capacitive heat demand has to be transported by the film ($\varepsilon = 1$). In order to raise the feed temperature by 1000 °C (1400 K - 400 K), the slag has to be overheated 20 times i.e. $x \cdot \vartheta = 20 \cdot 50°C$.

Heat Losses Heat losses arise from radiation through the furnace aperture, by heat conduction through the reactor walls and by gasing.

Radiation losses:

$$
\begin{aligned}
\dot{Q}_r &= 35 \quad KW \\
\dot{Q}_{R-r} &= 12 \quad KW
\end{aligned}
$$

Part of the radiation $\dot{Q}_{R-r}$ will be absorbed by the receiver at F_r. Otherwise the loss confers with the conduction loss of the radiation tube (section 4.3.4).

Thermal conduction losses:

$$
\begin{aligned}
\dot{Q}_\lambda &= \frac{\lambda}{\Delta r} \cdot \vartheta_2 \cdot F_M \\
&= 36 \quad kW \qquad \text{(see section 4.3.4)}
\end{aligned}
$$

Gassing losses:

$$\dot{Q}_g = c_p \cdot \dot{m} \cdot \vartheta_2$$
$$= 24 \; kW$$

The gasing losses are mainly for the air circulation through the furnace, which has to be superheated from receiver temperature to furnace temperature. An air velocity of 0,25 $\frac{m}{s}$ at the receiver was assumed.

Total losses:

$$\dot{Q}_\eta = \frac{1 - \eta}{\eta} \cdot \dot{Q}_r$$
$$= (1 - \eta) \cdot \dot{Q}_g$$
$$= 107 \quad kW$$
$$\eta = 0,76$$

The assessed values depend to a high degree on the reactor's solar-geometrical shape. By reducing the ratio $\frac{F_R}{F_W}$ the radiation loss $\dot{Q}_\sigma$ is decreased; at the same time, however, the wall temperature is highly increased with the result that the material transport by adhesion to the wall is reduced and that the liner material suffers damage.

The ratio $\frac{F_R}{F_W}$ can be reduced by means of a long radiation duct with the heliostat focus lying near F_r. This, however, entails an increase of the λ wall losses by enlarged surfaces.

The gas loss becomes very small if the radiation duct is led out of the tube receiver. The total heat balance, however, does not necessarily become worse.

4.3.4 Reactor Insulation

For the reactor brick lining and the insulation oxidic ceramics like Al_2O_3, MnO can be considered.

Full material	Working temperature	$\lambda_{(1000°C)}$	$\lambda_{(100°C)}$
Al_2O_2	1950	0,0061	0,0300
Zirconoxide	2500	0,0023	0,0020
SiO_2	1200	0,0021	0,0010
Mullite	1850	0,0039	0,0061

As the thermal conductivity of full materials is very high at low temperatures a two-layer-insulation with a porous sublining can reduce heat losses.

Apart from the thermal conductivity and the temperature stability the expansion factor, quenchresistance and chemical stability are to be observed.

An effective insulation is achieved by a pressure and quench resisting brick lining of the drum up to 300 °C and an outer heat insulation by mineral fibre.

$$
\begin{aligned}
\text{Silimanit} \quad & \vartheta_{max} & = & \quad 1600 & °C \\
\text{(10 \% porosity)} \quad & \Delta\vartheta & = & \quad 1100 - 1000 & °C \\
& \lambda_{1200°C} & = & \quad 0,0014 & \frac{kW}{mK} \\
& c & = & \quad 1 & \frac{kJ}{kgK} \\
& \rho & = & \quad 2400 & \frac{kg}{m^3} \\
& \gamma & = & \quad 4,6\cdot 10^{-6} & \frac{1}{K}
\end{aligned}
$$

good quench resistance

good chemical stability

$$
\begin{aligned}
\text{Silimanit} \quad & \vartheta_{max} & = & \quad 1800 & °C \\
\text{(50 \% porosity)} \quad & \Delta\vartheta & = & \quad 1000 - 400 & °C \\
& \lambda & = & \quad 0,0004 & \frac{kW}{mK} \\
& c & = & \quad 0,6 & \frac{kJ}{kgK} \\
& \rho & = & \quad 1500 & \frac{kg}{m^3} \\
& \gamma & = & \quad \text{very low} &
\end{aligned}
$$

$$
\begin{aligned}
\text{Asbestos or} \quad & \vartheta_{max} & = & \quad 400 & °C \\
\text{Mica} \quad & \Delta\vartheta & = & \quad 400 - 200 & °C \\
& \lambda & = & \quad 0,5\cdot 10^{-4} & \frac{kW}{mK} \\
& \rho & = & \quad 1000 & \frac{kg}{m^3}
\end{aligned}
$$

The heat passing through a cylindrical form is:

$$\lambda \cdot r \cdot \frac{d\vartheta}{dr} \Big|_{r_i}^{r_a} = q_i \qquad (35)$$

$$\ln \frac{r_a}{r_i} = \frac{2 \cdot \pi \cdot l}{Q_v} \cdot \sum \cdot \lambda \cdot \Delta \vartheta \qquad (36)$$

$$
\begin{aligned}
\Delta r \quad &= \quad 4{,}5 \quad \text{cm} \quad \text{Silimanit (20 \% porosity)} \\
&= \quad 10{,}5 \quad \text{cm} \quad \text{Silimanit (50 \% porosity)} \\
&= \quad 0{,}5 \quad \text{cm} \quad \text{Asbestos or Mica}
\end{aligned}
$$

The brick lining of 15 cm could also be carried out as composite construction.

For the radiation duct, the following insulation results:

$$\Delta r \quad = \quad 10 \quad \text{cm} \quad \text{Silimanit (50 \%porosity)}$$

The drum and the radiation duct have to be designed from heat resisting steel for operating temperatures of 300 °C.

4.3.5   Film Forming and Film Viscosity

The wall temperature is the outer material temperature decreasing over the adhering material layer in direction to the reactor wall. It increases over the circulation position and is cooled when passing through the bath. Thus, T_W is adjusted by the bath temperature or the bath viscosity by drawing a material stream suited to the insolation rate as a heat absorbing film on to the drum surface.

Film forming and film adhesion depend upon:

- Material adhesion

- Viscosity and temperature

- Flow direction or flow velocity

Oxidic slags create strong films or deposits. These deposits have to be avoided by the right material composition and temperature selection.

The flow velocity is changed during the drum's rotation. At the same time, viscosity is reduced by overheating the molten mass. At the vertical wall shear stress and gravitional force are in balance.

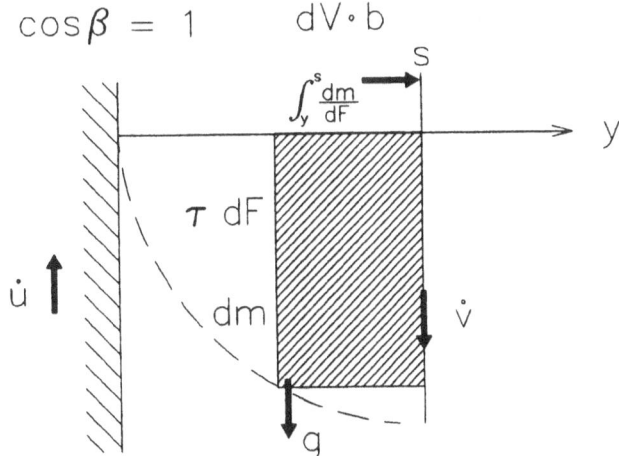

Figure 22: Mass Flow in the Melting Film

$\dot{u}$ = peripheral velocity
$\dot{v}$ = velocity of flow
s_v = vertical film thickness
s_w = horizontal film thickness

$$\tau = -\eta \frac{d\dot{v}}{dy} = g \int_y^s d\left(\frac{m}{F}\right) = g \int_y^s d\left(\frac{\dot{m}}{F}\right) \tag{37}$$

$$-\rho \cdot \nu \frac{d\dot{v}}{dy} = g \cdot \rho \int_y^s \frac{\dot{v} \cdot b \cdot dy}{\dot{v} \cdot b} = g \cdot \rho \left(s - y\right) \tag{38}$$

$$-\frac{\nu}{g} \int_{\dot{u}}^{\dot{v}} d\dot{v} = \int_0^s \left(s - y\right) dy \tag{39}$$

$$-\frac{\nu}{g} \cdot \left(\dot{v} - \dot{u}\right) = s \cdot y - \frac{y^2}{2} \tag{40}$$

$$\left[\frac{2 \cdot \nu(T) \cdot \dot{u}}{g}\right]_{\dot{v}=0}^{0.5} = s_v \tag{41}$$

The adhering vertical layer thickness s_v and the material transport at the drum wall is a function of the material viscosity at bath temperature and the peripheral velocity.

The viscosity of slags is to a high degree determined by the components melting at low temperatures. A change of the viscosity η with ϑ for glasses follows [8]:

$$\ln \frac{\eta_1}{\eta_2} = c \cdot \left(\frac{1}{T_1} - \frac{1}{T_2}\right) \tag{42}$$

This dependence on an e-function generally corresponds to the temperature function of Newtonian fluid. Compared with water, the factor c is about 30 times higher. Thus, viscosity is much more sensitive to temperature than water.

The composition of filter dusts assumes that the softening point lies at approx. 700 °C and that at 1000 °C a viscosity of $\eta = 1$ P corresponding to 100 times $\eta_{Water,20°C}$ is achieved.

$$\dot{m} = \rho \cdot \int d\dot{V} = \rho \cdot b \int_0^{s_v} \left(-\frac{g}{\nu}\left(s_v \cdot y - \frac{y^2}{2}\right) + \dot{u}\right) dy \tag{43}$$

$$\dot{m} = -\frac{\rho \cdot b \cdot g}{\nu}\left[-\frac{s_v^3}{3} + \frac{\nu}{g} \cdot \dot{u} \cdot s_v\right] = \frac{\rho \cdot b \cdot g}{\nu} \cdot \frac{s_v^3}{6} \tag{44}$$

The film thickness on the horizontal wall s_w is only $\frac{1}{3}$ of the vertical film thickness.

$$s_w = \frac{\dot{m}}{\rho \cdot b \cdot \dot{u}} = \frac{g}{\nu} \cdot \frac{2 \cdot \nu}{s_v^2 \cdot g} \cdot \frac{s_v^3}{6} = \frac{s_v}{3} \tag{45}$$

The viscosity ν is very sensitive to temperature and is determined by eutectic compositions and non-melting or melting portions of the slag flow respectively.

Apart from the slag composition the thermal conductivity in the molten mass, too, is important for the adhesion behaviour, that means the generation and melting of layers.

4.3.6 Thermal Conduction in the Film

The heat absorbed by the molten mass has to be conducted through the upper slag layer. In case of uniform rotations the heat flow is quasi-stationary.

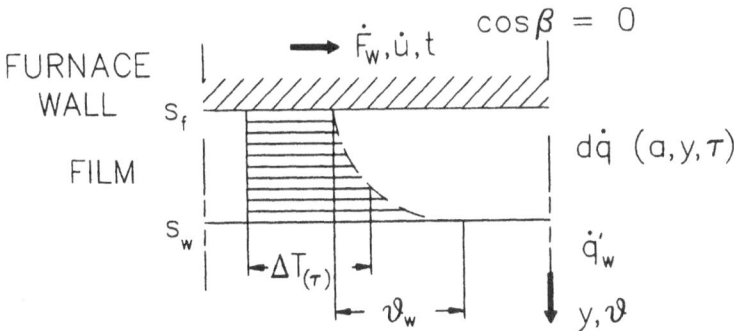

Figure 23: Thermal Conduction in the Film

$$
\begin{aligned}
\dot{u} &= \text{peripheral velocity} \\
I_r &= \text{reactor wall image} \\
\tau &= \sqrt{\frac{F_W}{\dot{u}}} \\
\tau_{max} &= \frac{l-r}{\dot{u}} \\
a &= \frac{\lambda}{c \cdot \rho}
\end{aligned}
$$

$$
d\dot{Q} = \dot{F} \cdot c \cdot \rho \cdot \frac{d\vartheta}{d\tau} \cdot dy = \frac{d}{dy} \cdot \left(\lambda \cdot F \cdot \frac{d\vartheta}{dy} \right) dy \tag{46}
$$

The differential equation for the film conductivity can be simplified for small values of $\frac{s_0}{F}$ with separate equations for $\frac{d\vartheta}{d\tau}$ and $\frac{d\vartheta}{dy}$. The average film temperature (without conduction) is:

$$
\dot{q}_w' \cdot F = \rho \cdot c \cdot F \cdot s_w \cdot \frac{\Delta\vartheta}{\Delta\tau} \tag{47}
$$

$$
\Delta\bar{T} = \dot{q}_w' \cdot \frac{\sqrt{F_W}}{\dot{u}} \cdot \frac{1}{\rho \cdot cs_w} \tag{48}
$$

$\dot{q}'_w$ refers to absorbed portion of the insulated energy $\dot{e}_s$. The surplus temperature of s_w against s_f is:

$$c \cdot \rho \cdot \left(\frac{d\vartheta}{\tau}\right)_c \cdot y \;=\; \dot{q} \qquad \text{heat capacity}$$

$$-\lambda \cdot \frac{d\vartheta}{y} \;=\; \dot{q}_0 - \dot{q} \qquad \text{heat conduction}$$

For $\left(\frac{d\vartheta}{\tau}\right) = constant$ is $c \cdot \rho \cdot \left(\frac{d\vartheta}{\tau}\right) \cdot s_w = \dot{q}'_w$, i.e. no temperature rise into the wall and quasi-stationary temperature profile.

$$\int_0^y \dot{q}_0 \cdot \left(\frac{y}{s_w} - 1\right) dy = \lambda \cdot \int_{\vartheta_w}^{\vartheta} d\vartheta \tag{49}$$

$$\Delta\vartheta = \frac{\dot{q}'_w}{\lambda} \cdot \left(\frac{y^2}{2 \cdot s_w} - y\right) \tag{50}$$

$$\Delta\vartheta_f = -\frac{\dot{q}_0}{\lambda} \cdot \frac{s_w}{2} = -\frac{\dot{q}_w}{\lambda} \cdot \frac{s_v}{6} \tag{51}$$

$$\Delta\vartheta_w = T_W - \bar{T} = \frac{2}{3} \cdot \Delta\vartheta_f \tag{52}$$

The temperature of the film against the bath temperature is therefore:

$$\Delta T_W = \Delta\bar{T} + \Delta\vartheta_w \tag{53}$$

$$\Delta T_W = \dot{q}'_w \cdot \left[\frac{g}{s \cdot \nu}\right]^{0,5} \cdot \left[\frac{\sqrt{F_W}}{\rho \cdot c} \cdot 3 \cdot \dot{u}^{-1,5} + \frac{2 \cdot \nu}{g \cdot \lambda \cdot g} \cdot \dot{u}^{0,5}\right] \tag{54}$$

The film thickness i.e. the material mass picked up by the rotating drum is determined by the viscosity at bath temperature. The bulk material is heated up as a function of the rotating velocity $\dot{u}^{-1,5}$ and of the material heat capacity c^{-1}. The lower the rotating speed, the higher the bulk temperature rise. The film wall temperature surpasses the bulk temperature with increasing rotating speed and in reverse proportion to the conductivity.

On the basis of the assumed reactor conditions the corresponding wall temperature with the following material values can be calculated for an assumed bath temperature:

$$F_w = 2,3 \ m^2$$
$$\dot{q}_w = 200 \ \frac{kW}{m^2 \cdot s}$$
$$\lambda_{bath} = 0,006 \ \frac{kW}{m \cdot K}$$
$$c_{bath} = 1,1 \ \frac{kJ}{kg \cdot K}$$
$$\rho_{bath} = 4000 \ \frac{kg}{m^3}$$

T_B	$°C$	1025	1000	975
η_B	$\frac{kg}{m \cdot s}$	0,40	1	3,4
$\Delta \bar{T}$	$°C$	50	70	80
$\dot{u}$	$\frac{m}{s}$	0,94	0,55	0,37
s_v	mm	3,9	5,3	8,0
s_w	mm	1,3	1,8	2,7
$\Delta \vartheta_f$	$°C$	21	29	45
ΔT_{wmax}	$°C$	64	89	110
T_W	K	1362	1362	1358
$\dot{m}$	$\frac{kg}{s}$	8,3	5,9	6,0
x	heating per charge	30	21	22

Wall temperature amounts to approx. 1360 K for varying bath temperatures. Bath viscosity decreases with 25 K of temperature increase by $\frac{\eta_1}{\eta_2} = 0,4$. At wall temperature the slag has a viscosity of $\eta = 0,1 \frac{kg}{m \cdot s}$ and drains off or drops off from the ceiling, a fact that can be counterbalanced by lowering the temperature. At lower temperatures the film pick up s_v becomes excessive and the heat conduction in the insulation zone does not allow a full melt-up. The operation in the narrow temperature range of suitable viscosities is very effectively counterbalanced by changing the drum rotation speed. Changes in the insulation, if not counterbalanced by heliostat movements, can be governed by feed variation as well as drum speed.

x gives the number of circulations per volume, that means the number of material overheatings at the reactor wall.

$$x = \frac{\frac{\dot{m}}{\rho}}{s \cdot \dot{u} \cdot L} \tag{55}$$

Therefore, only molten masses that do not generate mixed phases are problematic.

4.3.7   Operational Data

Weights of Installation   The assessment of the reactor dimensions results in the following values:

Radiant Tube:

Steel casing (300 °C)	
1,9 m $^\varnothing$* 5 mm * 2 m	0,5 t
Silimanit interior brick lining	
1,6 m $^\varnothing$* 15 cm * 2 m	2,5 t

Drum:

Steel casing	
2,40 / 1,80 m $^\varnothing$* 20 mm * 2,3 m	3,0 t
Silimanit interior brick lining	
2,1 / 1,5 m $^\varnothing$* 15 cm * 2,2 m	4,0 t

Bonds, Bearings:

3,0 m $^\varnothing$ * 0,02$^\square$* 2 m	3,0 t
4 bearings	1,0 t

Engine, Transmission, Control:

	1,0 t

Conveying Facilities:

Insertion hopper, feeder	1,0 t
slag quench, water	1,5 t
filling	1,0 t

Facilities:

Reactor lock during the night	1,0 t
Insertion - lift $2 * 1\frac{t}{h}$ @ 120 m	2,0 t
Crane	2,0 t

Production Control:

	0,5 t

The reactor mass assembled on the tower, plus of accessory aggregates, will amount to approx. 24 t exclusive of processing machinery that should be installed on the ground.

5   Operational Parameters

The process parameters as temperature, heat flux, material flow, exposure time etc. have to be channeled and regulated according to the process needs. This calls for a subordinate solar field sectioning control independent of the optimized energy performance of the overall solar plant. As stated in main section 4 the process requires also various energy levels for auxiliary and preheating operations.

It is therefore considered best to incorporate the process operation as part of a combined energy/process-plant concept.

5.1   Solar Field Sectioning

For the heat exchanger reactor with an indirect energy feed from the main receiver a field allocation does not apply.

For the radiation reactor a detailed analysis of field allocation and field control algorithm is given in section 4.2. Accordingly a drum reactor has been devised in section 4.3 based on the GAST-field with a tower height of 170 m and a heliostat focal length of 300 m aimed at the outer aperture of the tower. With this configuration the reactor aperture within the receiver of 1,1 m² receives 450 kW corresponding to a field sector of 2000 m² at the further field rim approx. 320 to 360 m from the tower base (field efficiency 0,6). Actually the participating field is distributed over twice the area but does only partially contribute to the aperture flux and is otherwise framed out by the pipe receiver.

If the heliostat focal length is adapted to the reactor aperture the allocated heliostat field can be reduced or positioned closer to the tower base corresponding to a lower tower height. This focal length adaption is of course necessary for a pure melting reactor field.

5.2   Solar Operating Time

In a combined field plant i.e. with melting reactor and pipe receiver, the time performance of the reactor can follow closely the general plant performance. At 700 $\frac{W}{m^2}$ design point for 100% production this means a 7-8 h production cycle in winter and a 9-10 h production cycle in summer. Since the process flux control necessitates the individual control of certain heliostat field sections, the production time of the reactor can easily maintained or even be prolonged by adding of focus-adapted heliostats though of course on the cost of other heat requirements. Further control means and control configurations are discussed in sections 4.2 and 4.3.

With a pure melting reactor the field efficiency is reduced since the limited angle of the reactor radiation duct means a field fairly remote to the tower base and therefore a sharp decline of field efficiency especially in the morning and evening hours.

This means that with a corresponding design point of 700 $\frac{W}{m^2}$ the heliostat field has to be larger or more precisely, the time use factor compared with the combined field is smaller. In winter though the field efficiency improves comparativly in line with the overall closer solar arc and with reduced heliostat shadowing of a small angle field.

In figure 24 and table 9 the field performances are compared based on assumed heliostat field efficiencies. As stated in the field calculation in section 4.2 the actual heliostat demands can only be computed for actual plant specifications.

Table 9: Heliostat Field Efficiency ([26] and Adapted Data)

Operation Time	Heliostat Field Efficiency	
GAST-Field	0,48 - 0,81	⊘ 0,71
GAST-Section	0,55 - 0,67	⊘ 0,62

Time		Solar Radiation $\left(\frac{W}{m^2}\right)$	Efficiency
June	12^{00}	850	∼ 0,75
	$7^{30}/17^{00}$	700	∼ 0,60
	$6^{30}/18^{00}$	550	(50% load)
Dec	12^{00}	950	∼ 0,72
	$8^{30}/15^{30}$	700	∼ 0,62
	$8^{00}/16^{00}$	500	(50% load)

5.3 Operation Parameters of a Melting Receiver/Reactor

The layout of the melting receiver/reactor is done for a throughput of 1 $\frac{t}{h}$ without preheating. In case of preheating to approx. 600 °C in a cyclone or fluidized bed the throug-

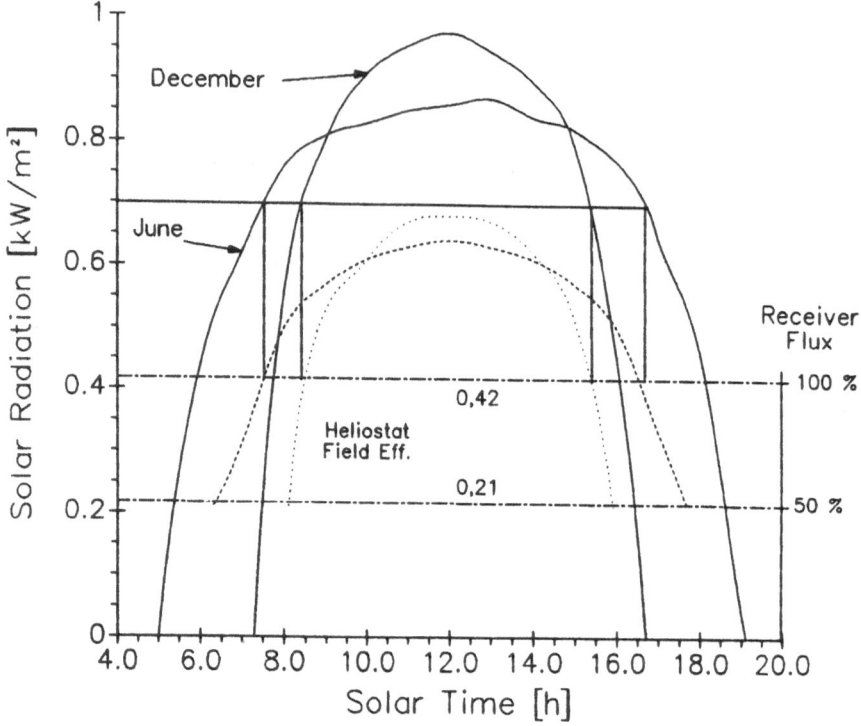

Figure 24: Solar Radiation Almeria - Heliostat Field/Receiver Efficiency for Melting Reactor Field Section

hput can more or less be doubled. In case of an enlargement of the reactor drum the material throughput increases proportionally to the adsorption area i.e. in line with the reactor weight. Higher specific performances can only be achieved by the concentration of the heliostat flux and correspondingly higher reaction temperatures.

The heat capacity of the reactor amounts to approx. $3,1 \cdot 10^6$ kJ. With a heat loss over the casing of about 8% corresponding to 40 kW the drum will lose about 50% of its heat in case of a stillstand of 14 h overnight. Heating up to the operating temperature will take approx. 1 h. This, however, is only possible in case of a good insolation cover of the reactor aperture overnight. Then, the production time of the reactor will amount to 9 h/day in summer and 7 h/day in winter. If heliostats can be switched variably on to the radiation reactor as discussed in section 4.2, 4.3 and 5.2 the heat flux can be achieved over longer times and at worse insolation conditions, too. This flexibility is necessary for the reactor-process control in any case. A combined solar plant for electrothermical and process-thermical use can achieve this flexibility best [2,25].

5.4   Necessary Further Examinations

As shown in section 4.3.1 the concentrated solar beam passes the receiver aperture and the radiation duct through the circular reactor seal and then meets the upper wall of the drum. The radiation energy is absorbed by the adhering material at the upper part of the drum. The thin molten layer on the ceiling wall is overheated during each drum rotation. The adhering layer thickness or the material transport at the drum wall respectively is a function of the material viscosity and the peripheral velocity. The viscosity is very sensitive to temperature and is determined by eutectic compositions and non-melting or melting portions of the slag flow respectively.

The material transport on the wall as well as the thermal conduction in this material layer is the main factor of a full working melting reactor/receiver. It is very important to get fundamental informations of the melting film under solar conditions.

6 Appendix

6.1 Nomenclature

Letters

c_p	heat capacity
d	diameter
d_h	diameter of ellipse of the receiver aperture
d_V	diameter of ellipse of the receiver aperture
D_h	rear diameter of reactor
D_R	reactor seal diameter
D_v	greatest reactor diameter
e_r	energy density at receiver aperture
e_R	energy density at reactor seal
e_s	insolated energy
f	focus of heliostats
F	surface
F_r	radiation duct cross section reactor seal
F_R	radiant tube cross section receiver aperture
F_S	radiant tube surface
F_W	insolated reactor wall
h	bed height
I_B	widest image on the reactor wall
I_w	reactor wall image
L_h	reactor insolation length
L_v	reactor extension length
$\dot{m}$	mass stream
Nu	Nusselt number
p	pressure
Pr	Prandtl number
$\dot{q}$	solar insolation
$\dot{Q}$	heat flux
Re	Reynolds number

s_v	vertical film thickness
s_w	horizontal film thickness
S	distance of receiver level
t_m	gas temperature
T	height of tower aperture
T	temperature
$\dot{u}$	gas velocity
$\dot{v}$	velocity
x	(apparent) dimension of heliostats
x	number of circulations per volume

Greek Symbols

α	aperture or reactor inclination
α	heat transmission coefficient
Δ	difference
η	dynamic viscosity
η	reactor heat efficiency
λ	caloric conductibility
ψ	division
ρ	density
Θ	angle
ε	proportionate convective heat transfer, free volume
φ	vertical field angle
$\partial\varphi$	angle expansion of the heliostat
ϑ	temperature
ϑ_{log}	average transmission temperature
τ	radiation time
ξ	horizontal field angle
ζ	radiant tube angle

6.2 References

[1] Aragon,J.M.,   Solar Heating of Granular Solids
     Segarra,M.A.,   in: Becker,M., (Ed.); Solar Thermal Central Re-
                                       ceiver Systems Volume 2: High Temperature
                                       Technology and its Applications
                                       Springer Verlag (1986) 975-980

[2] Babcock and Wilcox Co.   Experimental Process for Destruction of Hazar-
     Baberton (USA)   dous Waste Using Solar Energy, (1984)

[3] Becker,M.,   Ellgering,H.,   Construction Experience Report for the Cen-
     Stahl,D.,   tral Receiver System of the International Energy
                                       Agency Small Solar Power Systems Project
                                       DLR (1983)

[4] Boese,F.K.,   Konzept zur Weiterentwicklung von volumetri-
                                       schen Receivern
                                       DLR interner Bericht ET-3/88-1, 119-305

[5] Böhmer,M.,   Anlagen und Komponenten der Solarthermie
                                       in: Becker,M., Funken,K.-H.,(Ed.); Solarchemi-
                                       sche Technik Volume 2: Solare Detoxifizierung
                                       von Problemabfällen
                                       Springer Verlag (1989) 91-121

[6] Brown,C.T., Mackie,P.E.,   Advanced Component Research in the Solar
     Neale,D.H.,   Thermal Program
                                       Solar World Congress, Vol. 3 (1983) 1421-1425

[7] Chavez,J.M.,   Development and Testing of Advanced Central
                                       Receivers
                                       Proceedings of the 24th Intersociety Energy Con-
                                       version Engineering Conference
                                       Vol.4, (1989)

[8] D'Ans,J.,Lax,E.,   Taschenbuch für Chemiker und Physiker
                                       Springer-Verlag (1967)

[9]   DeLaquil,P.,   Feasibility Study of a High Temperature Direct
                                       Absorption Receiver
                                       in: Becker.M., (Ed.); Solar Thermal Central Re-
                                       ceiver Systems Volume 2: High Temperature
                                       Technology and its Applications
                                       Springer Verlag (1986) 869-873

[10] DLR   Solar Energy for High Temperature - Technology
                                       and Applications
                                       (1987)

[11] Drost,K.,   Evaluation of Solar Air Heating Central Receiver
                                       Concepts
                                       Progress in Solar Energy 5, Part 1 (1982) 293-298

[12] Epstein,M.,   Performance of Ceramic Tubes Solar Receiver
                                       in: Becker,M., (Ed.); Solar Thermal Central Re-
                                       ceiver Systems Volume 2: High Temperature
                                       Technology and its Applications
                                       Springer Verlag (1986) 635

[13] Erhardt,K., Vix,U.,   Direct Absorption of Concentrated Solar Radia-
                                       tion by Particle-Gas Suspensions
                                       in: Becker,M., (Ed.); Solar Thermal Central Re-
                                       ceiver Systems Volume 2: High Temperature
                                       Technology and its Applications
                                       Springer Verlag (1986) 853-868

[14] Fisk,W.J.,   Performance Analysis of a Windowed High Tem-
      Wroblewski,D.E.,   perature Gas Receiver using a Suspension of Ul-
      Hunt,A.J.,   trafine Carbon Particles as the Absorber
                                       Proceedings of American Section of the Inter-
                                       national Solar Energy Society Annual Meeting,
                                       Phoenix, June 2-6 (1980) 553-557

[15] Flamant,G., et al.   A 50 kW Fluidized Bed High-Temperature Solar
                                  Receiver
                                  in: Becker,M., (Ed.); Solar Thermal Central Re-
                                  ceiver Systems Volume 2: High Temperature
                                  Technology and its Applications
                                  Springer Verlag (1986) 843-852

[16] Fricker,H.W.,   Tests With a Small Volumetric Receiver
                                  in: Becker,M., (Ed.); Solar Thermal Central Re-
                                  ceiver Systems Volume 1: High Temperature
                                  Technology and its Applications
                                  Springer Verlag (1986) 596-612

[17] Grasse,W.,   Regenerative Energiequellen für den Einsatz in
                                  Kraftwerken, Möglichkeiten und Grenzen - Solar-
                                  thermie
                                  VDI-Berichte (1990)808,147-155

[18] Gupta,B.P., Bohn,M.S.,   Direct Absorption Receiver Research
                                  in: Becker,M., (Ed.); Solar Thermal Central Re-
                                  ceiver Systems Volume 2: High Temperature
                                  Technology and its Applications
                                  Springer Verlag (1986) 637-647

[19] Hedden,K., Vesper,D.E.,   Verhalten von Pigmenten und Füllstoffen bei der
                                  thermischen Behandlung von Lackschlämmen
                                  Chemie-Ingenieur-Technik MS 1772/89

[20] Hogan,R.E.,   A Direct Absorber Reactor/Receiver for Solar
      Skocypec,R.D., et al.   Thermal Applications
                                  Chemical Engineering Science 45(1990)8,2751-
                                  2758

[21] Hunt,A.J.,   A New Solar Thermal Receiver Utilizing a Small
                                  Particle Heat Exchanger
                                  Proceedings of the 14th Intersociety Energy Con-
                                  version Engineering Conference, Boston, August
                                  5-10 (1979)159-163

[22] Hunt,A.J.,   New Approaches to Receiver Design: Prospects and Technology of Using Particle Suspensions as Direct Solar Absorbers
in: Becker,M., (Ed.); Solar Thermal Central Receiver Systems Volume 2: High Temperature Technology and its Applications
Springer Verlag (1986) 835-842

[23] Hunt,A.J., Brown,C.T.,   Solar Test Results of an Advanced Direct Absorption High Temperature Gas Receiver (SPHER)
Proceedings of the Solar World Congress, International Solar Energy Society, Perth, Australia, August 15-19 (1983)959-963

[24] Karrais,B.,   Ultralight Modular Ceramic High-Flux Receiver
in: Becker,M., (Ed.); Solar Thermal Central Receiver Systems Volume 2: High Temperature Technology and its Applications
Springer Verlag (1986) 613-623

[25] Kesselring,P.,   Solar Thermal High Temperature Technology and its Applications
in: Becker,M., (Ed.); Solar Thermal Central Receiver Systems Volume 2: High Temperature Technology and its Applications
Springer Verlag (1986) 573-584

[26] Kiera,M.,   Computer Codes, Requirements, Comparison of Methods
in: Becker,M., Böhmer,M., GAST The Gas-Cooled Tower Technology Program
Springer Verlag (1989) 95-113

[27] Kolb,G.J., Chavez,J.M.,   An Economic Analysis of a Quad-Panel Direct Absorption Receiver for a Commercial-Scale Central Receiver Power Plant

[28] Melchior.E.,   Receiver Concepts and Design, Construction and Tests of Components
in: Becker.M., Böhmer,M., GAST The Gas-Cooled Tower Technology Program
Springer Verlag (1989) 193-209

[29] Pierrot,A., Olalde,G.,   High Temperature Honeycomb Solar Receiver for Gas Heating
in: Becker.M., (Ed.); Solar Thermal Central Receiver Systems Volume 2: High Temperature Technology and its Applications
Springer Verlag (1986) 625-634

[30] Skocypec,R.D., Romero,V., Thermal Modelling of Solar Central Receiver Cavities
Journal of Solar Energy Engineering 111 (1989) 2, 17-124

[31]   VDI-Wärmeatlas
Berechnungsblätter für den Wärmeübergang

[32] Vogelpohl,A.,   Trocknung fester Stoffe
   Schlünder.E.U.,   Ullmann, 4. Aufl., Bd.2, 698-721

[33] Barnett,C.O., Myers,J.E., Momentum, Heat and Mass Transfer
McGraw Hill

Study Relating to the Use of Solar Energy
for the Allothermal Gasification of Coal

H. Kubiak, H. Lohner,

Deutsche Montan Technologie (DMT),
Institut für Kokserzeugung und Kohlechemie

Contents

1. Introduction

2. Bases

3. Process schematic

4. Simulation

5. Dimensions of a pilot plant

6. References

1. Introduction

The use of solar energy in coal gasification can make an important contribution to economical, environmentally acceptable and resource-sparing energy supply. Consequently, it has been proposed to design a plant to the concept of using solar-produced high-temperature heat for allothermal coal gasification within the framework of the present research and development project. Main plant components are a GAST hot-gas circuit with solar receiver combined with an MBG plant (MAN Bergbau-Forschung Gaserzeugung). The MBG plant (Fig. 3) on which the project is based is being developed by DMT in cooperation with MAN-GHH and is to be demonstrated at a power plant. The plant has been designed to a concept that is the result of many years of development backed by small-scale and semi-technical operating experience /1/, /2/, /3/.

Considering the results of the GAST project the present study elaborates on an MGB process modified to suit the marginal conditions of the solar plant. In addition to optimizing the plant concept, the study is primarily concerned with investigations into various parameters with the object of achieving maximum conversion rates from solar energy into chemical energy.

2. Bases

When coal is converted into synthesis gas the following reactions chiefly take place /4/:

					Reaction enthalpy (298 K)
C	+	H_2O	=	$CO + H_2$	119 kJ/mol
CO	+	H_2O	=	$H_2 + CO_2$	-42 kJ/mol
CO	+	$3H_2$	=	$CH_4 + H_2O$	-206 kJ/mol

In summary the above reactions are endothermic. Therefore heat must constantly be fed into the process.

Depending on the type of heat supply used a distiction is made between autothermal and allothermal processes (Fig. 1). In autothermal processes, part of the coal input is burnt by oxygen being introduced into the reactor. In allothermal coal gasification, the necessary process heat is supplied by a heat carrier passing through a heat exchanger. Hence, only allothermal processes provide the possibility of utilizing solar energy in the form of process heat and process steam. One of the advantages afforded by this integration of solar energy is the fact that specifically more gas can be generated from a given coal input. Hence, the coal-specific emissions, particularly of CO_2, can be distinctly reduced.

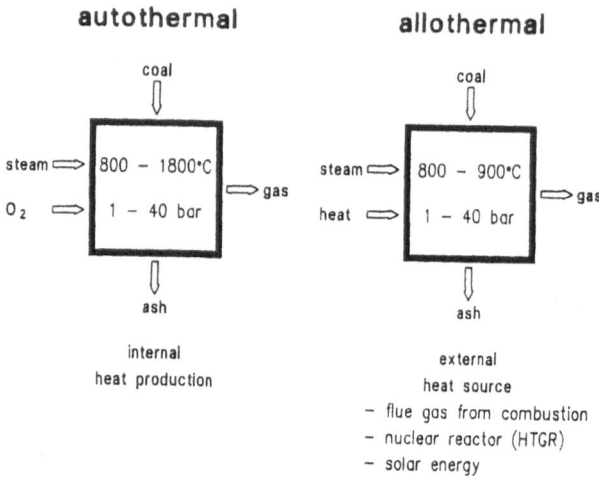

Fig. 1: Heat supply for gasifier

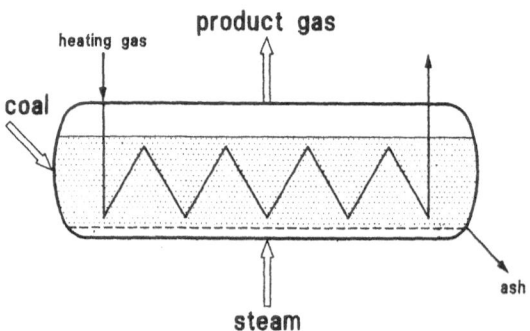

Fig. 2: Design concept of allothermal gasifier

Fig. 2 shows a gasifier working to the principle of the fluidized bed. In this process, fine-grain coal is injected using a special injection feeding system. This jet feeder allows to feed caking hard coal into a fluidized bed.

Steam, at the same time used as a reactand, serves as fluidizing agent. A tubular heat exchanger is immersed into the fluidized bed to introduce the necessary process heat into the fluidzed bed.

The results of the GAST project show that integration of solar energy into the MBG process is basically possible when proper pressure and temperature conditions exist. Air is a suitable heat carrier. This air must be available at a pressure of 9 bar and a temperature of 1000 °C maximum. During the months of April to September this availability exists over approx. 12 h/d. It diminishes to approx. 8 h/d during the winter months.

3. Process schematic

Fig. 3 shows the simplified process flow scheme of the MBG process /3/. The necessary process heat is supplied exclusively by the combustion of part of the product gas. Towards this end, air is initially compressed to operating pressure in a compressor and subsequently heated in this case to 950 °C. Heating is effected by the combustion of self-generated product gas in a combustion chamber. The combustion chamber is also used for

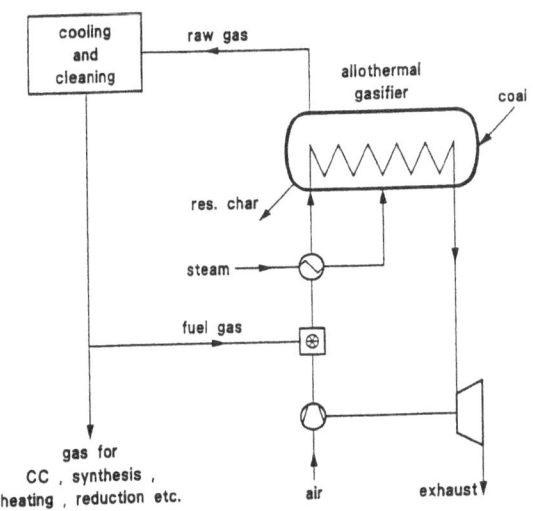

Fig. 3: Simplified process flow scheme of MBG process

superheating the process steam to approx. 850 °C. In the gasifier the flue gas generated in the combustion chamber is cooled to approx. 800 °C. At this temperature, it is passed to a turbine combined with the compressor before being relieved to ambient pressure. The heat contained in the raw gas coming from the gasifier is utilized for the generation of steam. Subsequently, the gas is cooled and conditioned.

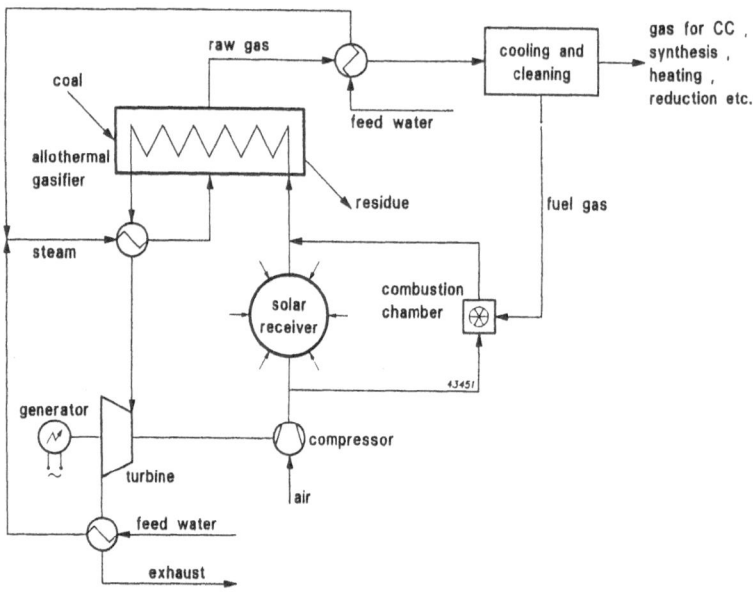

Fig. 4: Utilization of solar energy in the MBG process

To supply high-temperature solar energy to the MBG process, the flow scheme was changed according to Fig. 4.

Both the steam generator and the steam superheater are downstream of the gasifier to achieve better utilization of solar energy. With adequate insolation, the air may be completely heated my means of solar receivers. During periods of reduced insolation, a part of the air stream can be burnt in a combusiton chamber using part of the product gas. If the air is passed completely through the combustion chamber, the gasifier may also be operated during the night using practically unchanged operating conditions. Another possibility of reacting to fluctuations in insolation is the variation of operating conditions. For instance, the gasifier may be switched to part load operation when the heat supply is reduced and no additional gas is to be burnt to maintain coal throughput. Because of the good isolation of the gasifier and the heat capacity of the coal-bed solar transients may not have an influence to the process. Obviously the product yield decreases in these periods.

4. Simulation

The calculations carried out within the framework of this study are based on the "Aspen" simulation program. For a highly realistic assessment of the energy saving potential provided by the incorporation of solar energy into the gasification process, a total of 9 different total plant operation cases are analysed.

One variable is the temperature of the air used as a heat carrier. This temperature may vary between T = 900 °C, T = 950 °C and T = 1000 °C. Moreover, three different variants are simulated for air heating. In the 100 %-case, the total energy is provided by the utilitzation of solar energy. In the 0 %-case, air is heated solely in the combustion chamber. In the 70 %-case, the combustion chamber and the collector operate in parallel, i.e. 70 % of the air to be heated is passed through the collector. Table 1 summarizes some specific data from data thus gained. Fig. 5 to 7 show the energy yields for gas, heat and electricity. The different coal throughput rates at one and the same heat carrier temperature result from changes in the utilization of the combustion chamber. Since the fuel gases generated in the process are likewise passed into the reactor in the form of theat carriers, the heat carrier

air temperature		900°C			950°C			1000°C		
solar heat	MW	0	42.6	60.9	0	51.77	73.96	0	56.07	80.1
	%	0	70	100	0	70	100	0	70	100
coal feed	t/h	11.7	11.3	11.0	15.4	14.8	14.5	19.5	18.8	18.4
	MW	104.8	100.6	98.5	137.3	132.3	129.9	174.3	167.8	164.5
raw gas	$10^3 m^3/h$	33.42	32.12	31.43	43.79	42.19	41.43	55.59	53.53	52.45
	MW	119.8	114.9	112.6	156.6	150.9	148.2	198.9	191.5	187.6
fuel gas	$10^3 m^3/h$	21.69	6.6	-	26.78	8.04	-	29.28	8.78	-
	MW	77.7	23.6	0	95.8	28.7	0	104.8	31.4	0
products :										
product gas	$10^3 m^3/h$	11.73	25.52	31.43	17.01	34.15	41.43	26.31	44.75	52.45
	MW	42.1	91.3	112.6	60.8	122.2	148.2	94.1	160.1	187.6
current	MW	11.9	10.1	9.7	14.4	12.0	11.1	16.0	13.2	11.3
heat	MW	13.6	13.4	13.3	17.1	16.9	16.3	23.0	22.6	21.5
degree of integration	%	-	98.1	100.8	-	99.7	99.2	-	97.0	91.4

Tab. 1: Main data of solar-heated MBG gasification

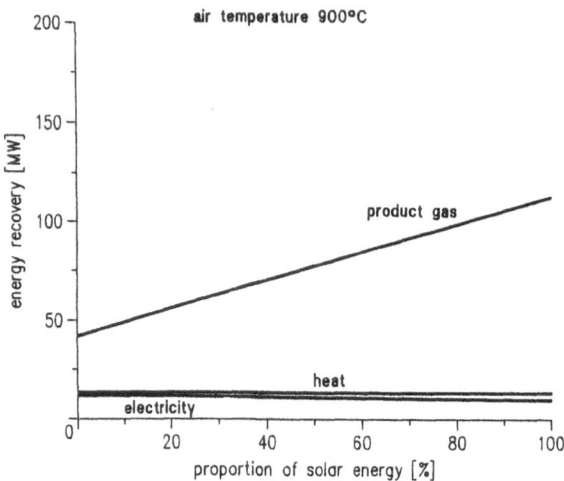

Fig. 5: Energy yields as a function of solar energy supply

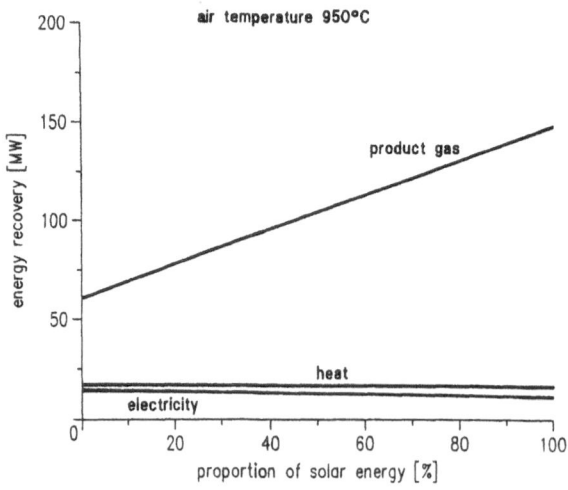

Fig. 6: Energy yields as a function of solar energy supply

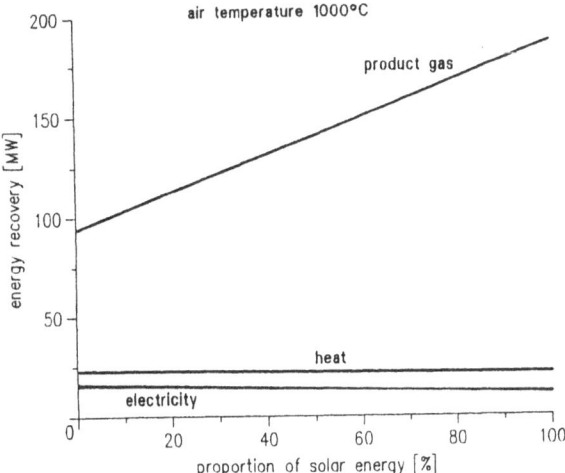

Fig. 7: Energy yields as a function of solar energy supply

mass flow will increase with increasing level of combustion chamber utilization. Interacting with the likewise changing heat capacity of the heat carrier this gives a varying energy feed into the reactor with a resultant effect on the gasification process.

It can be seen quite clearly that in all the cases analysed, the power gas recovery using solar energy integration will increase substantially while, at the same time, the yields of byproducts from the MBG process, electricity and heat remain practically on the same level. The degree of integration shown in Table 1 is calculated from the difference of product heats between the 100 %-case / 70 %-case and 0 %-case related to the solar heat used as follows:

$$y = \frac{\sum products\ (X\%\text{-case}) - \sum products\ (0\%\text{-case})}{solar\ heat\ (X\%\text{-case})}$$

Using the metereological data recorded in Almeria it was now possible to simulate all-year operations /5/ /6/. Based on the distribution of direct solar radiation over solar time for every month, the periods of solar energy utilization were supposed as shown in Fig. 8. It was expedient to break up the year into three periods, viz. February to May, June to September and October to January.

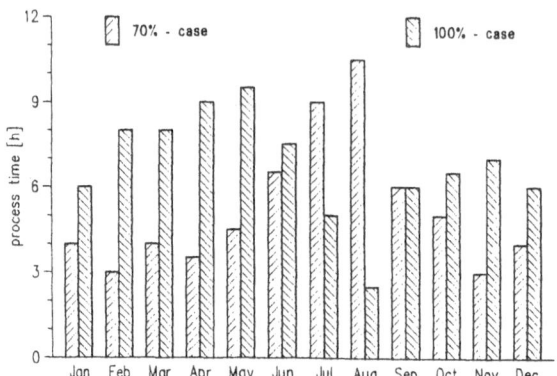

Fig. 8: Daily solar utilization periods

In the simulation of the all-year operations two different modes of operation were used. One variant provides the possibility of product recovery even during periods with no insolation (24-hour operation). Another mode of operation provides for the change-over of the plant to low profile operation. With this mode of operation (day-time operation) the plant is utilized only during periods of insolation. During low profile operation merely a steep temperature drop in the reactor and piping must be avoided. The energy used for this can be neglected because of the high heat content of the coal bed and proper insulation of the various plant components.

Table 2 shows the figures obtained from the calculations with a heat carrier temperature of T=900 °C. In the 70 %-case the coal savings are 5.05 t/h, in the 100 %-case they come to 6.22 t/h. Hence, the resulting savings are 73.07 t/d during period 1, 71.50 t/d during period 2 and 57.52 t/d during period 3.

If the plant is switched to low profile operation during hours without insolation, the annual mean coal savings come to approx. 49 %. This means that the coal-specific emissions are also reduced in the same order of magnitude. The latter effect is very important for the valuation of the process under environmental aspects. Another immense advantage is the use of solar energy to produce coal gas in areas with both coal resources and intensive insolation. In contrast to coal the generated gas can easily be transported even over long distances using pipelines.

operation periods		Feb. - May	June - Sep.	Oct. - Jan.
process time				
100% - case	h	8.5	5.0	6.0
70% - case	h	4.0	8.0	4.0
coal savings				
100% - case	t/h	6.22	6.22	6.22
70% - case	t/h	5.05	5.05	5.05
24h - operation				
coal feed with fully combustion chamber operation	t/d	280.8	280.8	280.8
coal savings	t/d	73.07	71.5	57.52
by solar energy	%	26.0	25.5	20.5
day-time operation				
coal feed with fully combustion chamber operation	t/d	146.3	152.1	117.0
coal savings	t/d	73.07	71.5	57.52
by solar energy	%	49.9	47.0	49.0

Tab. 2: Comparative data day-time operation / 24-hour operation

Based on the simulation of an all-year operation it is possible to estimate the process economy using the following datas:

rate of utilization : 90 %
depriciation time : 20 years
interest on equity : 8 %
interest on debt : 9 %
tax, insurance : 1,5 %
capitalization : 15,2 %

The increase of product yield of $6,53 * 10^7$ m^3/a means a profit increase of about 14,06 Mio. DM/a., if a production of electricity using a gas-turbine is assumed. Beside this, a coal saving of 2300 t/a occurs which means savings of 230 TDM/a.

The above calculations show that an additional investment of 94,1 Mio DM is possible to get the same electricity costs as in the original MBG-process.

5. Dimensions of a pilot plant

The technical feasibility of solar tower and MBG plant integration is to be demonstrated in the form of a small-scale pilot plant of the Plataforma de Almeria. The gasification reactor to be adopted has already been modelled with the help of a calculation program developed by DMT /7/. With an available air mass flow of 0.36 kg/s a coal throughput of 40 kg/h can be attained. At an air temperature of 1000 °C, a reactor volume of 0.42 m^3 and a pressure of 9 bar the coal is gasified at a gasification temperature of 859 °C to a conversion degree of 87 %. A vertical fluidized bed reactor of 58 m diameter and a fluidized bed depth of approx. 1.5 m will be used as a gasifier. Fig. 9 shows a flow diagram with the main data of the pilot plant. For the development of the semi-technical pilot plant we must take into account:

The design of the various plant components such as reactor, heat exchanger, steam generator etc. would appear to present no problems considering past experience. This also applies to the materials to be adopted. Only the linking of gasification plant and GAST hot gas circuit requires further notice.

Fig. 10 is a first design of the process schematic.

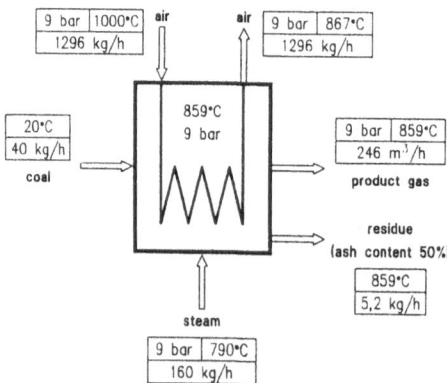

Fig. 9: Flow diagram of gasifier

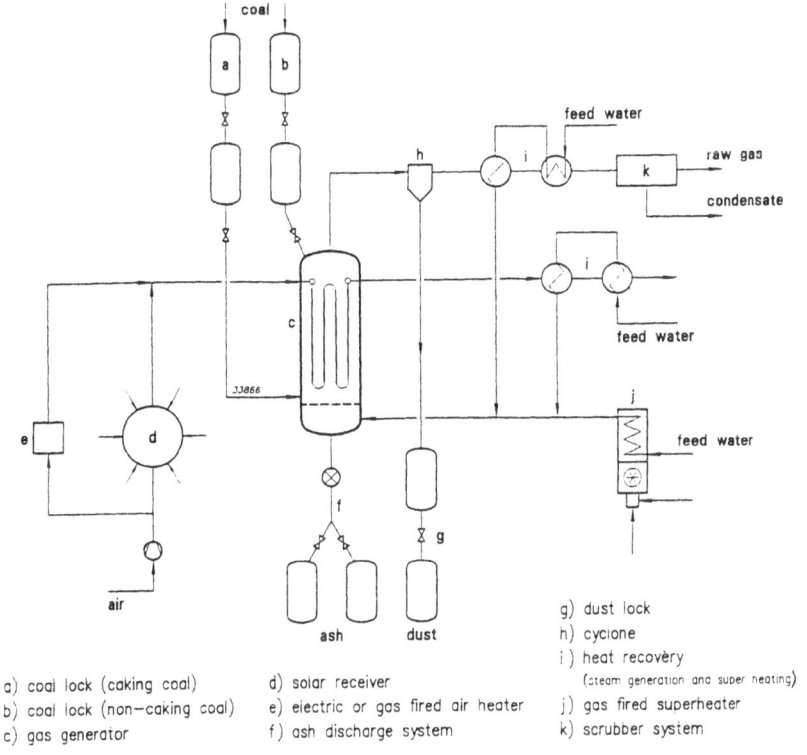

coal

a   b

feed water

h   i   raw gas

k

condensate

i   i

feed water

c

JJ866

e   d

j   feed water

f

air

g

ash   dust

a) coal lock (caking coal)   d) solar receiver   g) dust lock
b) coal lock (non–caking coal) e) electric or gas fired air heater h) cyclone
c) gas generator   f) ash discharge system   i) heat recovery
                                                               (steam generation and super heating)
                                                               j) gas fired superheater
                                                               k) scrubber system

Fig. 10: Process schematic of pilot plant

6. References

/1/ van Heek, K. H. et.al. "Wasserdampfvergasung von Kohle", Projekt Prototypanlage Nukleare Prozeßwärme, Final Report BMFT-FB-T85-153, Essen 1985

/2/ Knop, K., van Heek, K. H., Kubiak, H. "Entwicklungstand und Anwendungs-möglichkeiten des MBG-Kohlevergasungsverfahrens", Stahl und Eisen 1990, No. 8, p. 131-136

/3/ Wolters, G., Knop, K. "MBG - ein Verfahren der allothermen Kohle-Druckvergasung", MAN GHH Oberhausen

/4/ van Heek, K. H., Jüntgen, H. "Kohlevergasung", Thiemig-Verlag München, 1981

/5/ Mateos, J. et.al. "Adaption of the Cesa-1 Facility to the GAST Technology Program"

/6/ "Typical Direct Radiation and ambient Temperature Data" IEA-SSPS Doc.No. R. 18/89 E.Z.

/7/ Kubiak, H. "Modellierung einer allothermen Wasserdampfvergasung in einer Wirbelrinne", Dissertation Essen, 1982

/8/ Becker, M., Böhmer, M. "The Gas-Cooled Solar Tower Technology Program", Proceedings of the final presentation, May 30 - 31, Lahnstein, BRD

Solar Powered Energy Cycle with Coupled Biomass Production by Chemoautotrophic Bacteria

C.C. Bärtels, H. Tributsch,

Hahn-Meitner-Institut Berlin GmbH

Solar betriebener Energie Zyklus mit angekoppelter Biomasse Produktion durch chemoautotrophe Bakterien

Kurzfassung

Bakterielle Biomasse Produktion im Energiekreislauf— basierend auf einer solar thermischen Energieumwandlung — kann im Labormaßstab auf der Basis von Eisenverbindungen realisiert werden. Eisen (Fe^{2+})Verbindungen werden durch die Bakterien oxidiert und können elektrochemisch wieder reduziert (Fe3+) werden. Der Austausch des Fe2+ / Fe3+ -Zyklus gegen einen Sulfid / Sulfat -Zyklus hat das Ziel eine höhere Energiedichte für die Bakterienzellen zu liefern. Dies fordert die Untersuchung zweier hauptsächlicher Problemstellungen: a) die Ausnutzung einer heterogenen sulfidischen Energiequelle durch die Bakterien und b) die solar-thermische elektrochemische Reduktion des Sulfates. Beide Gebiete wurden während dieses Projektes untersucht.

In Ergänzung unserer Untersuchungen wurden im Jahr 1990 verschiedene Bakterienstämme auf unterschiedlichen Energiesubstraten (Schwefel-Verbindungen) eingesetzt. Dabei konnte die unterschiedliche Aktivität verschiedener Stämme miteinander verglichen und der Einfluß der Substratbeschaffenheit abgeschätzt werden. Das Anheften der Bakterien an die Substratoberfläche wurde unter dem Einfluß von Chemikalien getestet. Zum ersten Mal konnten die fließenden lebenden Prozesse kontinuierlich über Zeiträume von Tagen mit Videotechnik aufgezeichnet werden. Die Ergebnisse lassen die Entwicklung eines Modells zu, das die Fähigkeit der Zellen zur Substraterkennung und Auflösung beschreibt.

In einer parallelen Initiative wurden in unserer Abteilung die Abscheidungstechniken zur Herstellung dünner Pyritschichten weiter optimiert. Eine genaue Charakterisierung von Eisensulfidmodifikationen (Markasit / Pyrit) lieferte Ergebnisse, die bisherige schwerverständliche Resultate bezüglich des bakteriellen Verhaltens erklären. Die Untersuchungsmethode des bakteriellen Verhaltens in Kombination mit dünnen Pyritschichten eignet sich in der jetzigen Form als meßbarer Parameter für den bakteriologischen Erzbergbau einerseits, als auch für die Optimierung der Kinetik, zum Zwecke der Biomasseproduktion und der Kohlendioxidfixierung andererseits. Zur Vervollständigung des propagierten Brennstoffzyklus wude eine völlig neue Zelle zur Sulfatreduktion aus Kunststoff gebaut, die in einen extra angefertigten Metallmantel paßt, der über integrierte Widerstände den Korpus gleichmäßig aufheizt. Ein integriertes Quarzfenster ermöglicht daß Einstrahlen von Laserlicht.

Solar energy cycle with simultaneus biomass production by chemoautotrophic bacteria

Abstract

On the basis of iron-II-sulphide compounds, a bacteriological biomass producton can be realized via an energy cycle — based on solar thermal energy conversion. Iron-II-compounds are oxidized by the bacteria and then again reduced using solar energy. The object of exchanging the Fe^{2+}/Fe^{3+} cycle for sulphide / sulfate cycle is to yield a higher energy density for a technological installation. This requires the analysis of two main problems: 1) The use of a heterogeneous sulphide source of energy and 2) the solar thermal electrochemical reduction of the sulfate. Both fields were investigated in the course of this project. In 1990, as a supplement to our investigation, various bacterial strains were cultured on different energy substrates (sulfur compounds). In this process, it was possible to compare the different activities of various bacterial strains and to estimate the influence of the constitution of the substrates. The attachment of bacteria to the substrate surface was tested under the influence of various chemical substances. It was possible for the first time to record on video the dynamic heterogeneous processes continuously over periods of several days. The results permit us to develop a model, describing the capability of the cells to recognize and dissolve substrates. In a parallel initiative of our department the depostion conditions for the production of thin pyrite films were further optimized. A detailed characterization of iron sulfide modifications (pyrite/marcasite) yielded results, which allow the interpretation of up to now unexplained results on bacterial behaviour. In its present form, the method of analyzing bacterial behaviour combined with thin pyrite layers is both suitable as a measurable parameter of bacteriological ore mining and for optimizing kinetics serving the purpose of biomass production, as well as carbon dioxide fixation. For the completion of the described fuel cycle, a completely new synthetic cell for the sulfate reduction was constructed. It fits a specially made metal envelope that guarantees a stable heating of the body via integrated resistors. An integrated quartz window facilitates irradiation of Laser light.

Table of contents

Final aim

We understand the final aim of this project to be the technical development of an energy cycle which nature already demonstrates for hundreds of millions of years. Different from the natural situation the energy source should not be geothermal energy but solar energy.
In aquaeous environment bacteria are supplied with a sulfur compound as energy substrat so that they can fix carbon dioxid from the environment to produce energy rich molecules. The energy substate donates electrons which are introduced by the bacteria into their respiration chain. In this way the substrat, the sulfurcompound is oxidised. The sulfur compound should in the oxidised state be transfered in to a chamber supplied with solar energy where it will be reduced by a solar technological process. This means electrons are donated again to the oxidised sulfur compound which is in its reduced form again transfered to the bacterial suspension. The energy for this reduction process should come from solar power. Photon energy is collected and converted in to electricity which is why a catalyst is used for the reduction of the substrate. Solar heat is used for heating the reaction chamber. In this way from solar energy via a sulfur compound as an energy source, biomass is produced by bacterial carbondioxid fixation.

This system is only dependant on solar power the excess of carbon dioxid and a few essentials mineral soles. It is not fixed to a particular place.

Introduction

For the aim of this project a special method for the insitu observation of chemoauthotrophic bacteria (Thiobacilli; Ø 0,5 x 1µm) on sulfid surfaces have been developed. The combination of microscope, videocamara, videorecorder and videoprinter permit clear, high-resolution images of individual cells up to a 2500 times magnification. These pictures assume a key position during the investigation of bacterial behavior and bacterial oxidation activity. On the basis of observed movement of cells and locally the rival changes of the substrate surface it is possible to draw conclusions on both, the behavior of cells and the property of the substrate. The understanding of bacterial behavior is as much a precondition for the technical realisation of a fuelcycle with additional biomassproduction as the development of a reduction process for the cyclic preparation of an energysubstrate for the bacteria. The scheme for the cycle has already been illustrated several times /Bärtels/. The technique involving a modification of the energysubstrate "Thin Pyrite Filmes" has been selected because this substrate can be produced and modified in our department. In relation to the bacteria the substrate is always present excessively. The investigation of alive cells in the natural environment is a straight forward method which provides the possibility to influence the system cell / substrate and which allows direct observation of any changes.

Scientific and technical aims

The oxidation mechanism of sulfur compounds by Thiobaccilly is largely unexplained. There is a series of hypotheses of which none is experimentally verified. The following questions have to be posed:

- How can a cell which contains up to 90 % of water disolve a rigid cristal structure?
- How is information for recognition of the energysubstrate coded?
- What parameters on the surface of the substrate have significance for the bacteria?
- What processes are the oxidation of sulfursulfide and disulfide based on (proteinchemical, enzymatical, biophysical)?
Some economical aspects can be added of which only the following four should be shortly mentioned:
1.   Bacterial ore leaching
2.   The distruction of calciumbased architecture (concrete) by Thiobacilly (historical monuments, sewage canals)
3.   The desulferisation of carbon
4.   The cleaning of sulfurcontaining exhaust gases

to 1.
Bacterological mining of ores uses the ability of cells to disolve metalsulfides. In this way noble metals are separated from accompaniing minerals. This occures on a large scale in open pits in leaching piles and in specially prepared orebearing geological layers.
For the time being no simple method exists which allows an insitu controle of cellactivity for the purpose of its controle.
The technique developed during this project which combines the use of thin sulfide layers with a video observation technique allows a direct controle and with further development even a calibration for the purpose of controle of cell activity.

to 2.
The widely present Thiobacilly can dispurse themselves via the dropplet infection. In our environment the cells find in rough and porous surfaces a microclimate which provides them adequate living conditions. Monuments and canalsystems frequently fulfill these conditions. Air and water provide CO_2 for the build up of chemicals and sulfur compounds from the environment provide energy for their respiration cycle. Oxidation of sulfur compounds in humid environment produces sulfuric acid. The acid is directly involved in the disolution of calcium carbonate of concrete and building rocks. /Bock/Sand/. Via this process the action of acid rain is multiplied.

to 3.
The presence of sulfide in the coal which is used as fuel in powerplants for energy conversion is an essential reason for the pollution of the atmosphere. The desulfurisation of coal by thiobacilly is intensively investigated in the USA and Sweden /Davis/Klubek/Ollson/.

to 4.
In Japan the possibilities investigated to clean waste gas from sewage from H_2S inpurities via imobilized Thiobacilly /Fukuyama/Kanagawa/Nakamura/Tanji/. This concerns not only the general airpollution but also the industrial healthcare. In Austria substrates for bioreactors have been developped in order to decrease the emmision of H_2S and CS_2 with the help of microorganisms. The wastegases stream through a particulate substrate where Thiobacilly adhere to a very large surface. This process is especially interesting for the viscose industry /Windsperger/.

Optical analysis and information processing via video

The lifecycles of bacteria are recorded via a video camera and can be documented via a video printer on thermo paper. With this method, in-situ observations are possible up to a magnification of 1250 times.
The cells are cultivated in tiny glas chambers. Up to now no other in-situ studies of Thiobacilly on substrate surfaces have become known.

Fig.1

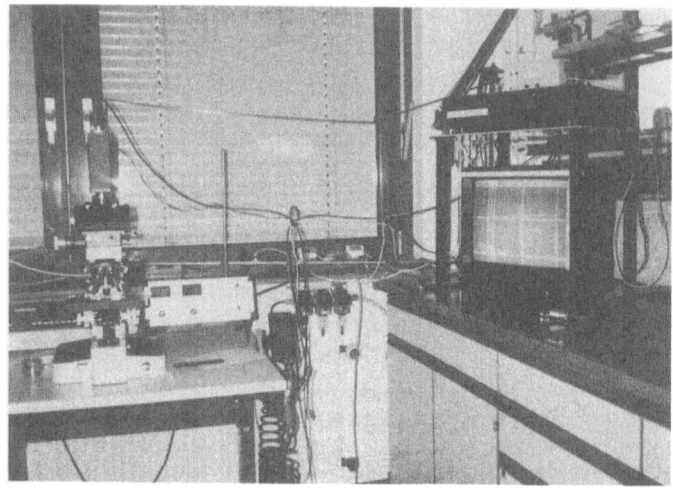

The photo shows the microscopic video unit at the time of surface research in fall 1990.

Experimental work

Sulfur compounds
Thin pyrite layers, approximatly 1000 Å were deposited on glas via the MOCVD
method (Molecular Organic Chemical Vapor Deposition).
Also the deposition of sulfur would be possible, however, the product turned out not to
be suitable for the observation of the cells because it shows a too pronounced
roughness of its surface (Fig. 2; photo MOCVD-sulfur). Bacterial modifications of the
substrate can therefore not be recognized. For the direct observation of Thiobacilly on
sulfur another method has therefore been developed (sulfur droplet in a flat chamber).
An essential progress in the development of the MOCVD technique brought systematic
X-ray diffraction studies. From these studies, it was learned that up to now many
MOCVD-layers which have been exposed to bacteria contained only few or no pyrite
fractions. This explains to a large extent the wide variation of results on cell behavior
on the supplied substrates.
Many supposed pyrite-layers contained FeS compounds or FeS_2 as marcasite. Now
layers with homogener surface without pinholes and impurities (deposits of
pentacarbonyl) in the rival thicknesses can be produced.
The layerthickness is at the time a parameter which cannot yet be reproduced 100%.
Because of turbulences in the depositionreactor the layerthickness of two
simultaneously produced layers can vary up to 50%.

Fig. 2
Photo: MOCVD sulfur

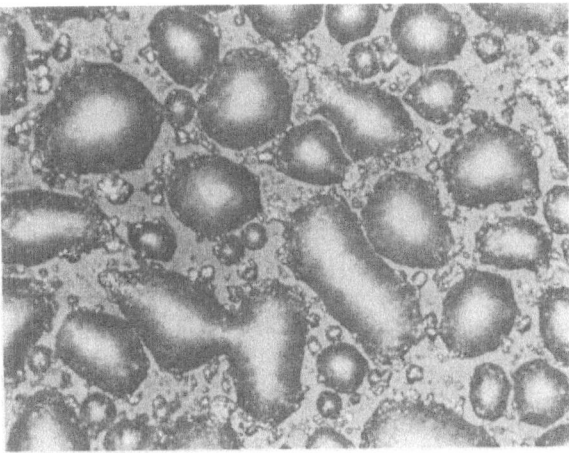

Sulfur-surface
MOCVD-sulfur-depostition
Magnification 1000 times
Individual drop formations are recognizable; no homogeneous surface is present.

X-ray diffractometry

The diagrams of MOCVD-layers (compare capital evaluation and discussion) illustrate the development during the production of thin pyrite films. They further document difficulties during the interpretation of proceeding investigations. The ration marcasite/pyrite appears to have significant influence on the bacterial behavior.
The techniques:
Diffractometer: Siemens D500 with Cu-K-alpha-radiation and graphite monochromator.
Exitation voltage: 45KV; current flux: 30mA, Cu-anode
Stepwidth: 2-4 x 100^{-1} degree; measurement period: 2-200 sec.
The total measurement period results from the measurement period per step in deppendance of layer thickness; it varies from 3-72 hours.
Pyrit and marcasite are marked in the diagrams with a line perpendicular to the abse: Pyrite shows in the used experimental setup the characteristic signal at 28.522 (2-theta); marcasite at 25.879 (2-theta).

IR measurements on pyrite layers

IR Spectroscopy was applied with natural pyrite and marcasite cristals. The aim was, to investigate how the spectroscopic pattern of cristals which originate from Italy, Elba and in the case of marcasite from CSSR and Bohemia behave in relation to literature data before and after application of bacteria. It is expected that by bacterial activity the cristal surfaces change in their composition to such an extent that these changes can be detected IR spectroscopy. Ironoxides and sulfides could form. In addition organic substances from cells and membranfragments should be deposited on the surface.
After treatment with bacteria the cristal surface is carefully rinsed with destilled water in order to disolve soluble soles from the nutrient solution. Cells which are not fixed are washed away.

Experimental:
Equipment: Fourier-transform Infraredspectrometer IFS 113v, make: Bruker.
Configuration: Glovar-rod as sightsource; KBr-Radiationconductor, TGS-detector for the medium IR, Resolution 4cm-1, 256 or 512 scans per spectrum.
Sampelholder: standard-reflectionunit for variable angleadjustment with KRS-5-Polarisator (Specac Co.). For the fixation of the samples black ridgid paper was used.
The reference as far as not otherwise indicated: aluminum mirror.
The obtained spectra were registered under the insidentsangle of 45°. The %-values of reflection should be understood as tendancy and not as absolute values since they depend on the sample size which is not further elaboratet here. Spectra have not been correted carefully with respect to the basis.

Up to now marcasite could only be studied in the untreated condition. The cristal shows a typical spectra for marcasite; in the investigated orientation the orthorombique system shows coincidence with literature values /Lutz/ Wäschenbach/. On the measured cristal surface also after several cleaning procedures organic compounds were still detectable. The cristal has for this reason not yet been used for the planed investigations.
Pyrite was investigated before and after treatment with bacteria the sketch shows the sectioning and the inoculation of the cristal surface.

Fig. 3
Sketch:

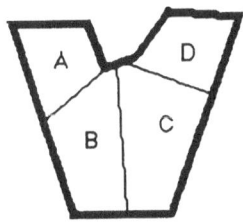 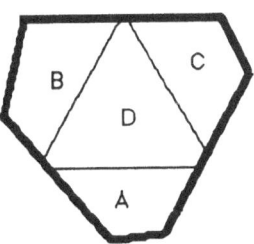

Sectioning of cristal surface of a natural pyrite cristal with polished surface inoculated with two Thiobacilly ferrooxidans cultures, of 60 µl volume containing 240.000 cells each.

A: R2 / 1406; pH 1,5   (R2 is a wild strain from Chile, Rancagua Mine)
B: DSM 583; pH 1,5   (Deutsche Sammlung für Mikroorganismen)
C: R2 / 1406; pH 2,0
D: DSM 583; pH 2,0

On the position "A" of the V-surface no significant changes could be found.

Fig.4:
V-surface of the position "A"

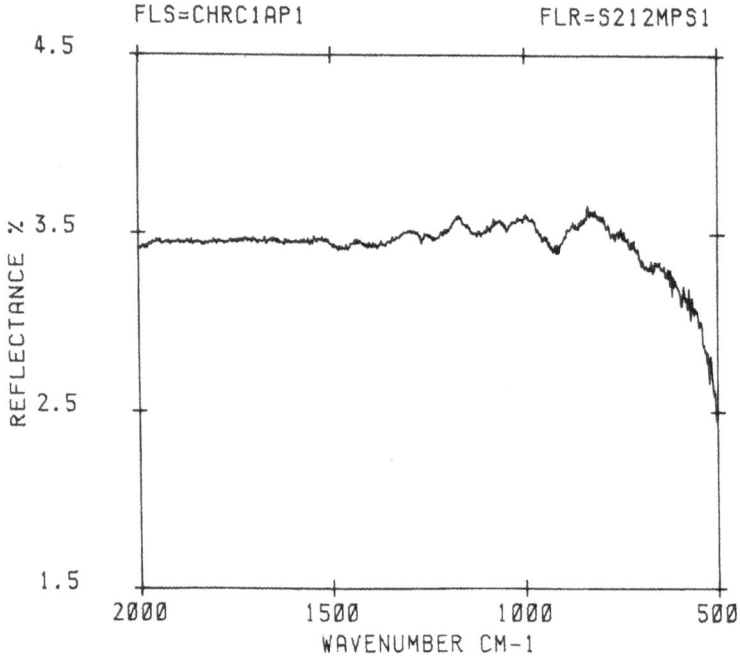

The position "D" of the V-surface (Fig. DV5) shows 1100 cm-1 an absorption which points to SO_4^{2-}./Nakamoto/

Fig.5:
V-surface of the position "D"

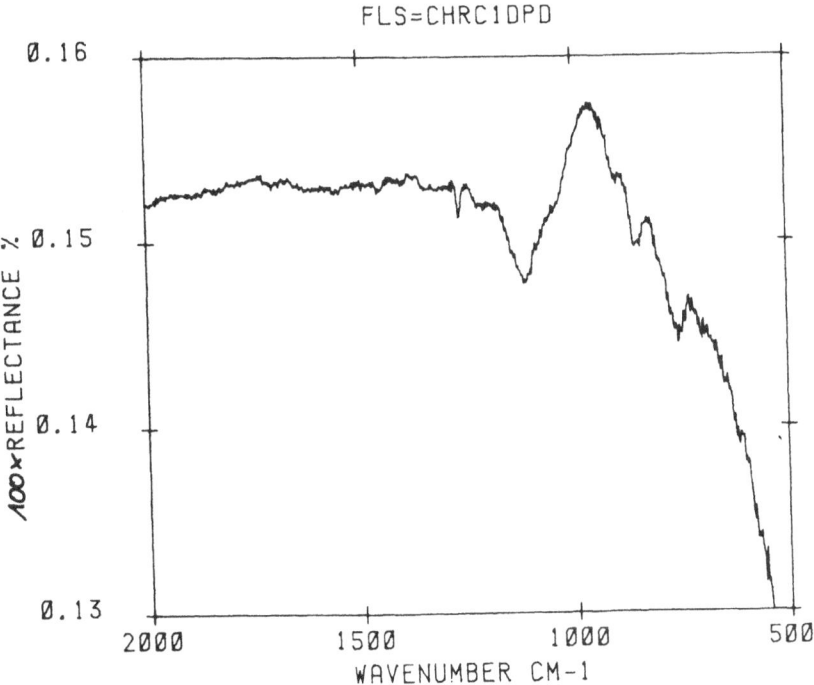

FLS=CHRC1DPD

Fig.6:
Area: "triangle"; position: "D"

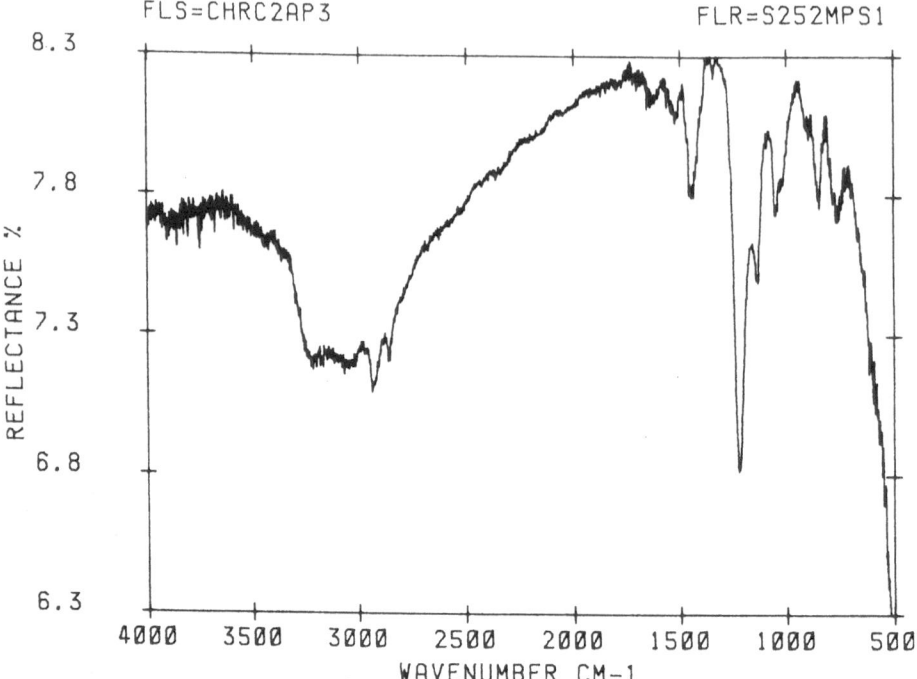

FLS=CHRC2AP3   FLR=S252MPS1

In the parallel experiment on another surface of the same cristal, changes were
observed with respect to the untreated area. The evaluation showed a bandpattern
which points to its organic groups (CH, CO, NH, OH) (Fig. 6 = pos. "A", triangle area;
Fig. 7 = pos. "D", triangle area). A reliable identification is not possible at the time. Pos.
"A" shows a somewhat different spectrum compared to pos. "D"; different intensities
are recognizable.

Fig.7:
Area: "triangle"; position: "A"

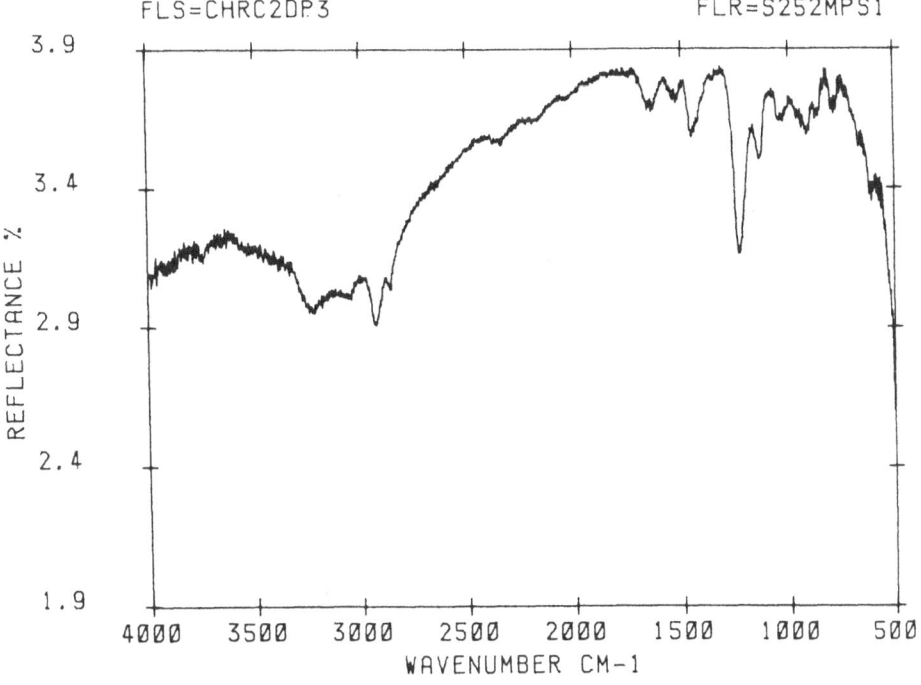

Closed Flat Chamber

For the observation of the lifecycle and the oxidation behavior of the cells a flat chamber of glas was constructed. In one case the MOCVD-layer provides the basis (Fig. 9; Photography) in the other case an object-carrier (Fig. 8; Illustration). The upper cover is a coverglas (20x20 mm) as it is standard in microscope technology. As sealing and contactmaterial, silicon rubber as is used. The solidification time should amount to 24 hours. The chamber is perforated at the silicon rim with a syringe and inoculated. A second syringe serves for ventilation. The chamber height corresponds to the free working distance of the objective (0.3 to 1.2 mm). The chamber contains 100 µl and a sufficient air volume. The problem of evaporation of small volumina does not occur with this construction.

Fig: 8
Illustration: Closed flat chamber

Objectholder with cover glas and a syringe

Fig: 9
Photo

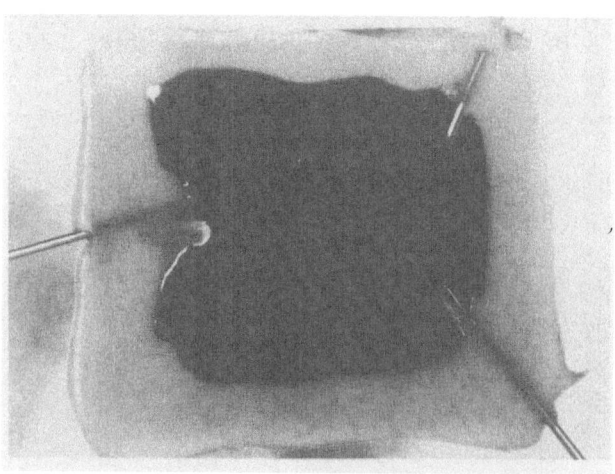

Closed flat chamber
In this model the three syringes serve for supply and distance holding

Different cultures on MOCVD-layers

On various MOCVD-pyrite layers different cultures were investigated for comparison. The following three MOCVD-layers (UK483, ENN6, MK348) were inoculated in the same way: Inoculation occurred clockwise. It was started above on the left side.
A = "TM" inoculation medium, pH 1.5
B = R2/1406 Thiobacillus ferrooxidans (wild strain) pH 1.5
C = Thiobacillus thiooxidans ATCC 8085 ph 4 - 5 during inoculation
D = FTO-strain; laboratory cultiva of Thiobacillus ferrooxidans; cultivated in APH medium at pH 3.0; the medium corresponds in its composition to the DSM medium 271.
Always the same amount of cells per volume and field were inoculated. As expected, the sterile solutions and bacteria of the species Thiobacillus thiooxidans do not show any activity in comparison with Thiobacillus ferrooxidans. Interestingly the FTO culture showed a stronger oxidation activity compared to the R2 culture. This is interesting because a more acid pH should support an additional chemical oxidation.
The series:
UK483 (Figs: 10 - 14), ENN6 (Figs.: 15 - 19) and MK348 (Figs.: 20 - 24) were photographed after 24 and 48 hours and after 6, 8 and 10 days each.

Fig: 10
Series UK483
Photograph: Magnification 60 times after 24 hours

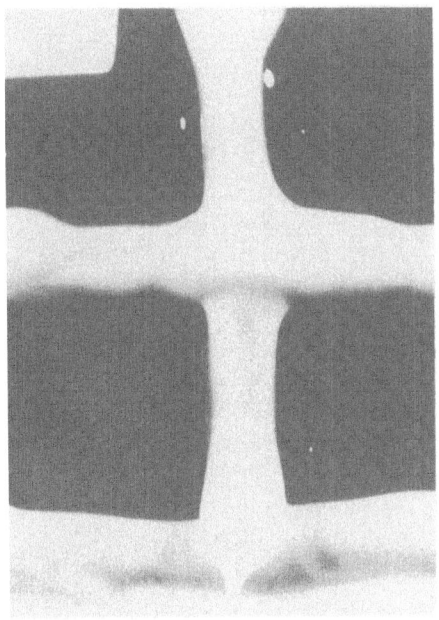

Figures: 11 + 12
Photo-Series UK483
above after 48 hours
Magnification 60 times
Below, 40 times magnification
after 6 days

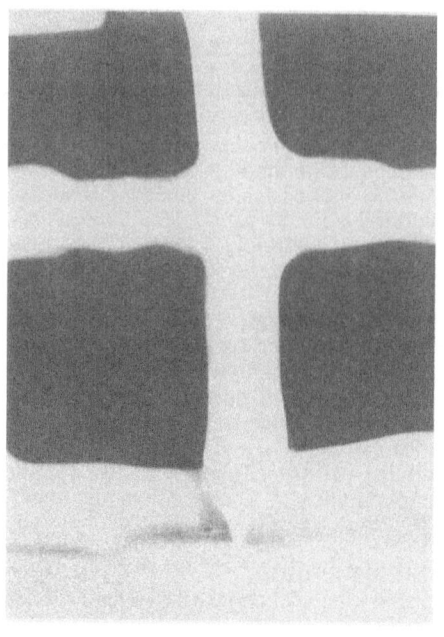

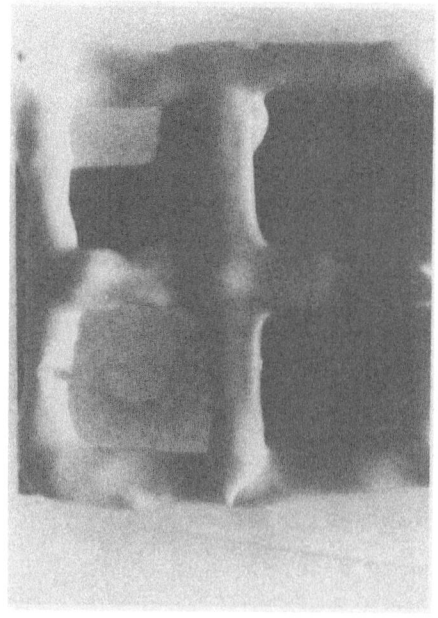

Figures: 13 + 14
Photo-Series UK483
above, after 8 days,
with30 times magnification
and below after 10 days,
magnification 36 times

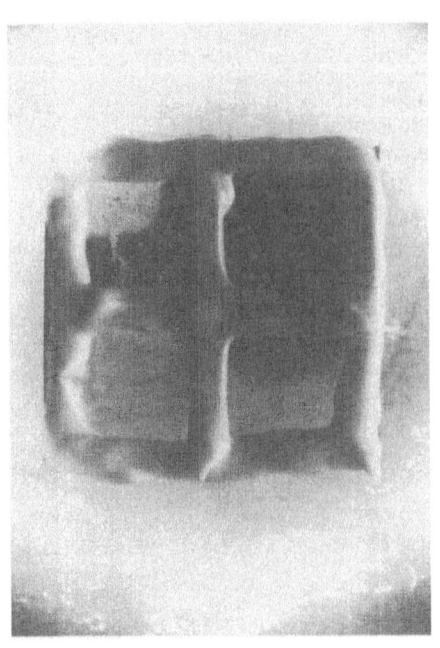

Figures: 15 + 16
Photo-Series ENN6
Magnification 60 times
above after 24
below after 48 hours

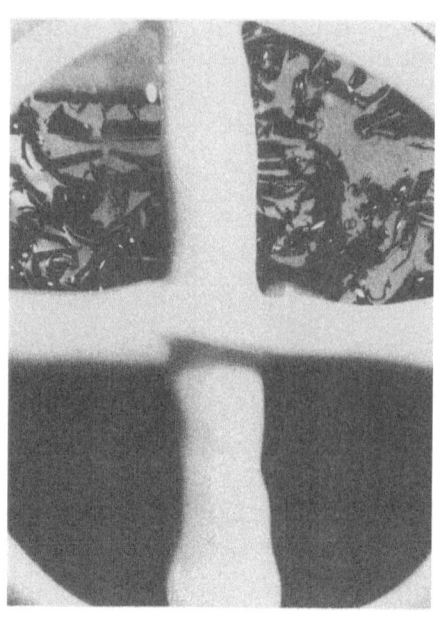

Figures: 17 + 18
Photo-Series ENN6
above 36 times magnification
after 6 days
below 30 times magnification
after 8 days

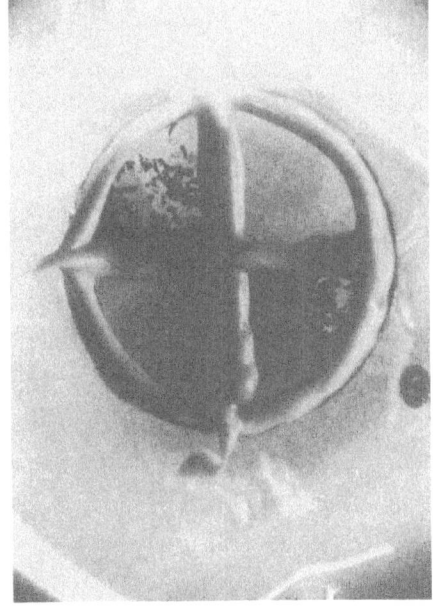

Fig: 19
Photo-Series ENN6
after 10 days
Magnification 36 times

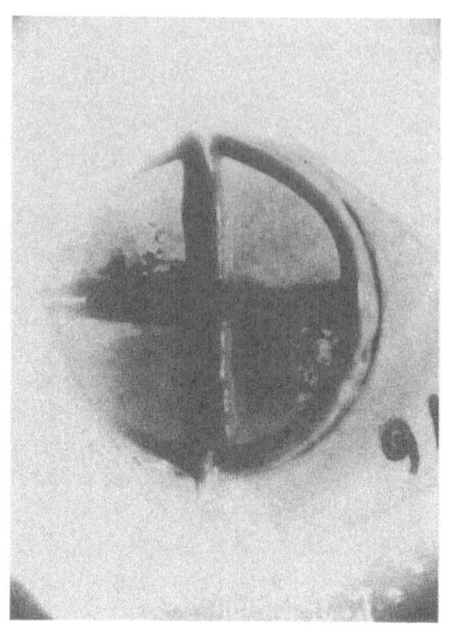

Fig: 20
Photo-Series MK348
Magnification 60 times
after 24 hours

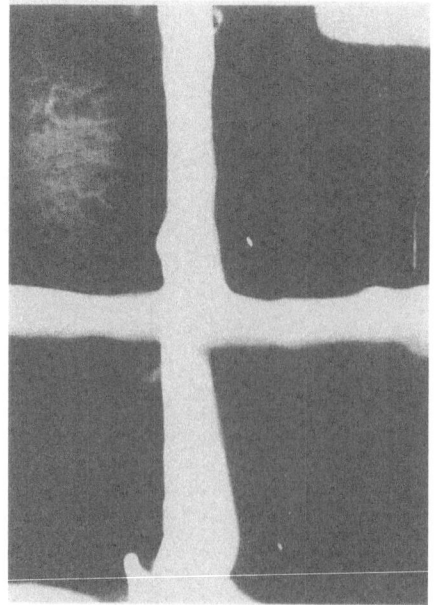

Figures: 21 + 22
Photo-Series MK348
Magnification 60 times
above after 48 hours
below after 6 days
Magnification 40 times

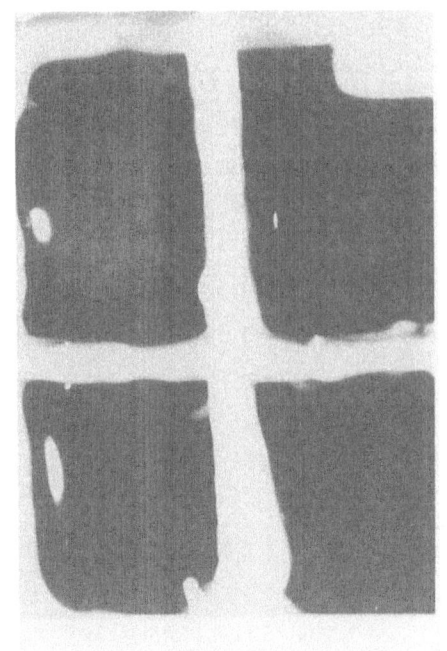

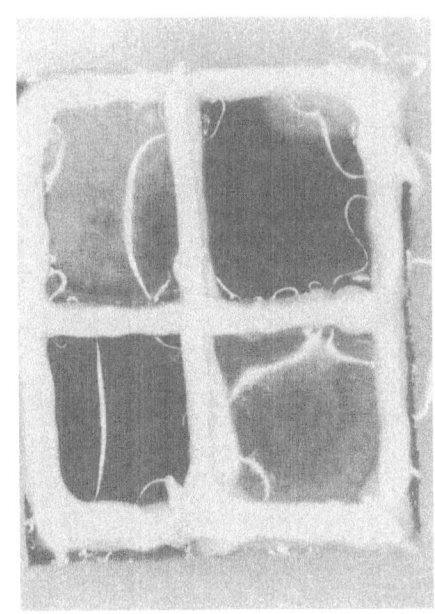

Figures: 23 + 24
Photo-Series MK348
above 30 times magnification
after 8 days
below 36 times magnification
after 10 days

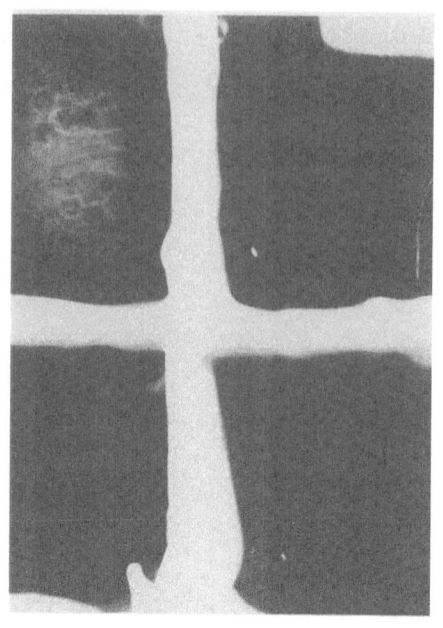

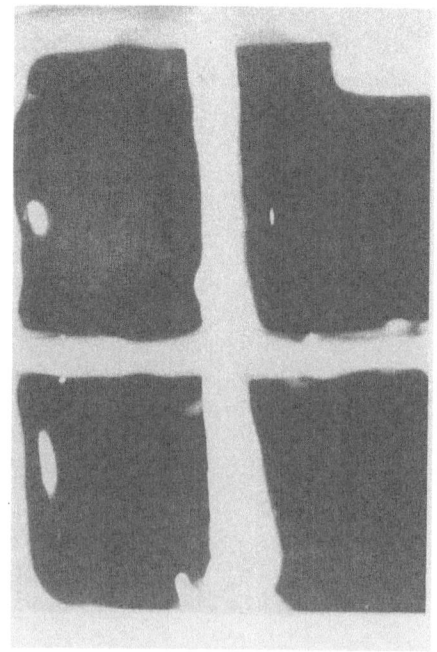

Adhesionbehaviour

The adhesion of cells to their substrate which provides the energy for CO_2-fixation determines critically the amount of oxidation. From the mechanism of bacterial substrate identification and the beginning of oxidation we may receive information relevant to the mechanism of substrate degradation. The adhesion of cells should be dependent on various parameters and be influenced by them /Grudev/. In addition it is known that the oxidation activity of Thiobacilli can change during their growth /Kodama/.

1) The cells must recognize the substrate as a suitable energy source and adhere to its surface. A change of the surface should make substrate recognition more difficult or easier.

2) From 1) follows that it should theoretically be possible to influence substrate recognition. Precondition is a knowlegde of the mechanism of recognition and adhesion by cells. Should a hypothetical complex of a substance X on the substrate be recognized specifically by the cells, an accumulation of this substance on the surface should cause more cells to attack the substrate faster. On the other hand, covering this substrate X should cause fewer cells to attack the substrate slower.

3) In the presence of similar substrates with different surfaces (FeS_2 as pyrite, MOCVD pyrite, Marcasite) the cells could show different attachment.

4) In the presence of different substrate surfaces the cells should show different behaviour.

5) Different cultures of Thiobacillus could show a different behaviour on the same substrate.

Evaluation and Discussion

When bacterial cells from an aquaeous medium attach to a surface and alter this surface during a degradation process, the visual technique is needed for adequate recording and monitoring of any changes. Such information should be available and documentable at any time.

For the time being two functional systems exist which can document bacterial activity: 1) Bacterial activity is visually detectable via a comparative observation of oxidation of MOCVD-pyrite layers which are sub-divided into fields exposed to different bacteria.in various concentration. One field which is only inoculated with culture medium without cells serves as a reference. The different oxidation speed of the sulfide fields is a measure for the different degradation velocities.

It can depent on:

a) the cell density; b) the cell activity; c) the adaptation of the culture to the substrate; d) the pH value, the medium, temperature

2) Microscopically the cell activity is detectable and comparable via a) the degradation of the sulfur droplets (sulfur flower melted on glas) in a glas chamber and b) through the degradation of MOCVD-pyrite layers deposited on glas and inbuilt as the floor of a glas chamber (see illustration and photograph; chapter "Closed flat chamber") and c) by investigation of natural pyrite and marcasite crystals. These studies are not yet terminated.

to 1) Two T. ferrooxidans strains (FTO + R2) were repeatedly tested in comparison to each other and to a T. thiooxidans strain on MOCVD-pyrite layers. Strain FTO lives in an environment of pH 3.0; R2 in pH 1.5. For the culture medium and T. thiooxidans the lowest degradation rates were observed as to be expected (Series UK483; ENN6; MK348) From this it follows:

1) Two different T. ferrooxidans strains are independently of each other able to recognize and to utilize the exposed substrate (MOCVD-pyrite) as an energy source.

2) The substrate can not be used by T. thiooxidans.

3) The substrate is only insignificantly attacked by the culture medium. The chemical oxidation can be neglected in comparison to the bacterial oxidation.

4) The pH did not have an affect which increased the oxidation at pH 1.5 beyond the bacterial fact at pH 3.0.

5) The FTO culture showed as compared to the R2 culture an accellerated degradation of the substrate. This may be explained by the higher vitality of the culture. This has filled to be confirmed.

MOCVD-Pyrite-Layers

A thin (1000 - 2000 Å thickness) layer of pyrite is deposited on a glas substrate in a reactor / Chatzitheodoru / from the gaseous compound ironpentacarbonyl and from sulfur. Besides of the standard conditions (temperature, pressure, stoichiometry, gasflux) also difficult controllable parameters such as stream behaviour, turbulences and area dependent cooling of the substrate influence the deposition of pyrite on glas. In spite of these difficulties the observation of thin layers of pyrite with the microscope video technique is for the time being the only procedure which allows an in-situ study of cell activity directly on the substrate.
For this reason this system was steadily developed and improved. A comparison of different MOCVD-layers produced partially results which were difficult to understand. In part the layers were not attacked at all by the Thiobacilli, in part they were only slightly oxidized and others were already detached from the substrate during addition of cell culture. By systematic analysis it was found that a large part of the used supposed pyrite layers did not contain pyrite at all (Fig.: 25; Sample WKH 108) or only in small concentrations (Fig.: 26; Sample ENN9A). Some layers contained only marcasite (Fig.: 27; Sample ENN10), others neither marcasite nor pyrite. In other samples the FeS-compound present was not classifiable (Fig.: 28; Sample MK390 RAW + Fig.: 29; Sample UK439).
This explains part of the results which were difficult to understand up to now. Only recently it was possible to produce MOCVD-layers with a nearly 100% pyrite portion (Fig.: 30; Sample ENN12A)
The destruction of the layer after inoculation with the culture medium is independent of the cell kind, the cell density, the pH value and the nutrient medium. Layers were partially also destroid by simply adding water (aqua. dest., Layers were partially also destroyed by simply adding water (aqua. dest., tab water). This phenomena depends as a recent investigation showed only from the interaction of MOCVD-pyrite with the glas substrate. The dissolution phenomenum was investigated and turned out to be reproducable when pyrite on glas was exposed to aquaeous media. The composition of the glasses turned out to be an important parameter (quartz / AF 45-standard glas).

Fig.: 25
X-ray diffractogram
WKH 108
The sample contains only marcasite.

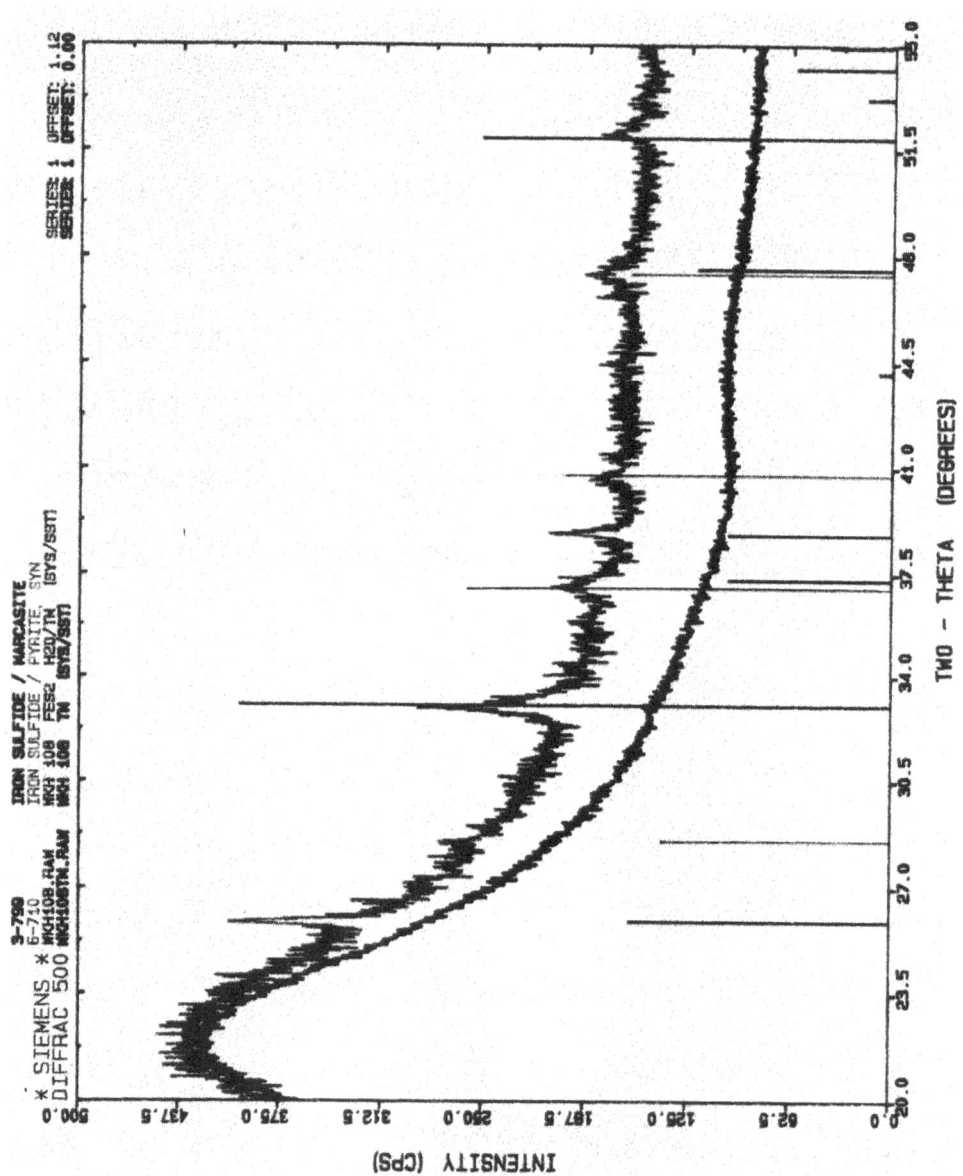

Fig.: 26
X-ray diffractogram
ENN9A
The sample contains traces of pyrite.

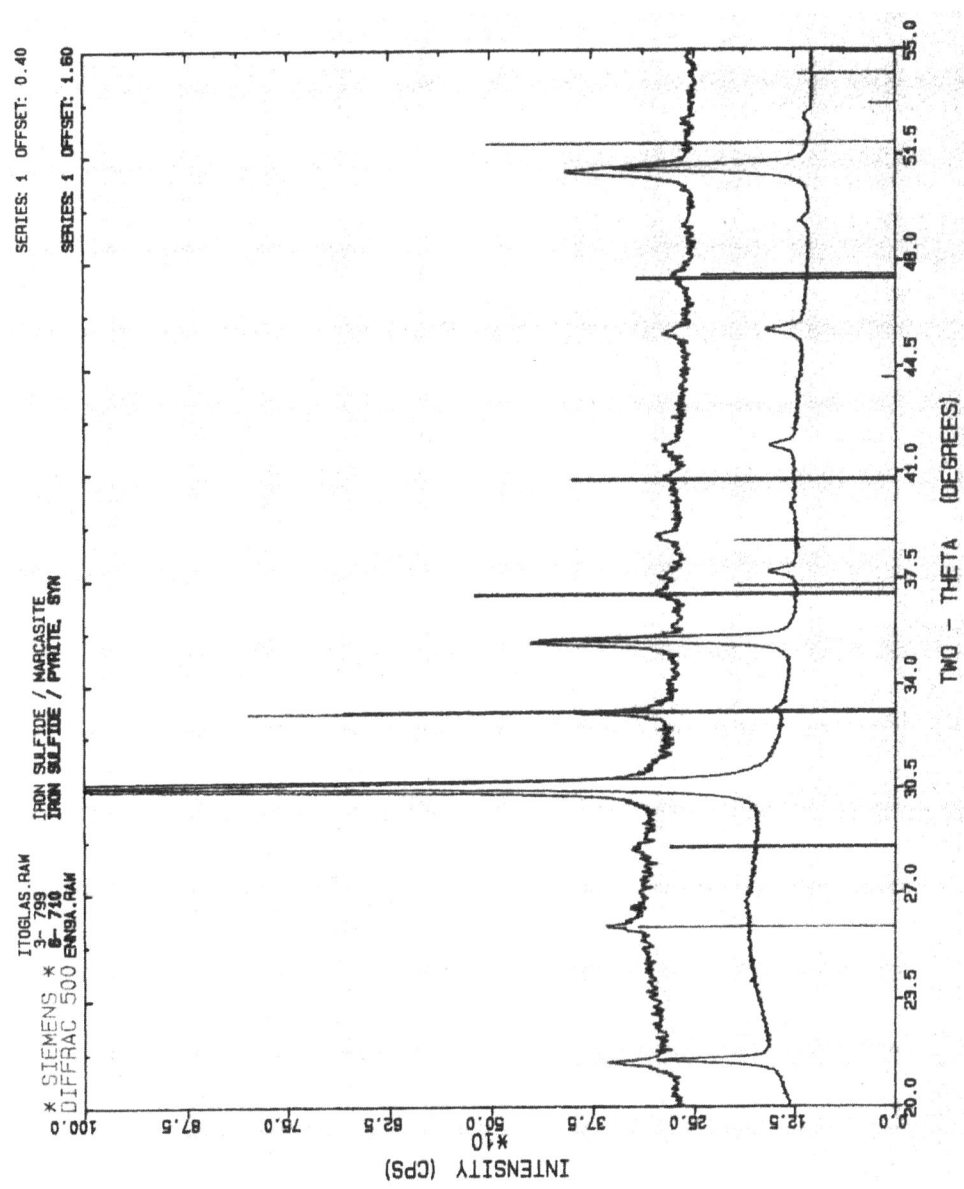

Fig.: 27
X-ray diffractogram
The sample contains mainly marcasite.

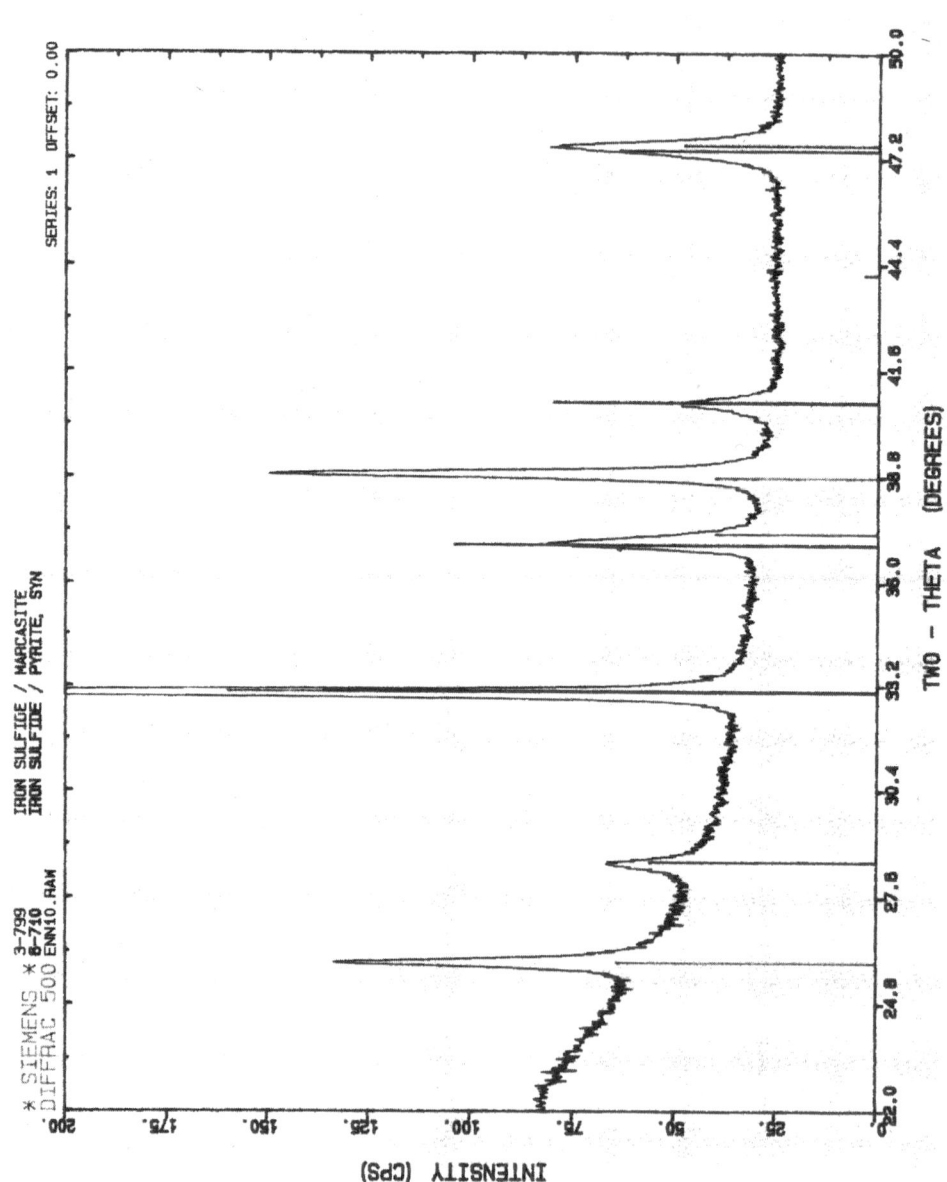

Fig.: 28
X-ray diffractogram
MK390
The FeS compound cannot be classified.

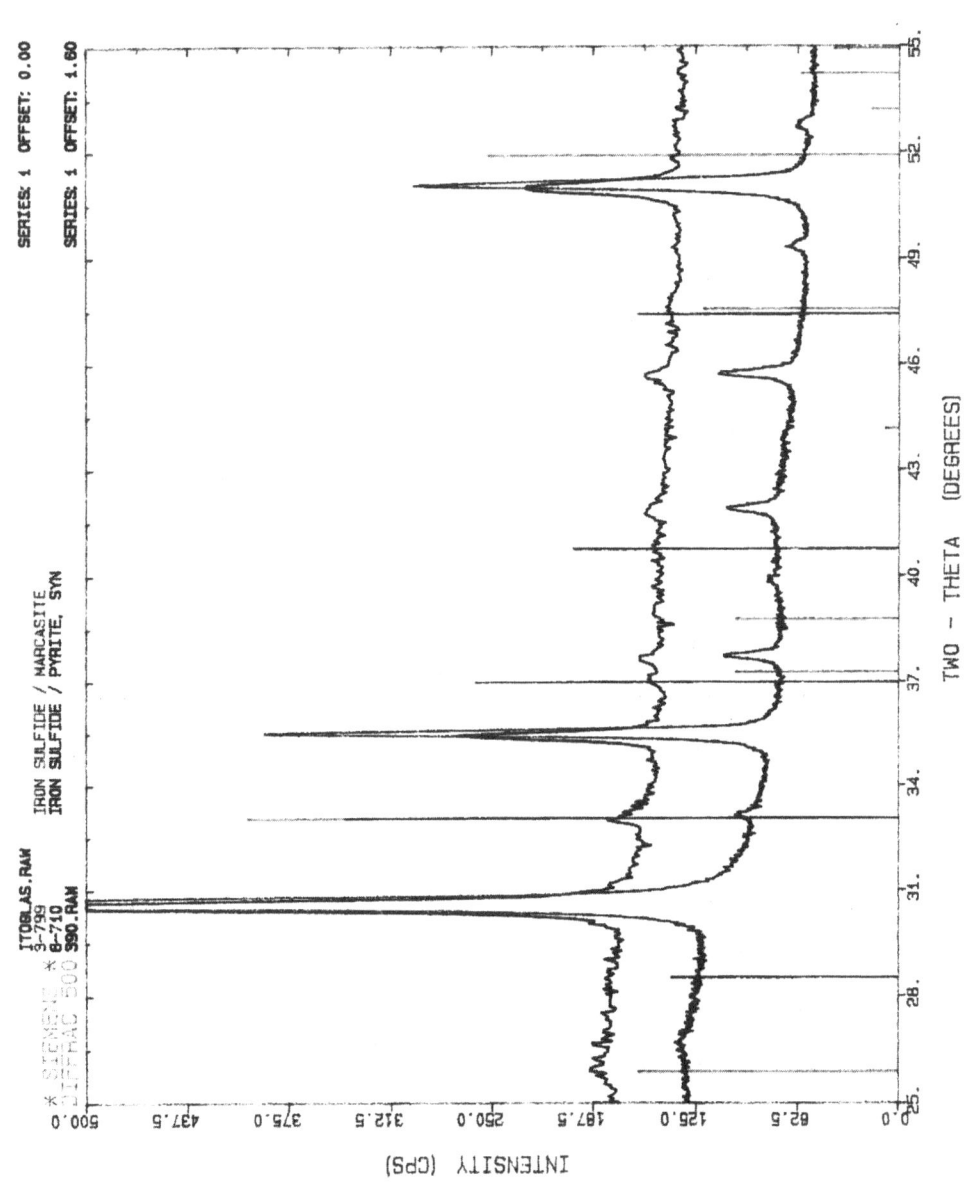

Fig.: 29
X-ray diffractogram
UK439
The sample contains neither marcasite nor pyrite.

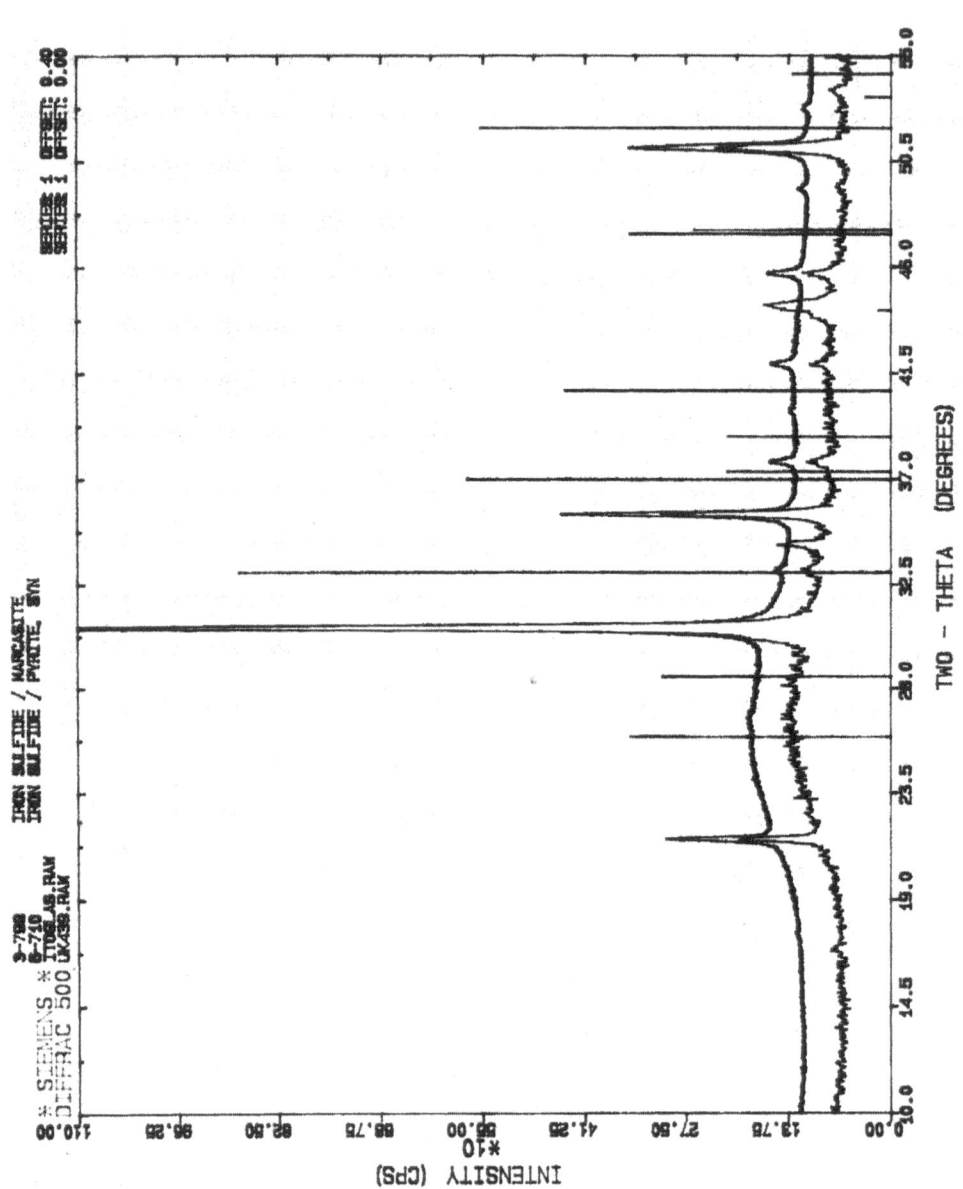

X-ray diffractogram
ENN12A
The sample contains nearly 100% pyrite.

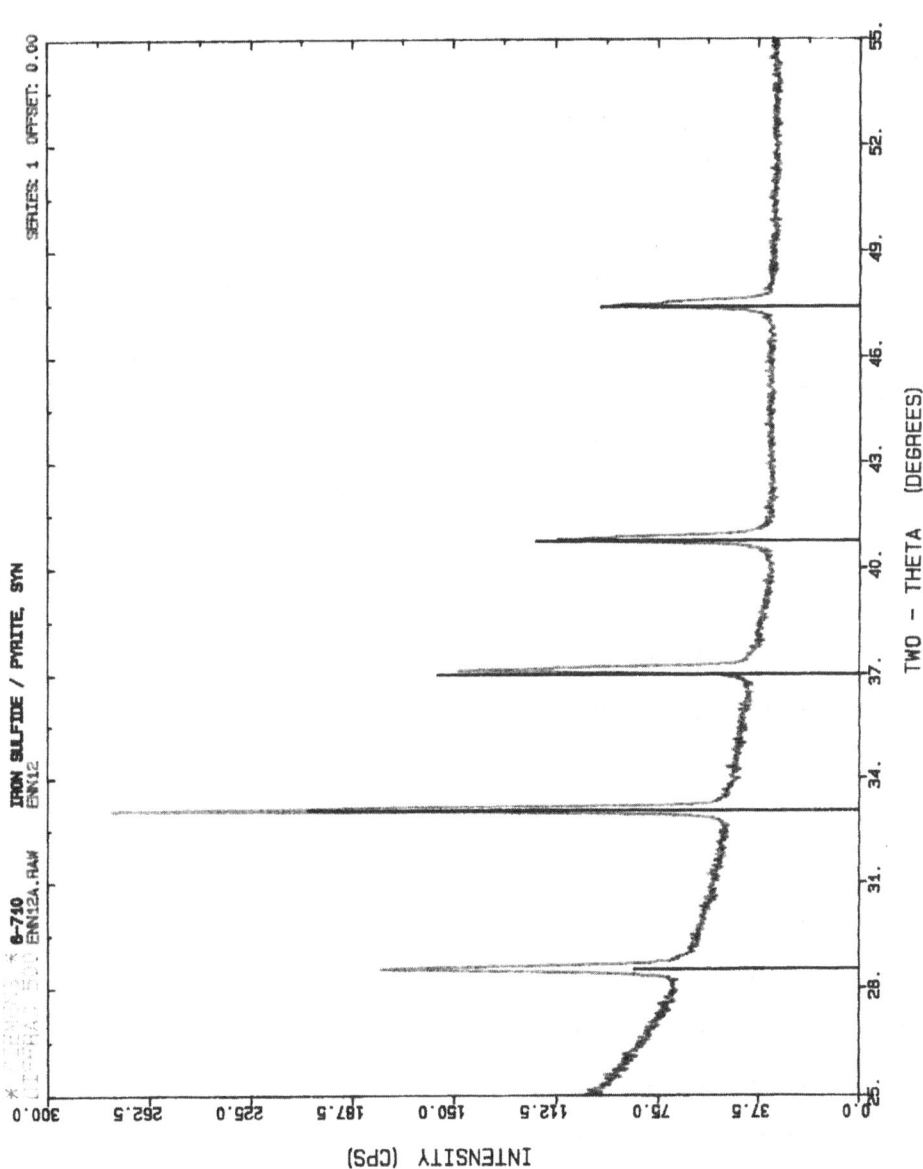

Control unit

For a continous long-term observation of degradation processes of pyrite and sulfur by bacteria, it was necessary to construct a system which in programable periods records the process of bacterial activity in a well defined way. This is in principle possible with a microscope and an attached video camera and a video recorder and could also be done by hand. But for a continous long-term observation a system had to be developed which is able to record periodically day and night. The periods for recording and interuption should be adaptable to the special samples to be investigated. The technical parameters which are relevant are a) the heating of the samples by the light source of the microscope and b) the material turnover of the bacteria known from the experiments. The recording should start at a time x and with some delay considering the reaction time of the recorder light sources (reflection and transmission) of the microscope had to be adjusted to the desired light intensity (thin layers require less light intensity for transmission as compared to thicker ones). A too rapid increase of light intensity (100 watt miniature halogen lamp) would lead to a premature destruction of the light source. During turn-off the recording has also to be interrupted. A programable regulation unit for the microscope was not available on the market and had to be constructed.
Now it is possible to record in a continous way the oxidation processes occuring in the miniature glas chamber during a period of several days (1 to 960 cycles of 15 sec. duration / Recording period: 15 sec. to 30 min. / Interruption time: 1 to 256 min.). The video camera reacts very sensitive with respect to a strong light intensity and the halogen lamps are very sensitive when the power access is too fast. The control unit for the light sources solves this problem. The lamps can for transmission and reflection studies be controlled continously and independently.

<u>Chemical alterations of oxidation of MOCVD-layers</u>

When bacteria recognize the surface of a substrate as a suitable energy source , it could be concluded that the cells use a chemical signal for detection. The pyrite surfaces can chemically be modified by chemical agents. In our case we assumed that an oxidized surface is a non-interesting energy source for cells which is therefore avoided.
Surfaces of natural pyrite crystals (cut and polished) and of MOCVD-pyrite layers were treated with the following chemicals ($HgCl_2$; H_2O_2; $KMnO_4$; NEM "N-Ethyl-Maleimid") in different concentrations (10 to 40 mM) and exposure times (20 to 30 min.). As a measurable parameter the decrease of a defined cell density from the liquid of the cell suspension is taken which was added onto the surface of a crystal or a layer. The surface amounted in average to approximately 20mm^2. A rim of silicon rubber fabricated around the layer form a reservoir which contained 100 - 400 µl of liquid. Then the decrease of cell density in the liquid was determined. As a control the decrease of cell density above an untreated parallel surface of the same sample was determined. After 2, 4, 6, 8, 10, 15, 20 and 30 minutes 10 ml were taken from the liquid and the density of cells was determined with a counting chamber. The extracted amount of liquid was replaced with a medium without cells. It is expected that after treatment of the substrate surface a change of the adhesion behaviour is observed with respect to the untreated surface. This reaction would show itself in the change of the cell density in the liquid. It is necessary that for a reliable conclusion the changes in the cell density of the treated surface must clearly be different from that of the untreated surface. The initial cell density is said to be 100% and the decrease of cell density as a function of time is recorded. When the selected chemicals influence the surface of the energy substrate for cells in a negative way the cell density in the liquid should clearly be higher as compared to the untreated control. The first count starts at 2 minutes; t = 0 corresponds to 100 %, t = 2 minutes corresponds to x %. The following plots show the behaviour of cells exposed to differently treated surfaces:
(Figures: 31 - 40; Decrease of cell density in the liquid)

<u>Figures: 31 + 32</u>

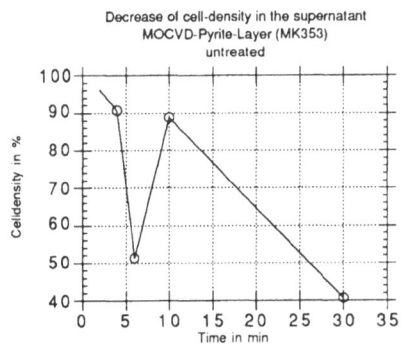

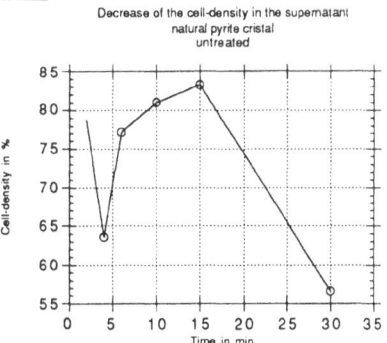

Figures: 33 + 34

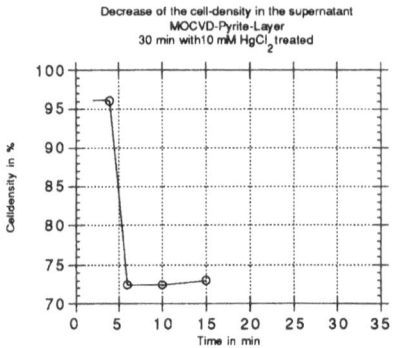

Decrease of the cell-density in the supernatant
MOCVD-Pyrite-Layer
30 min with10 mM HgCl$_2$ treated

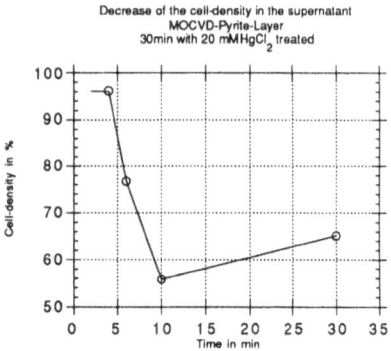

Decrease of the cell-density in the supernatant
MOCVD-Pyrite-Layer
30min with 20 mMHgCl$_2$ treated

Figures: 35 + 36

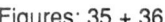

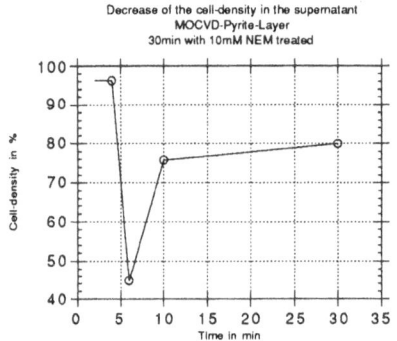

Decrease of the cell-density in the supernatant
MOCVD-Pyrite-Layer
30min with 10mM NEM treated

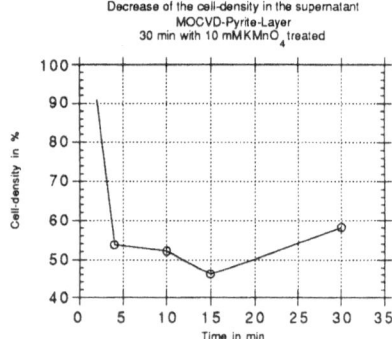

Decrease of the cell-density in the supernatant
MOCVD-Pyrite-Layer
30 min with 10 mMKMnO$_4$ treated

Figures: 37 + 38

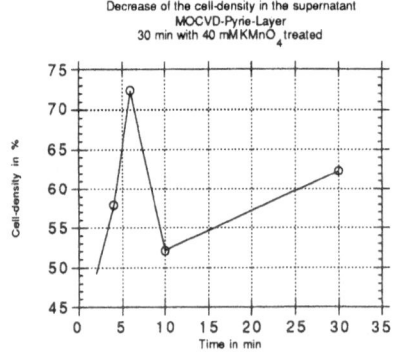

Decrease of the cell-density in the supernatant
MOCVD-Pyrie-Layer
30 min with 40 mMKMnO$_4$ treated

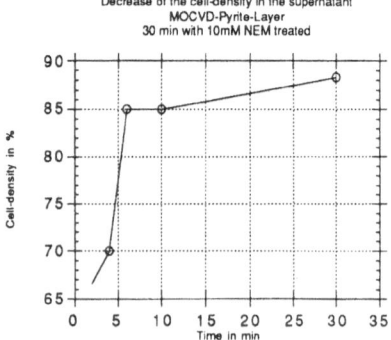

Decrease of the cell-density in the supernatant
MOCVD-Pyrite-Layer
30 min with 10mM NEM treated

Figures: 39 + 40

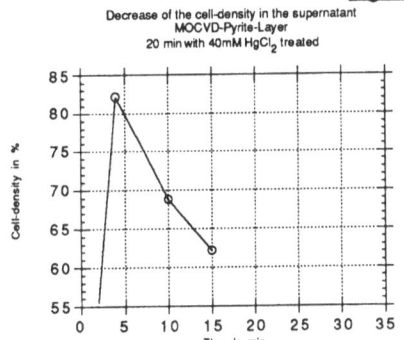

Decrease of the cell-density in the supernatant
MOCVD-Pyrite-Layer
20 min with 40mM HgCl$_2$ treated

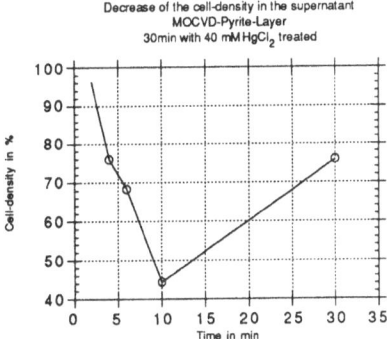

Decrease of the cell-density in the supernatant
MOCVD-Pyrite-Layer
30min with 40 mM HgCl$_2$ treated

In eight out of ten cases after 4 to 6 minute repeatedly a decrease of cell density to 50% was recorded. This is also seen with the controls and is not considered as a significant phenomenon relevant to utilization of the sulfid substrate. In the case of the HgCl$_2$, a change was expected with a 10 mM concentration. This was not observed so there the the concentration was stepwise increased to 40 mM / 30 min. Also here a significant difference could not be observed. The chemical KMnO$_4$ showed similar results.

As the most important result of this series of experiments it can be said that a surface treatment with NEM causes the cell density in the supernatant to increase again after 4 to 6 minutes. This is insofar remarkable as the cells have to move away from the substrate against the force of gravity. This indicates a blockage of the hypothetical recognition parameter of the cells. The decrease of cell density within the first five minutes can be caused by not further determined effects like gravity. These effects have also been observed during subsequent investigations of the adhesion behaviour.

At this point a result of subsequent investigations has to be mentioned. The presented results are based on the assumption that the adhesion behaviour is a reliable parameter. The possibilities to interprete the results should, however, not be overestimated. Here, only the behaviour of the cells in presence of one chemical is given. Comparable conclusions with respect to other chemicals are not reasonable. All conclusions have a qualitative character. These conclusions can be supported by the following control experiment..

Control-experiment for adhesion behaviour

The adhesion behaviour of T. ferrooxidans was investigated with different substrates (Figures: 41 - 45). With adhesion behaviour ("attachment") we understand in our experiments the adhesion of bacterial cells from the liquid medium to the surface of the solid energy source (which at the same time is an interface solid / liquid ; another interface is the substrate boundary in contact with the liquid medium). Adhesion can occur by arbitrary movements or by directed movements. As a first control serves the adhesion of T. ferrooxidans R2 species on glas in order to recognize purely physical influences (physical adhesion, sedimentation). As a second control T. thiooxidans (ATCC 80 85) was deposited on a natural pyrite crystal with polished surface. This experiment serves for the pure purpose of estimating the effect of the investigated pyrite material with a bacterium which is very similar to T. ferrooxidans but cannot use ironsulfite as an energy source. The higher mobility of T. thiooxidans as compared to T. ferooxidans was not considered in this experiment. But it is expected that T. thiooxidans is not attracted by pyrite. In the experiment 100 µl of bacterial suspension were added onto the surface to be investigated and in time distances of 2, 4, 6, 8, 10, 15, 30 minutes, 10 µl of the supernatant were taken and the cell density determined in a counting chamber. After each removal of a liquid sample the liquid volume was replenished to 100 µl.

After removal of seven samples (= 30 minutes) the theoretical dilution of the supernatant can be calculated to 55% (referring to the initial cell conzentration). A directed adhesion behaviour should - if measurable - amount a decreased cell density of less than 50% within the first six minutes. Investigated was T. ferooxidans R2-culture on natural pyrite, on MOCVD-pyrite and on MOCVD-marcasite.

Figures: 41 + 42

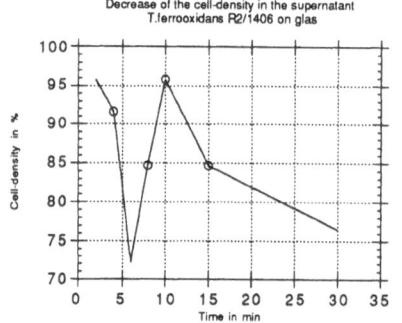

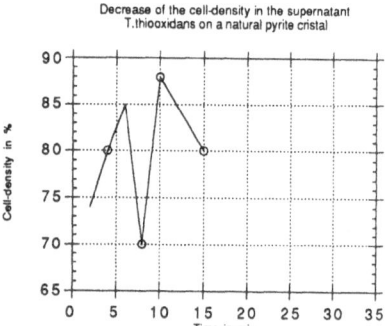

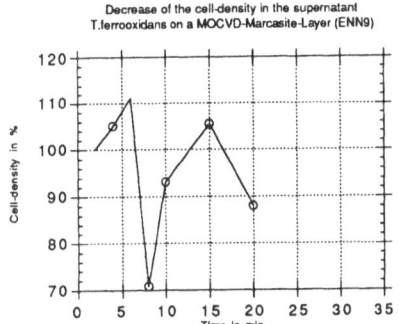

Decrease of the cell-density in the supernatant
T.ferrooxidans on a MOCVD-Marcasite-Layer (ENN9)

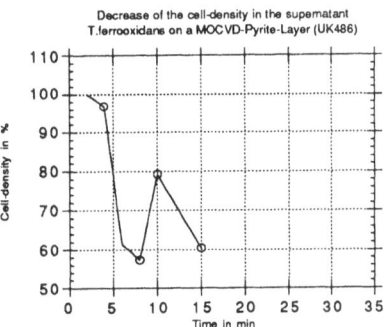

Decrease of the cell-density in the supernatant
T.ferrooxidans on a MOCVD-Pyrite-Layer (UK486)

Fig: 45

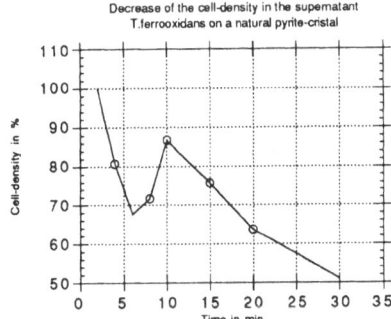

Decrease of the cell-density in the supernatant
T.ferrooxidans on a natural pyrite-cristal

In none of the cases it was possible to observe a clear adhesion behaviour which is significantly different from possible side mechanisms (gravity, dilution, physical adhesion) within 30 minutes. From our investigation the following observations and models can be used:

1)   Observations:

T. thiooxidans and T. ferrooxidans need time for information recognition (recognition of the energy source) and information processing (adaption? movement of the cells to the energy source). This amounts for an observable number of bacteria, which is a significant fraction of the overall number of bacteria, hours or days.

2)   Model: Adhesion behaviour of T. thiooxidans

Phase 1: Because of the high mobility of this bacterial species numerous cells hit undirectionally (by chance) the substrate (drop of sulfur).
Phase 2: These cells oxidize sulfur to sulfuric acid and thus lower the pH value in the closed environment of the substrate (change of micro-environment, production of a proton-gradient).
Phase 3: The increase of proton concentration is a signal for the bacteria which are freely moving in the liquid. As a result, these migrate following increasing acidity towards the substrate and adhere to its surface.
Phase 4: While cells which come in touch with elemental sulfur oxidize the sulfur to sulfate [via $(S_2O_3^{2-})_x$] those which do directly adhere to the sulfur (because they don´t find sufficient space) may oxidize the soluble intermediate products - as it is also possible with Thiobacillus ferrooxidans / Hazeu / which they oxidize further to sulfate. This seems to occur because of the extremely high cell density (80 cells / 0.0025 mm^2) up to 5 mm away from the solid energy source. An exchange of cells between those on the sulfur surface and those in the liquid was not observed. A bacterium once adhering to the substrate seems to remain fixed. Therefore the continuation of oxidation assumed in phase 4 has to be concluded. Via this directional movement towards the energy source, i.e. by a concentration of cells and by a possible splitting of work by the bacterial collectives an acceleration of the oxidation of the sulfur drop occurs. This happens independently of the increased substrate turn-over caused by an increasing cell concentration in the medium by cell growth.

3)   Model for T. ferrooxidans

Preconditions:
A) T. ferrooxidans does not find a sufficient concentration of dissolved Fe^{2+} or sulfurions in its environment.
B) The energy supplying substrate is solid (pyrite).
C) T. ferrooxidans follows the steps described in phase 1 to phase 3.
Note:
This is not possible for a T. ferrooxidans culture at pH 1.5. The model is valid for the R2 strain only with restrictions. At a similar cell concentration R2 oxidizes the substrate slower than the T. ferrooxidans FTO-culture. It is possible that cells only adhere arbitrarily to the substrate (gravity-supported) and that there is no chemical signal (pH-reduction) acting as information for other cells. These models can be tested experimentally for FTO and R2 cultures.

Cells adhere head-on on the substrate

In the model experiment for T. thiooxidans cells were observed as they approach a sulfur droplet. The technique based on a closed chamber of glas has been optimized to an extend that makes it possible to observe the cell behavior at a magnification up to 1000 times over a period of several weeks. A tiny amount of sulfur flower was given onto an object carrier and shortly heated to melt the sulfur. In this way a round form with a relatively sharp edge was created. The droplet in this way adhered also better to the glas support.

Fig: 46

Photography: Sulfur flower as a droplet in a flat closed chamber
Magnification: 500 times

The cells (T. thiooxidans) approach head-on the sulfur droplet and stay in this position at the rim of the round droplet . (Fig: 47 - 48).
Within a period of 12 to 24 hours chains of two bacteria form. With chains of more than three cells, it was frequently observed that these also adhere head-on to the substrate with the last chain member (Fig: 47 - 48). In general, clear movement of the cells towards the substrate was recorded. This speaks for a chemotactical behavior, i.e. for a programed substrate recognition.

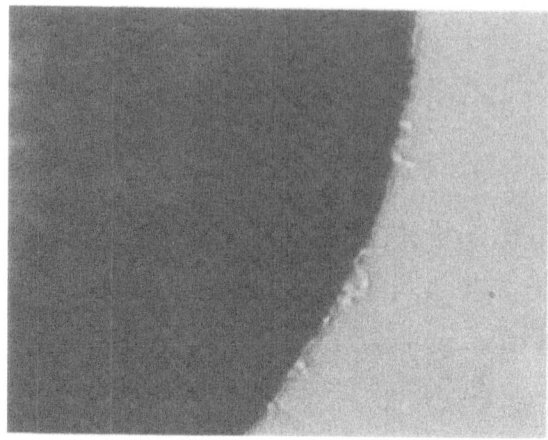

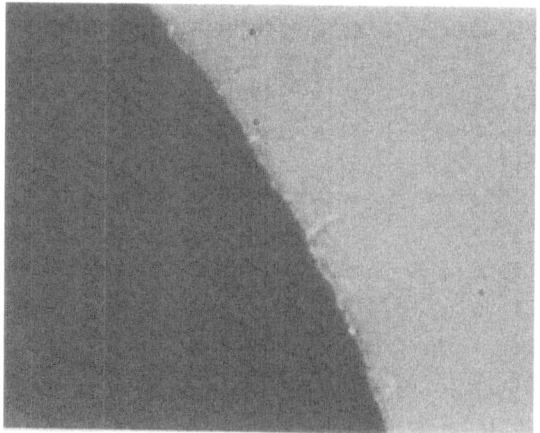

Pictures of a sulfur droplet rim in a chamber with cells which stick head-on to the substrate.
Magnification: 1000 times

Sulfur drops were - as described - supplied as substrate to the cell in a chamber (Figure). Here it is useful for the investigation that Thiobacillus thiooxidans oxidizes colloidal sulfur better than sulfur in powdered form /Tano /. Sulfur powder (sulfur flower) is used in "batch"-cultures. Over a period of several (2 to 9) weeks the sulfur droplets were oxidized according to their size to 70 to 100%.
The sulfur droplets show strong morphological changes at the edge of the droplets (and at the interface towards the liquid) during significant accumulation of cells in that region.

Fig: 49

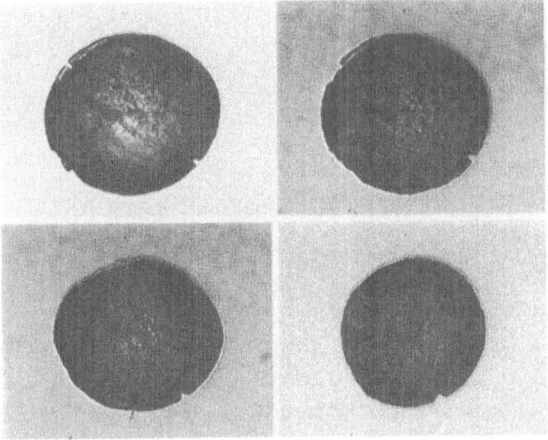

Photography: Sulfur droplets with morphological changes as a function of time
Magnification: 200 times

Fig: 50

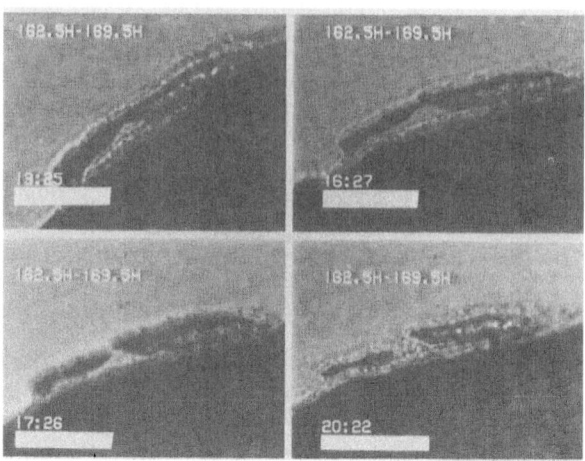

Photography: Sulfur droplet with changes at the rim - detail
Magnification: 500 times

Fig: 51

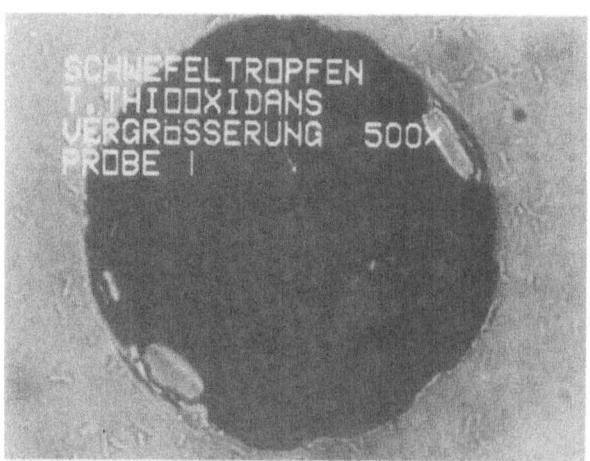

Photography: Sulfur droplets after 7 weeks
Magnification: 500 times

An increase of cell density towards the substrate could clearly be documented. Through a subsequent investigation it should be tested to what extent the pH in the neighbourhood of the sulfur droplets is changed.

Figures: 52 + 53

Photography: cell distribution at the sulfur droplet
The cell density decreases with the distance from the substrate.

It is intended to use pH-indicators thymolblue (pH-conversionpoint 1.2 - 2.8) and methylviolet (pH-conversionpoint 0.1 - 3.2). The sensitivity of the cells towards the substances has still to be tested. A change in colour in the neighbourhood of the droplets would indicate a pH-change caused by the cells (H_2SO_4 produced) . This parameter could act as a signal for cells further away.

Energy access for Thiobacillus ferrooxidans by electrochemical reduction of iron III + ($Fe_2^{III+} SO_4^{II-})_3$)

Bacteria of the genus thiobacillus ferrooxidans are able to deduce energy from Fe^{2+} SO_4^{2-} for their metabolism. This way $Fe^{3+} SO_4^{2-}$ developes. It is known that Fe^{3+} can electrochemically be reduced to Fe^{2+} (+770mV / NHE). Cultures of Thiobacillus ferrooxidans were already with good success and high cell density cultivated in the laboratory. A reaction chamber with electrodes was constructed . This chamber consists two platinum wires ($\varnothing$ 0.025 mm) as working and as counterelectrode, respectively. As a reference electrode a silver wire ($\varnothing$ 0.05 mm) was used which covers itself in acid environment in a silversulfat-layer, so that a defined potential ($Ag/AgSO_4$) developes.
In a preliminary experiment it was observed and recorded that the cells moved towards the cathode (A working electrode) after application of a potential (-70 $mV/AgSO_4$) which is sufficient to reduce Fe^{3+} to Fe^{2+}.
A reaction chamber is in the state of development, where the silver wire will be replaced by an external reference elctrode, what will be directly connected with the reaction chamber.

Highpressure hightemperature (HDHT) cell / reaction cylinder

An entirely new type of cell was developed. For this cell typ the nomination reaction cylinder (Figures: 56 + 57) was used. The newly constructed reaction cylinder has the following characteristic properties as to compared to the HDHT-cell (Figures: 54 + 55):
1) It consists of plastic
2) It has a copper cover
3) It has an integrated quartz window
4) It works - for the moment - to up to 230°C
5) It contains a new kind of photoelectrodes that are stable through a series of experiments
6) It has a volume increased to 5 ml in the reaction chamber (HDHT-cell approximately 1 ml)
7) It has a "real" reference electrode
8) It has a new kind of temperature sensors
The external pumping control system remains the same. As with the HDHT-cell, measurements are possible while the medium is flowing through the system. This is intended for chromatographic analysis of reaction products produced at the working electrode.
The concept of the HDHT-cell (Figures: 54 + 55) has been abandoned in favour of changes 1 - 8. The entire system has been installed on an optical bench (Figures: 56 + 57). For several reasons (danger of explosion, thermal isolation) the reaction cylinder with copper cover has been put into a metal cylinder, which was housed in an aluminum box.
By the changes 1 - 8 the following advantages were achieved
1) In the plastic housing with connections of the same material (PEEK, poly-ether-ether-keton) aggressive solutions (acids, bases) can be investigated without the generation of corrosion as in the HDHT-cell which were falsifying results.
2) The copper cover contains a resistance heating which permits a homogeneous heating of the entire reaction cylinder.
3) The quartz window is placed in the optical axis laser - working electrode, so that a direct illumination is possible.
4) At normal conditions (environmental temperature, 760 Torr) 230°C were obtained in the reaction chamber.
5) The photoelectrodes consist of a) a PEEK-cover. In this layer there is b) a hightemperature epoxidglue-cylinder, this contains c) a platinum wire, on which d) the electrode (for example RuS_2-crystal) can be mounted.
6) A larger volume allows more accurate measurements and higher product turnovers.
7) The reference electrode allows exact photoelectrochemical investigations (potential measurements, kinetical investigations)
8) The temperature sensor is directly installed besides the working electrode. This construction allows an accurate temperature measurement at the working electrode.

Fig: 54

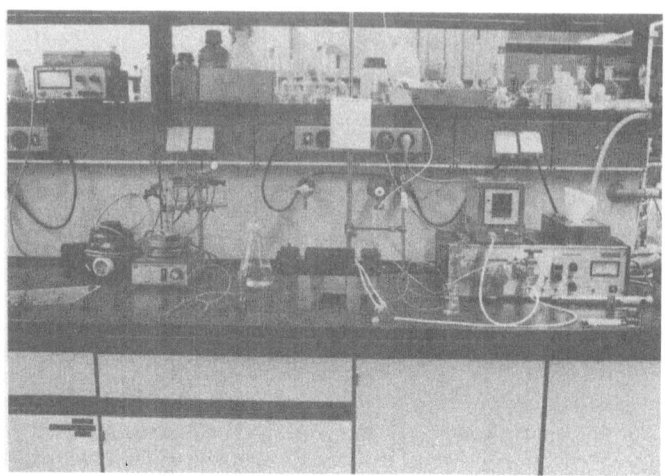

Photography HDHT-cell 1989
Overview over the experimental set-up

Fig: 55

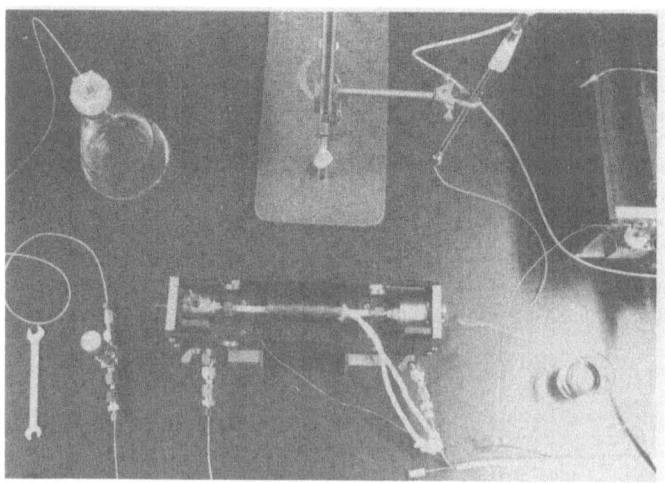

Photography: HDHT-cell 1989
Detailed picture of the reaction cylinder of Al_2O_3

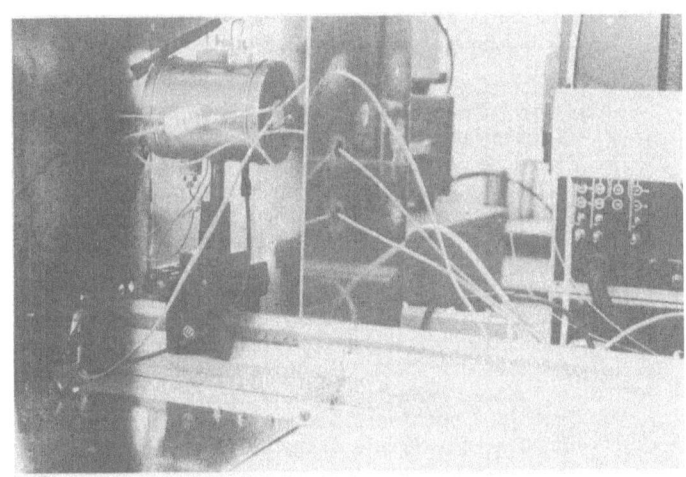

Photography: Reaction cylinder (1990) with metal cylinder and aluminum security box
At the left side of the picture a tiny bright spot of the laserbeam is visable, which
illuminates through a hole in the aluminum box the cylinder through a quartz window.

Fig: 57

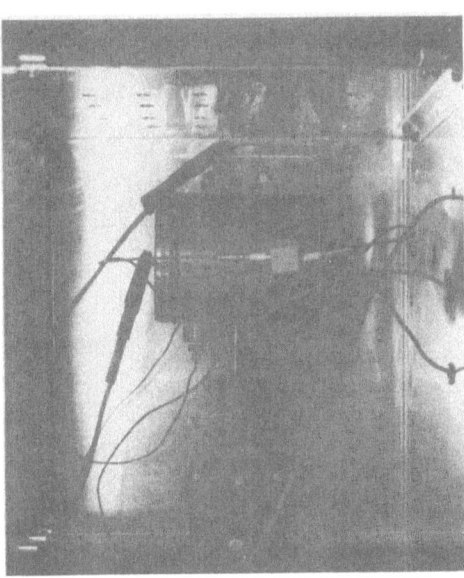

Photography: Detail picture of the reaction cylinder (1990) of PEEK with metal cover
and wiring.

References

/Bärtels/   Bärtels, C.C. Tributsch, H.
Solar Thermal Energy Cycle Based on Sulfur and Sulfide
Oxidizing Bacteria
Deutsche Forschungsanstalt für Luft und Raumfahrt E.V. (DLR)
"Solar Thermal Energy Utilization"; German Studies on
Technology and Application, Vol.5: Final Reports 1989; Edts:
M.Becker, K.-H- Funken, G. Schneider, Springer Verlag
1991

/Bock/   Bock, Eberhard; Sand, Wolfgang
Applied elctron microscopy on the biogenic destruction of
concrete and blocks. Use of the transmission electron
microscope for identification of mineral acid producing bacteria
Proc. Int. Conf. Cem. Microsc. 8th , 285 - 302; Edited by:
Bayles,J.; Gouda,G.R.; Nisperos,A. ; Int.Cem. Microsc. Assoc.:
Duncanville, Tex.
1986

/Chatzitheodoru/   Chatzitheodoru, G. et al.
Thin Photoactive Pyrite Films
Mat .Res. Bull., Vol.21, No.12, pp. 1481 - 87
1986

/Davis/   Davis, A.J.III. , Yen,T.F.
Feasibility studies of a biochemical desulfurization method
Adv. Chem. Ser. 151
(Shale Oil, Tar Sands, Relat. Fuel Sources, Symp. , 1974), 137-
43
1976

/Fukuyama/   Fukuyama,Yoji; et al.
Deodoration Method
Jpn. Kokai Tokkyo Koho, 3pp
1990

/Grudev/   Grudev,S. et al.
Adhesion of the bacterium Thiobacillus ferrooxidans to mineral
surfaces
God. Vissh., Minno.-Geol. Inst., Sofia, 24 (Pt.3), 352-9
1978

/Hazeu/   Hazeu,W. et al.
The production and utilization of intermediary elemental sulfur
during the oxidation of reduced sulfur compounds by
Thiobacillus ferrooxidans
Arch.Microbiol., 150 (6), 547-49
1988

/Kanagawa/	Kanagawa, Takahiro Biological deodorization of air containig sulfur compounds Yosui to Haisui, 31 (5), 397 - 404 1989
/Klubek/	Klubek, B. ; Clark, D. Microbiol removal of organic sulfur compounds from coal (bacterial degradation of sulfur-containing heterocyclic compounds): final report, March 1. - December 31., 1987 Report, DOE/PC/79863-T17; Order No DE 89008368, 21pp. Avail. NTIS from: Energy Res. Abstr. 1989, 14 (11), Abstr.No. 23206 1988
/Kodama/	Kodama, Akido; Mori, Takeshi Metabolism of a sulfur oxidizing bacterium IV. - Growth and oxidation of sulfur compounds in Thiobacillus thioooxidans Plant Cell Physiol., 9 (4), 709-23 1968
/Koenig/	Koenig, W.A.et al. Identification of volatile organic sulfur compounds in municipal sewage systems by GC / MS HRC CC, J. High Resolution Chromotogr. Chromatogr. Commun. 3 (8) , 415-16 1980
/Lutz/	Lutz, H.D.; Wäschenbach, G. Infrared Reflection Spectra, Directional Dispersion of the Phonon Modes and Dynamical Effective Charges of FeS_2 - Marcasite Phys. Chem. Min.; 12, 155-60
/Nakamoto/	Nakamoto, K. Infrared and Raman Spectra of Inorganic and Coordinaton Compounds Wiley & Sons; New York; III.Auflage, S. 239 1978
/Nakamura/	Nakamura Kanji; Nagami, Takashi Apparatus and method for biological deodorization of waste gases containing sulfur compounds Jpn. Kokai Tokkyo Koho, 6pp 1989
/Ollson/	Ollson, Gunnel; et al. Biological technique for removal of sulfur from coal Kem. Tidskr., 100 (5), 54-6 1988

/Tanji/   Tanji, Yasunori et al.
Removal of dimethylsulfide, methylmercaptan, and hydrogen
sulfide by immobilized Thiobazillus thioparus
Ferment. Res. Inst.
1989

/Tano/   Tano, Tatsuo; Imai kazutami
Physiological Studies on Thiobacillus thiooxidans
Hakko Kyokaishi; 26 (7), 322-27
1968

/Sand/   Sand, Wolfgang; Bock, Eberhard
Biogenic sulfuric acid attack in sewage systems
Biodeterior, [Sel.Pap. Int. Biodeterior Symp] 7th, Meeting date
1987, 113 - 17 Edited by: Houghton, D.R.; Smith, R.N.; Eggins,
H.O.W.;
Elsevier: London, U.K.
1988

/Sand/   Sand, Wolfgang
Importance of hydrogen sulfide, thiosulfate, and
methylmercaptan for growth of thiobacilli during simulation of
concrete corrosion
Appl. Environ. Microbiol., 53 (7), 1645-8
1987

/Windsperger/   Windsperger, Andreas
Eignung verschiedener Füllkörper als Träger für immobilisierte
Mikroorganismen zur biologischen Abluftreinigung
Chem. Ing. Tech. 62 Nr.11, 962 - 963
VCH Verlagsgesellschaft mbH , D- 6940 Weinheim
1990

Appendix

Culture Conditions

Each culture was grown under thermostatic (30° C) conditions in the dark. The cultures were shaken with 60 rpm. The amount of liquid was in every case 100 ml, the inoculum 100 µl.

The wild strain of Thiobacillus ferrooxidans R2, 1406 grew in the DSM Media 70 (Deutsche Sammlung für Mikroorganismen).

KH_2PO_4	0.4 g
$MgSO_4$ x $7H_2O$	0.4 g
$(NH_4)_2SO_4$	0.4 g
$FeSO_4$ x $7H_2O$	33.3 g

adjusted to pH 1.5 with concentrated sulfuric acid; ad to 1000 ml with distilled water.

The strains of FTO and FA, both Thiobacillus ferrooxidans (FA is identical with the ATCC 13661 strain) were grown in the medium DSM 271

$(NH_4)_2SO_4$	2 g
K_2HPO_4	0.5 g
$MgSO_4$ x $7H_2O$	0.5 g
KCl	0.1 g
$Ca(NO_3)_2$	0.01 g
$FeSO_4$ x $7H_2O$	40.0 g

adjusted to pH 3.0 with concentrated sulfuric acid; ad to 1000 ml with distilled water.

Thiobacillus thiooxidans TI and TA (TA is identical with the ATCC strain 8085 and has been used as a control for the medium; it was not taken for the data shown) were grown in the listed Medium.

KH_2PO_4	3.0 g
$(NH_4)_2SO_4$	0.2 g
$MgSO_4$ x $7H_2O$	0.5 g
$CaCl_2$	0.25 g

adjusted to pH 4.4 - 4.7 with H_2SO_4 or KOH respectively; ad to 1000 ml with distilled water.

The cells in a chamber were grown at room temperature, exposed to daylight and night. During video-recording (15 - 30 sec.) the chamber was exposed to the light beams of the transmission and reflection light sources. The recording time was interrupted for 128 - 256 minutes, depending on the status of cell-activity.

MOCVD (Metal Organic Chemical Vapor Deposition) Pyrite Layer

Chemical Vapor Deposition Technique has been used to generate thin pyrite films on glass, ITO (Indium Tin Oxide), Quartz, Si, GaAs, GaP, TiO_2 etc /Ennaoui/.

The production of thin pyrite layers on ITO coated glass is the basis to observe in situ the behavior of the bacteria attacking the pyrite surface.

Due to the oxidation of pyrite by the bacteria, the layer becomes thinner in time, and the thinner it becomes, the more light gets through. This change of the transmission-light intensity becomes visible during microscopical observation.

Partly, little holes emerge in the pyrite layer under the bacteria. The density of the holes rises in time; the number of cells increases proportionally.

To recognize these expected changes clearly, a layer with a surface as homogeneous and smooth as possible is necessary. This layer should have neither holes nor any roughness, so that bacterial influences are recognizable, without a doubt.

The surface quality is strictly coupled to the growth of the pyrite and its crystallization on the ITO-coated glass.

A variety of parameters influence the quality of the layers' surfaces:

1) The geometry of the reactor
   The inside of the reactor was enlarged in the area of the substrate holder and scaled down in other areas in order to yield different flow velocities and flow patterns.

2) The geometry of the substrate holder and its position in the reactor
   The surface of the substrate holder was varied from circular (diameter 1 cm) to rectangular (2 cm^2), the latter form delivered the best results up to now.

3) The nature and surface of the glass
   -a-   coated or uncoated
   -b-   the cleaning procedures
   -c-   the material itself
         contamination of the material with other elements, thermal expansion, stability of the material (in some cases Na, which is present in glass, was found in the pyrite layer).

4) The gasflow
   -a-   Flow speed of the carrier gas for the singular reactive components (iron / sulfur)
   -b-   Flow resistance.

5) Heating / Cooling
   Reaction and glass surface temperature (250 - 350° C).

6) Aerodynamics
   The flow patterns of the gas with the reactive components influence the growth and crystallinity of the thin pyrite layer.
   A very important change is the arrangement of the substrate holder itself. The new construction held the glasses into the gasstream at angulars from 30 - 60°.

7) Temperprocess
   Iron-Di-Sulfide crystallizes into marcasite and pyrite. The two phases are stable at 250-400 °C. Additional heating in S_x atmosphere after deposition converts marcasite into pyrite.

The most difficult problem is to find a connection between flow patterns, flow resistance, flow velocity and deposition. On the one hand the gas stream should be a laminar flow when it hits the glass surface. But on the other hand a good turbulence is necessary to mix the two reactive components before the meet the glass surface.

These parameters are not measurable by now.

For the covering of the glass-substrate ITO seems to be the best material.

The cleaning of the glass surface is another problem because up to now it is not possible to work under sterile conditions.

A tempering procedure after the depostion process (under S_x atmosphere, 450 - 550°C) turned out to be a very successful step /Ennaoui/.

References

/Ennaoui/   Ennaoui, A., Fiechter, S., Schroetter,S., Tributsch,H.
                 Infrared Spectroscopy and X-Ray Diffraction Characterisation of
                 Iron-Disulfide Prepared by Metal Organic Chemical Vapour
                 Deposition (MOCVD)
                 Journal of Materials Science Letters
                 submitted, 1991

/Ennaoui/   Ennaoui, A. et al.
                 Photoelectrochemical Energy Conversion Obtained with
                 Ultrathin MOCVD-Layers of FeS_2 (Pyrite) on TiO_2
                 Journal of Electrochemical Society
                 to be published, 1991

Sulfate reduction as a problem of multi-electron catalysis

The reduction of sulfate to sulfide is accomplished by sulfate reducing bacteria at the expense of organic energy sources at ambient temperature and with optimal rate at 35°C. This demonstrates that sulfate reduction is basically a problem of efficient catalysis. Up to now this process has not been in the center of interst mainly do to the lack of immediate technical applications. Applications are, however, immaginable, for example in the area of sulfuric acid waste disposal and in energy cycles.

Thermodynamically, the standard redox potential of the SO_4^{2-}/S^{2-} reaction is near +0.3V vs SHE, that is more positive than the potential for hydrogen production. Kinetically the sulfate reduction is however strongly inhibited, so that the reduction of an aqueous solution containing sulfate, with typical metal electrodes and in the absence of siutable catalysts, only yields hydrogen. Reduction of waterfree sulfuric acid, on the other hand, yields S^0, H_2S and SO_2. The relative concentration of these products depends largely on the H_2O and SO_3 contamination of the sulfuric acid and the temperature /Parshin,Beck/ SO_4^{2-} reduction to H_2S in presence of parafines has been observed at temperatures between 175 and 350°C /Toland, Anisimov, Kiyosu/. Also in presence of Fe^{2+} /Shanks/ or CO_2 sulfate is reduced between 200 and 300 °C. This corresponds to the conditions of sulfate reduction on the sea bottom. Typically, the reaction only starts after very low partial pressures of oxygen have been reached. With increasing temperatures the redox equilibrium shifts towards the reduced sufur species plus oxygen. As intermediate sulfur species thiosulfate $S_2O_3^{2-}$, sulfite SO_3^{2-}, elemental sulfur S , polysulfide S_nS^{2-}, the S_2^-, hydrogen sulfide H_2S , HS^- and at higher pH S^{2-} are to be expected /Giggenbach/. The most inhibited step will be the reduction of sulfate (S^{6+}) to sulfite (S^{4+}) while the steps leading from sulfite to thiosulfate, to elemental sulfur, to polysulfide and sulfide are much faster although they still require catalysis and elevated temperatures to proceed reasonably fast. They may react at 150°C with half times of less than one hour while the sulfate reduction at 300 °C may still need several days /Ohmoto/.

At pH 2 the stability limits of a 0.1 M solution of total sulfur species shift between 25 and 250°C from a log p_{O_2} of -62 to -32. As a consequence very high temperatures would be required to yield significant outputs of sulfate reduction by purely thermal reduction. In presence of an oxidizable compound such as CO_2 sulfate reduction is terminated at 300°C within 8 h. The same reaction at 240°C, on the other hand, takes 6720 h /Möller /. On the other hand, significant kinetic rates should be expected during electrolytic reduction at elevated temperature (250°C). In this case a reverse reaction of the sulfide would be suppressed by separating oxidation and reduction. The most attractive way to develope a technically solar powered sulfide reduction would be a photoelectrochemical reduction or an electrochemical reduction using solar electricity at elevated temperature (250-500°C).

Sulfate reduction under such conditions is presently explored using the described specially adapted electrochemical cell integrated into a high pressure liquid chromatography system. A mayor difficulty is the selection of the electrode for sulfate reduction which not only has to be stable but also highly catalytic for an electron transfer process involving 6 - 8 electrons. The technical problem of catalysis of sulfate reduction cannot be expected to be solved in a straightforward way. Fundamental studies are necessary to reveal the requirements of this complicated heterogeneous process.
In natural environments the dominant mechanism is the reduction by the iron species. The ratio Mg/Ca in the environment is supposed to exert a significant influence on the rate /McDuff/. Sulfate reduction at 350 °C with the minerals fayalite and magnetite has been reported to be quite fast /Shanks/.

For technical sulfate reduction at temperatures below 500°C catalytic electrodes will have to be developed. They must be efficient for the 6 - 8 electron transfer to sulfate, but inefficient for the competitive hydrogen evolution. At the same time these electrodes have to be resistant to corrosion. Two different classes of catalytic materials were explored :
a) Electrodes of the composition $(Mo,M)_6X_8$ which contain bimetallic octahedric transition metal clusters, turned out to be stable at elevated temperatures (up to 400°C) and efficient catalysts for multielectron transfer. For example $Mo_4Ru_2S_8$ or $Mo_2Re_4S_6$ bimetallic clusters have proved to function as electrodes for oxygen reduction to water which qualitatively can be compared with platinum, which is presently the best catalyst for oxygen reduction in acid fuel cells /Alonso-Vante/ . Like with platinum only 4% H_2O_2 occur as a side reaction. However, they also turned out to be efficient catalysts for the evolution of hydrogen.
Experience gained up to now suggests that efficient catalysis of multielectron transfer requires the cooperative action of a bimetallic transition metal cluster. Hydrogen evolution, on the other hand, is catalyzed by the catalytically most active individual metal atom /Schubert/. In the case of bimetallic clusters like $Mo_4Ru_2S_8$ or $Mo_2Re_4S_6$ hydrogen evolution is catalyzed by Ru and Re respectively which are efficient

catalysts too. Other element combinations with less catalytic metals in the transition metal cluster (e.g., Fe) will have to be identified for an efficient catalysis of sulfate reduction under simultaneous suppression of hydrogen evolution from aqueous electrolytes.

b) As an alternative strategy rhodium has been tried as a catalyst for electron exchange with sulfur species . This metal catalyst has been successfully applied as a catalyst for the photocatalytic cleavage of H_2S to H_2 and S on the surface of semiconducting CdS /Borgarello/ Quantum efficiencies of up to 45 % have been reported. In our laboratory an aqueos rhodium complex, $[Rh(OH)_6^{2-}]$aq in an aqeous electrolyte containing sulfuric acid was electrochemically deposited as on an illuminated p-InP electrode to form a metallic rhodium film consisting of many small grains. In presence of this film and the $[Rh(OH)_6^{2-}]$aq complex in the electrolyte reduction of SO^{4-} was observed /Bogdanoff/. This is a very interesting first success and the katalyst will systematically be explored in subsequent work.

References

/Alonso-Vante/ Alonso-Vante, N. , Schubert, B., and Tributsch, H.
Transition metal cluster materials for multi-electron transfer
catalysis; Material Chemistry and Physics, 22 , 281-307, 1989

/Anisimov/ Anisimov, L.A.
Conditions of abiogenic reduction of sulfates in oil- and gas-
bearing basins. Geokhimia, 11, 1692-1702, 1978

/Beck/ Beck, F.,
Electrochim. Acta, 17, 2317-2331, 1972

/Bogdanoff/ Bogdanoff, P.
Hahn Meitner Institut Berlin, unpublished results,1991

/Borgarello/ Borgarello, N. Serpone, E. Pelizetti, M. Barbeni
J. Photochem., 33, 35, 1986

/Giggenbach/ Giggenbach, W.F.,
Equilibria involving polysulfide ions in aqueous sulfide solutions
up to 240°C, Inorg. Chem., 13, 1724-1730, 1974

/Kiyosu/ Kiyosu, Y.,
Chemical reduction and sulfur-isotope effects of sulfate by
organic matter under hydrothermal conditions. Chem. Geol., 30,
47-56, 1980

/McDuff/ McDuff, R. and Edmond, J.M.,
On the fate of sulfate during hydrothermal circulation at mid-
ocean ridges, Earth and Planetary Science Letters, 57, 117-132,
1982

/Möller/	Möller et al, personal communication
/Ohmoto/	Ohmoto, H., Lasaga, A.C., Kinetics of reactions between aqueous sulfates and sulfides in hydrothermal systems, Geochem. Cosmochem. Acta , 46, 1727-1745, 1982
/Schubert/	Schubert, B. and Tributsch, H. to be published
/Shanks/	Shanks, W.C., Bischoff, J.L. and Rosenbauer, R.J., Seawater sulfate reduction and sulfur-isotope fractionation in basaltic systems: interaction with fayalite and magnetite at 200-350°C. Geochim. Cosmochim. Acta , 47, 1977-1995, 1981

We would like to express our gratitude:

A part of this work was financially supported by the DLR.

The MOCVD-technique was improved by Dr. Ahmed Ennaoui.

Thiobacillus thiooxidans was a present from Patricia Boskovsky (Österreichisches Forschungszentrum Seibersdorf), who also investigated the behavior of T. thiooxidans in the environment of sulfur droplets in a chamber.

The new prototype of the HDHT - cell was constructed by the group of Dr. Abdel-Latif Bilal and Henrik Collel.

Preparation and Characterization
of Novel Mixed Titanium/Iron Oxide Photocatalysts
for the Detoxification of Polluted Aquifers

D. Bockelmann, R. Goslich, D. Bahnemann,

Institut für Solarenergieforschung GmbH,
Hannover

Abstract.

The removal and decomposition of toxic halogenated hydrocarbons in waste water effluents and ground water wells is a problem of increasing importance for our industrial society. Since conventional methods such as chemical oxidation or microbial treatment are often not efficient for the destruction of these toxins, alternative routes for detoxification are required. It has been shown that semiconductor particles can be used as photocatalysts which are capable of inducing the complete mineralization of many of these hazardous compounds.

Here we present our investigations concerning the use of new Ti/Fe mixed oxide particles as potential photocatalysts for solar applications.

Colloidal Ti/Fe mixed oxide particles were synthesized in aqueous solution with the Fe^{3+} content varying from 0.05 up to 50 at%. The absorption spectra of these colloidal Ti/Fe-mixed oxide particles could be shifted into the visible part of the solar spectrum. This effect was more pronounced with increasing iron content. The photophysical properties of the newly synthesized particles were examined in detail. Steady state illumination experiments of mixed oxide colloids (50 at% Fe) showed a reversible bleaching of the absorption spectrum in the presence of a hole scavenger (1 M ethanol) indicating that a negative excess charge could be stored. The origin of the bleaching was studied in detail using time-resolved flash photolysis measurements.

The photocatalytic activity of the Ti/Fe mixed oxide particles was compared to that of pure colloidal TiO_2 by studying the photooxidation of dichloroacetic acid in acidic and alkaline media. It could be demonstrated that the incorporation of iron into TiO_2 leads to an enhancement of the degradation efficiency, especially in acidic solutions.

1. Introduction

Persistent organic chemicals are present as pollutants in waste water effluents from industrial manufacturers, dry cleaning facilities or even normal households. They can be found in ground water wells and surface waters where they have to be removed to achieve drinking water quality [1, 2]. Therefore many processes have been proposed in the last couple of years and are currently being employed to destroy these toxins.

Biodegradation is probably the most frequently used technique [3-7]. However, many toxic mixtures are often lethal to microorganisms. This limits the applicability of the method to cases where well-defined pollutant mixtures can be expected. Chemical oxidants, e.g. chlorine, ozone or hydrogen peroxide, do not suffer from these limitations. But often the formation of toxic side products is encountered in those processes and they are generally too expensive for the treatment of vast volumes [8].

Molecular oxygen is not a good oxidant since due to the triplet character of its ground state reactions with singlet state molecules are extremely slow. Catalysis of the oxidation by O_2, e.g. homogeneously by transition metal ions or heterogeneously on particle (for example metal oxide) surfaces, is also applied for detoxification processes but restricted to special applications [9,10]. Free radicals, such as the hydroxyl radical, are much more powerful oxidants and can in principle induce the mineralization of nearly all toxic organic chemicals. They can be generated in water by γ-radiolysis using a ^{60}Co-irradiation source. The applicability of the method has been demonstrated [11-15], but it is not applicable in technical scale due to the high expenses for associated safety measures.

Free radical reactions can also be induced by photochemical principles. These include the direct conversion of toxic molecules following the absorption of photons as well as indirect methods, where $^\bullet$OH radicals are generated by the UV-photolysis of hydrogen peroxide or ozone [16,17]. The potential of homogeneous phase photochemistry for the destruction of halogenated hydrocarbons has been reviewed recently [18-20].

The combination of two or more of the above mentioned methods can sometimes be beneficial to achieve complete destruction of all hazardous compounds present in a polluted water. For example, photochemical treatment can lead to compounds which are not toxic to microorganisms, thus a consecutive biodegradation step will result in clean water [21, 22].

2. Photocatalytic Detoxification

A novel method for the complete oxidation of polluted aquifers, the photocatalytic detoxification has been discussed in the scientific literature since 1976 [23]. Considerable public attention has been focussed on this possibility of combining heterogeneous catalysis with solar technologies to achieve the mineralization of toxins present in water [24-27].

2.1 Principles of Photocatalysis

A photocatalytic system consists of semiconductor particles suspended in a solvent. In all cases mentioned in this paper the solvent will be referred to as water, but numerous examples about other solutes as well as photocatalysis of gas phase reactions have been reported in the literature [28].

A single spherical semiconductor particle including its energetic solid state properties is depicted in figure 1. It is well-known that a semiconductor consists of a valence band (VB) filled with electrons and a conduction band (CB) which contains empty energy states.

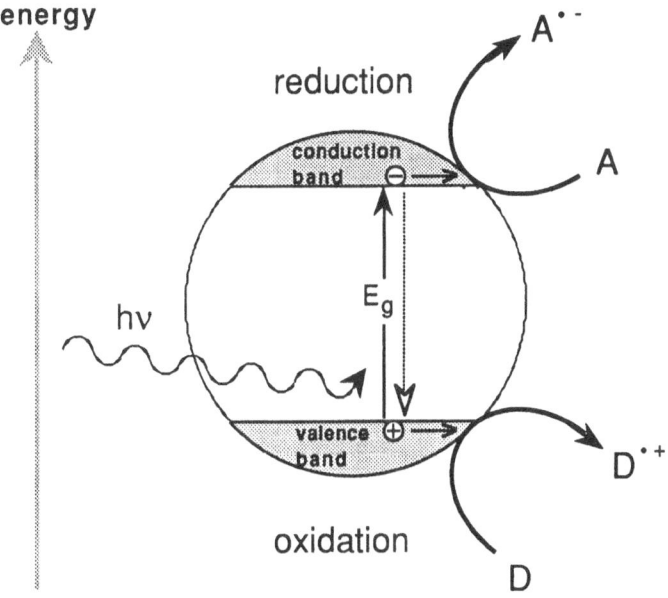

Figure 1: The semiconductor particle as a redox-catalyst (schematic presentation: energetic featues and geometric picture ar superimposed).

Upon illumination with photons the energy hv of which exceeds the band gap energy E_g of the semiconductor, electron/hole (e^-_{CB}/h^+_{VB}) pairs are formed within the particle. In the presence of appropriate redox couples these charge carriers are able to induce reductions

$$A \quad + \quad e^-_{CB} \quad \longrightarrow \quad A^{\bullet -} \quad \quad (a)$$

or oxidations, respectively.

$$D \quad + \quad \overset{\smile}{h}^+{}_{VB} \quad \longrightarrow \quad D^{\bullet +} \qquad (b)$$

It is a striking feature of semiconducting metal oxides that $h^+{}_{VB}$ possess extremely positive oxidation potentials and should thus be able to oxidize almost all chemicals [29,30]. Even the one-electron oxidation of water resulting in the formation of hydroxyl radicals

$$H_2O \quad + \quad h^+{}_{VB} \quad \longrightarrow \quad {}^{\bullet}OH \quad + \quad H^+{}_{aq} \qquad (c)$$

should be energetically feasible in illuminated semiconductor suspensions [29]. Numerous observations can indeed be explained by the intermediacy of $^{\bullet}OH$, however, due to the short lifetime and high reactivity of this radical the ultimate unambiguous experimental evidence for the formation of hydroxyl radicals in suspensions of metal oxide semiconductors (which clearly exhibit the most positive valence band edges as compared with other semiconducting materials) has so far not been given [31].

Potentials and limitations of microheterogeneous systems for photocatalysis and photosynthesis have been reviewed recently [31,32]. It should be noted that a large number of critical parameters exist which can influence photocatalyzed redox processes qualitatively and quantitatively. For example, an enlargement of the diameter of the semiconductor particle can result in a transition from free radical reaction routes to almost exclusive two-electron transfer processes [31]. Often, yields of reaction products are found to increase only with the square-root of the increasing illumination intensity [33]. Ollis has recently presented a generalized model to account for the influence of the light intensity on photocatalytic mineralizations in a solar parabolic trough reactor [34]. While a linear dependence of the reaction rate on the intensity (i. e. a constant quantum yield) is to be expected at low fluxes, the described square-root law is valid in an intermediate regime. Finally, at very high fluxes ($\geq$ 100 suns), transport limitation (i. e. a reactor design parameter) will cause the rate to remain constant upon further increase of the illumination intensity.

2.2 Photocatalytic Mineralization of Pollutants: A Review

A wide variety of toxic chemicals have been treated by photocatalytic processes to test the general applicability of the method. Halogenated hydrocarbons are readily mineralized in aqueous suspensions of titanium dioxide [35-42]. The following overall stoichiometry has been observed for the quantitative degradation of chlorinated hydrocarbon molecules $C_xH_yCl_z$ [43]:

$$C_xH_yCl_z + \left(x + \frac{y-z}{4}\right)O_2 \longrightarrow x\,CO_2 + z\,HCl + \left(\frac{y-z}{2}\right)H_2O \quad (d)$$

Aromatic molecules such as phenol are also quantitatively oxidized photocatalytically [45-48]. While the overall stoichiometry appears to be [45]

$$C_6H_5OH + 7\,O_2 \longrightarrow 6\,CO_2 + 3\,H_2O \quad (e)$$

details of the complex reaction mechanism are still not understood. The presence of halogen substituents does not limit the applicability of the method to aromatic compounds. In fact, chlorinated phenols, biphenols and even dioxins are apparently also completely oxidized yielding CO_2 and HCl as final products [49-59]. A few stable intermediates of these oxidation processes have been identified (e. g. hydroquinone [54]), however, an unambiguous proof that pertinent toxic side-products are not even formed in trace amounts is still not available in the literature. The intermediacy of free radicals formed upon the initial attack of h^+_{VB} or surface bound hydroxyl radicals ($^\bullet OH_S$, cf. reaction (c)) onto halogenated phenols has been discussed [60], but until now not proven experimentally.

Dyes [61,62], phthalates [63], DDT [64] and surfactants [65,66] represent other classes of problem chemicals the mineralization of which has been achieved with TiO_2 as the photocatalyst.

Wherever different semiconductor materials have been tested under comparable conditions for the degradation of the same compound, TiO_2 generally exhibited the highest activity [52,64,67,68]. Only ZnO possesses an activity similar to that of TiO_2 which has been observed for the photocatalytic degradation of pentachlorophenol [67], chloro-fluoromethanes [69] and chloroaromatic derivatives [70-72]. However, zinc oxides dissolves in acidic solutions [73] and can therefore not be used for technical applications.

The anatase modification of titanium dioxide shows much higher photocatalytic activity than the rutile form [47,75], which has been explained by a faster recombination of charge carriers in rutile

$$e^-_{CB} + h^+_{VB} \longrightarrow heat \quad (f)$$

and also by a considerably lower amount of reactants adsorbed on the surface of a rutile particle [47]. The latter effect can be rationalized by a lower number of surface hydroxyl groups present on these particles [74]. Indeed, it has been shown that adsorption of the toxic molecules onto the catalyst particle is a prerequisite for a highly efficient detoxification [48,75]. The importance of the acid/base chemistry of the surface hydroxyl groups present on metal oxide particles dispersed in water (cf. chapter 2.3) for the adsorption properties has been discussed in detail elsewhere [31]. It should be pointed out here that only relative values for the efficiency are given in the literature cited above. Absolute determinations of quantum yields are usually not presented by the authors, and available experimental details are normally not sufficient to estimate yields. Therefore it is almost impossible to compare data from different laboratories.

Titanium dioxide, the material with the highest photocatalytic detoxification efficiency is a wide bandgap semiconductor ($E_g \approx 3.2$ eV[29]). Thus, only light below 400 nm is absorbed and capable of forming e^-/h^+ pairs [76]. Therefore only 5% of the solar energy reaching the surface of the earth could in principle be utilized when TiO_2 is used as photocatalyst. Hence it is evident that for solar applications other materials have to be found or developed which exhibit similar efficiencies as anatase, but possess spectral properties more closely adopted to the terrestrial solar spectrum.

It has already been pointed out that the absorption of photons by semiconductors leads to the formation of oppositely charged charge carriers. While the fate of the hole which induces the desired oxidation process has been studied in detail, most authors did not examine the role of the cathodic process, i. e. the reactions of e^-_{CB}. It is generally assumed that molecular oxygen [43-45,47,49,53-55,62,77] acts as the oxidant :

$$e^-_{CB} \quad + \quad O_2 \quad \longrightarrow \quad O_2^{\bullet-} \qquad (g)$$

however, hydrogen peroxide which should be formed as a stable product by the subsequent reaction of superoxide radicals generated in reaction (g)

$$2\,O_2^{\bullet-} \quad + \quad 2\,H^+_{aq} \quad \longrightarrow \quad H_2O_2 + O_2 \qquad (h)$$

is only found in trace amounts when TiO_2 is used as the photocatalyst [78]. Further reduction of H_2O_2 leads to the formation of $^{\bullet}OH$ radicals:

$$H_2O_2 \quad + \quad e^-_{CB} \quad \longrightarrow \quad {}^{\bullet}OH \quad + \quad OH^-_{aq} \qquad (i)$$

It has in fact been shown that the rate of photodegradation can be considerably enhanced when H_2O_2 is used as oxidant [79].

Separation of the anodic and cathodic process is in principle not possible in microheterogeneous photocatalytic systems containing semiconductor particles. Hence, it cannot be decided whether hydroxyl radicals are formed via the oxidation of water (cf. reaction (c)) or the reduction of molecular oxygen (reactions (g)-(i)), i. e. whether electrons or holes are more important for the initial step of pollutant degradation which is generally believed to be the reaction of $^{\bullet}OH$ with the substrate molecule S:

$$^{\bullet}OH \quad + \quad S \quad \longrightarrow \quad S^{\bullet}-OH \quad \longrightarrow \quad products \qquad (j)$$

Since the efficiency of a complex process is always limited by the slowest reaction step, it will be necessary to distinguish between the two possibilities discussed above and to study them separately. Photoelectrochemical investigations

with separated anode and cathode are thus required to be able to optimize particulate photocatalytic detoxification systems.

Various different reactor designs have been tested in the laboratory studies
reported above [35,39,51,55,65]. Chemical engineering problems characteristic
for the different reactor types such as mass-transfer limitations have already
been mentioned [34,80,81] and are not subject of this paper.

Photocatalysis has been compared with other methods for the destruction of
toxic chemicals in aqueous solutions on economical [8] and technical [82]
grounds. The authors agree that the method bears considerable potential and
should be pursued.

Thinking of a solar application of the photocatalytic detoxification it is essential
that the incident sunlight is effectively utilized. So the investigations presented in
this paper concentrate on the synthesis and characterization of photocatalysts
which absorb the visible part of the solar spectrum to a greater extent with the
simultaneous improvement of their photocatalytic detoxification properties.

2.3 Particles and Colloids

For a detailed understanding of photocatalytical processes it is important to have
a certain knowledge about the photophysical and photochemical properties of the
photocatalyst itself. Most of the photocatalytic detoxification experiments
performed worldwide used suspensions of semiconductor particles. The
experiments described in this paper have been carried out with colloidal
solutions of mixed oxide semiconductor particles. Colloidal particles have a
diameter which is smaller than the wavelength of the incident light, so their
aqueous solutions are completely transparent. Thus it is possible to obtain
absorption spectra of the material and therefore the kinetic of processes taking
place on the surface can be recorded, if they were reflected in a change of the
absorption.

In order to get stable colloidal solutions it is important that the surface of each
particle is sufficiently charged, so that electrostatic repulsion prevents an
agglomeration of the particles. In the presence of water all mixed oxide particles
are covered with surface hydroxyl groups [83]. According to the pH-value of the
solution these groups are able to add or substract protons:

$$\equiv TiOH \ + \ H^+ \ \longleftrightarrow \ \equiv TiOH_2^+ \qquad (K^s_1) \qquad (k)$$

$$\equiv TiO^- \ + \ H^+ \ \longleftrightarrow \ \equiv TiOH \qquad (K^s_2) \qquad (l)$$

Because of these acid-base equilibria surface charges can be introduced very easily by variation of the $[H^+]$ or $[OH^-]$-concentration. Considering the equilibrium constants K^s_1 and K^s_2 the surface charge of a metal oxide particles will be positive at low pH-values and negative at high pH-values. For each metal oxide a specific pH-value can be found where the net charge of all surface groups are not changed. This pH-value (»zero point of charge«, pH_{zpc}) can be derived from the acid-base equilibrium constants according to [83]:

$$pH_{zpc} \quad = \quad \tfrac{1}{2} \quad \left(pK^s_1 \quad + \quad pK^s_2 \right) \tag{m}$$

The acid-base equilibrium constants and thus the pH_{zpc} can be experimentally obtained by titration measurements.

3. Experimental Procedures:

The Ti/Fe-mixed oxide colloids with different iron content are prepared as powders, which are stable at room temperature for several months and can be resuspended in water or water/ethanol mixtures yielding transparent colloidal solutions.

For the preparation of Ti/Fe-mixed oxide colloids freshly distilled $TiCl_4$ cooled to -20°C was added slowly to cold (~ 0°C) freshly prepared $FeCl_3$-solution under vigorous stirring . The concentration of the $FeCl_3$-solution is varied in order to get colloids with different iron content. It is the intention of this preparation to thoroughly incorporate Fe^{3+} into the growing particles. In order to prove this samples of colloidal solutions have been precipitated and were tested colorimetrically for Fe^{3+}-ions after removal of the solid material. The detection limit was below 5 ppm.

The resulting colloidal suspension is stable for several hours at temperatures below 5 °C. To increase the stability of the colloids and to facilitate powder formation during evaporation of the solvent, the ionic strength was subsequently reduced by dialyzing against pure water until a final pH between 2.5 - 3 is reached [76]. Aliquots of 200 ml are dried with the aid of a rotary evaporator (25 mbar, 30 °C). The residue is dried under higher vacuum (1 mbar) for one minute, resulting in crystalline yellowish to brownish powders.

Detailed photophysical and photochemical investigations of colloidal Ti/Fe mixed oxide suspensions have been performed using flash photolysis apparatus [84] for time-resolved measurements. The experimental setup is shown in figure 2.

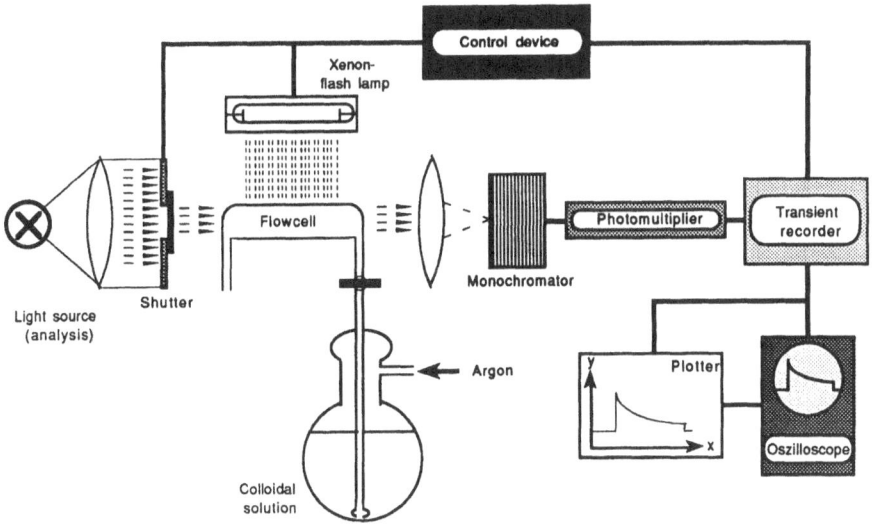

Figure 2:   Schematical drawing of the UV-flash photolysis experiment.

The flow cell had an optical path length of 6 cm and a thickness of 0.7 mm. The xenon flash lamp and the cell were located at the two focuses of an elliptical cavity. The lamp was triggered by a spark-gap using a 4 µF/10 kV capacitance. A 100 W quartz halogen lamp was used as analyzing light source. Signals were detected with a photomultiplier EMI 9558 and recorded with a transient recorder DL905 (Data Lab., England) in a range of 200 ms and had therefore a time resolution of ~2 ms. The excitation flash had a half-time of 9 µs and an energy of approximately 100 J/pulse. All flash photolysis experiments have been performed in 20% methanol/water solutions with 0.16 g/l colloidal Ti/Fe mixed oxides. The solution also contained 10^{-3} M $HClO_4$ to adjust the pH-value to ~2.3. For experiments in the alkaline pH-regime (~11.3) an appropriate amount of 1 M NaOH was added quickly to the colloidal solution. Then the solution was deaerated with argon for about one hour.

The photocatalytic activity of the Ti/Fe mixed oxide particles was tested with detoxification measurements using dichloroacetic acid as the probe molecule. According to eqn. (n), the oxidation of one dichloroacetic acid molecule leads to the formation of one proton. Therefore we used a pH-stat technique [85] which allows the in-situ measurement of H^+_{aq} formed during the photolysis experiments with extremely high sensitivity. The data from the autotitration system were transferred to a computer which calculates the concentration of the generated protons from the amount of the added base with respect to the elapsed time. Corrections due to the dissociation equilibtria of simultaneous formed H_2CO_3 (as HCO_3^- at $6.3 <$ pH < 10.3 and as CO_3^{2-} at pH > 10.3) have been considered in the computer program.

The autotitration system (Metrohm) was connected to a ROSS semimicro combination pH electrode 81-15 SC. The titrant solution (0.02 or 0.1 N NaOH) was kept under Ar and calibrated weekly with 0.1 N HCl (Titrisol/Merck). The photochemical reactor was made of quartz glass and filled with 50 ml colloidal solution which was thermostated and vigorously stirred by a magnetic stirring bar.

Each colloidal solution contained 0.5 g/l of the Ti/Fe-mixed oxide, 2.5 mM dichloroacetic acid and originally showed a pH value beween 2.7 and 2.4. The adjustment of pH~2.6 had been performed by adding appropriate amounts of 1 M HCl or NaOH. According to the preparation of the Ti/Fe-mixed oxides the colloidal solution contains 2-3 mM chloride. For experiments in the alkaline pH-regime (~11.3) an appropriate amount of 1 M NaOH was added quickly to the colloidal solution.

Illuminations were carried out with a mercury doped high pressure Xe-lamp (OSRAM HBO 500W). Short-wavelength UV-radiation was eliminated by a 320 nm band pass filter and the IR radiation was prevented by a water filter filled with 10 cm of water.

Actinometry was performed using Aberchrome 540 [86] in order to determine the light intensity in the wavelength region between 310 and 370 nm. The determination of the light intensity is essential in order to calculate quantum yields. The quantum yields i.e. the ratio of reaction rate and corresponding light intensity has been determined to characterize the efficiency of the degradation process.

A Dionex 4500i ion chromatograph equipped with a HPIC-AS4A separator column was used to determine chloride and dichloroacetic acid concentrations. Detailed conditions are given in the literature [87].

Absorption spectra of the colloids have been performed using a Bruins Omega 10 UV-VIS spectrometer.

Chemicals and solvents were of reagent grade and used without further purification. The water employed in all preparations was purified by a Milli-Q/RO system (Millipore) resulting in a resistivity > 18 MΩ.

4. Results and Discussion:

It is important for a solar driven chemical process that the incoming sunlight can be effectively utilized. Therefore the catalyst should exhibit an absorption over a broad range of the solar spectrum and the separation of the photogenerated charge carriers should be effective.

Until now titanium dioxide (TiO_2) has been the catalyst which was most frequently employed in photocatalytic detoxifications. In comparison with other semiconductors it could always be shown that TiO_2 is the most active and stable [52,64,67,68] photocatalyst. Unfortunately, due to its large bandgap of 3.2 eV TiO_2 absorbs photons below 380 nm. Further investigations have shown [88] that TiO_2 containing 0.5 at% iron exhibits a redshifted absorption spectrum and also a suppressed recombination of charge carriers compared to pure TiO_2. Therefore we started to synthesize colloidal TiO_2 particles with different iron content (0.05 - 50 at%) and examined their photophysical and photocatalytic properties.

4.1 Photophysical properties of Ti/Fe-colloids

From X-ray-fluorescence data it could be derived that the proportion of titanium to iron in the synthesized materials corresponds to the rough preliminary estimations within an uncertainty of approximately 2%. Particle size analysis with a transmission electron microscope (TEM) showed that freshly prepared Ti/Fe-colloids are as big as 5 nm in diameter. They agglomerate within some hours to long rolls that are still 5 nm in diameter and approximately 8-100 nm in length.

Absorption spectra of some Ti/Fe mixed oxides are given in figure 3. It can be seen that the absorption edge is shifted towards the visible region of the spectrum as the iron content is increased. Additionally a new absorption shoulder around 480 nm appears in the spectra when the particles contained more than 5 at% iron. Colloids containing 25 at% or more iron exhibit a significant absorption even at 550 nm. This can be interpreted as a decrease of the effective bandgap by more than 1 eV. Indeed the shape of the absorption spectrum of colloidal particles containing more than 25 at% iron is very similar to the spectrum known from pure colloidal Fe_2O_3 [89].

An increasing amount of iron can therefore be utilized to enhance the absorption properties of TiO_2.

Changes in the spectral characteristics upon illumination with near UV light (λ_{ex}=360 nm) in the presence of hole-scavengers such as alcohols are shown in the case of Ti/Fe-colloids with 50 at% iron content (fig. 4). The absorption onset is shifted towards shorter wavelength following an UV-irradiation for

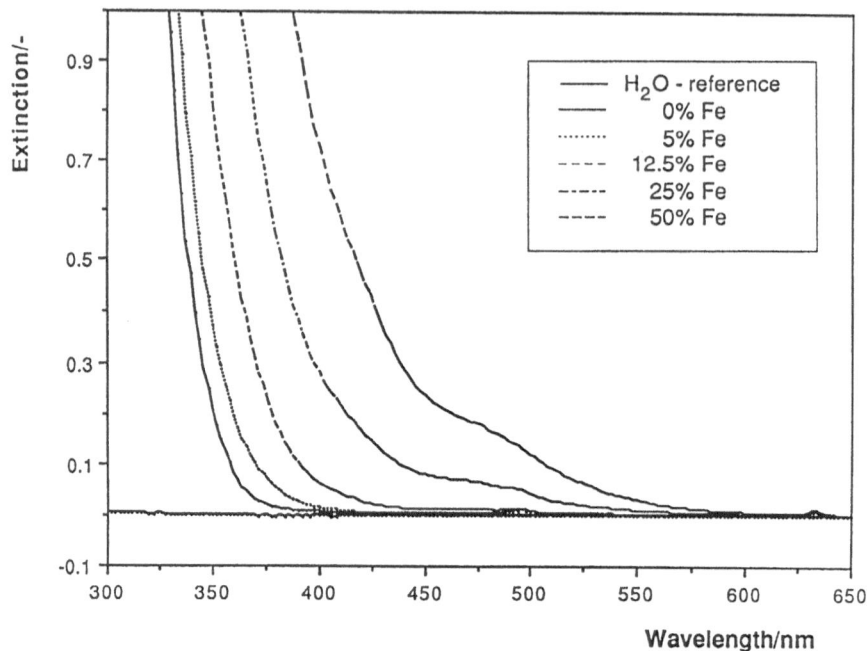

Figure 3: Absorption spectra of Ti/Fe-mixed oxide colloids with different iron content. The solutions contained always 0.5 g/l of the colloid (pH 11.3).

15 minutes. Subsequent addition of an oxidant (e. g. H_2O_2) leads to complete recovery of the spectrum. In pure quantum-size TiO_2-colloids a similar effect has already been reported. A slightly blue-shifted absorption spectrum can be observed after anoxic UV-illumination [76]. Also it has been reported that such an effect can be observed when quantum-size ZnO or CdS particles contain excess negative charges [90-92, 73].

This spectral blue-shift is explained by an increased confinement of the exiton arising from trapped negative charges (e. g. as Ti^{3+}). In the case of quantum sized TiO_2 the formation of Ti^{3+} can also be observed. The lifetime of the trapped electrons in TiO_2 is very long provided an oxidant (e. g. oxygen) is absent. Addition of O_2 leads to complete recovery of the spectrum [76].

This is different with the Ti/Fe-colloids. Even in the presence of oxygen the absorption spectrum of the Ti/Fe-colloids is bleached without showing the absorption band of Ti^{3+}-centers. Therefore we envisage that photogenerated electrons are not mainly captured by Ti^{4+}-centers but can in fact reduce Fe^{3+}-centers within the microcrystals to Fe^{2+}. Recovery of the spectrum is observed only for oxidants with an oxidation potential more positive than that of oxygen, such as H_2O_2 (*cf.* fig. 4).

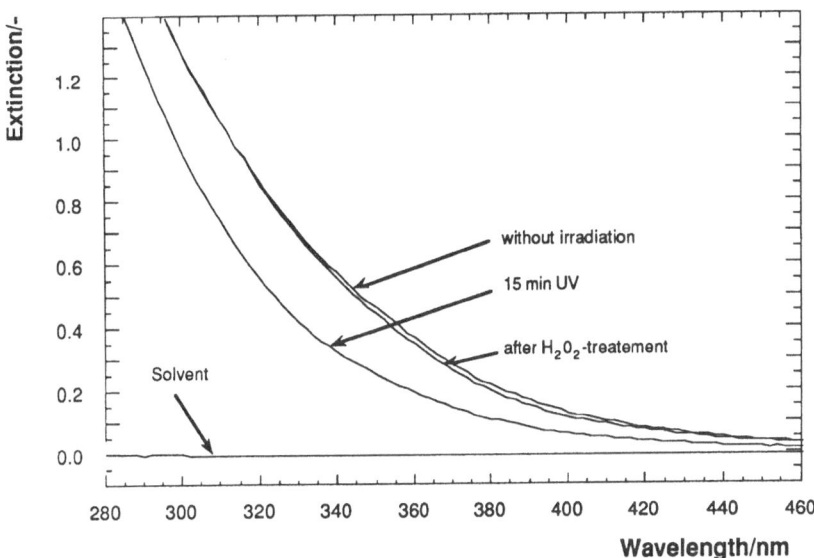

Figure 4:   Absorption spectrum of a Ti/Fe-mixed oxide colloid (50 at% Fe) at pH 3 in 1 M Ethanol and subsequent addition of H_2O_2.

For a more detailed understanding of this bleaching effect time-resolved studies have been carried out using an UV-flash-photolysis-apparatus. Figure 5 shows the difference of absorption (ΔE) before and about 2 ms after the UV-flash versus recordered wavelength for representative materials at pH 3. Pure TiO_2 colloids (▲) exhibit a broad transient absorption following the UV-flash starting at 350 nm and peaking around 600 nm. Similar absorption changes have previously reported by Kormann et al. and attributed to the formation of Ti^{3+}-centers [76]. The lifetime of these transients can be estimated from figure 6; even 200 ms after the UV-flash the intensity of the transient band has hardly changed. In the absence of molecular oxygen or other oxidants photogenerated electrons can obviously be stored in the TiO_2-particles where they are localized at Ti^{3+}-centers.

Mixed Ti/Fe oxide particles with an iron content of 0.5 at% were also investigated. Following the UV light flash a similar broad absorption band is observed with an increased intensity as compared to pure TiO_2.

Colloids with iron contents of 5 at% (○) exhibit a transient absorption with a broad band starting above 400 nm. This indicates that excess photogenerated electrons in these mixed oxide semiconductors might have a different chemical environment than the Ti^{3+}-centers observed in pure TiO_2.

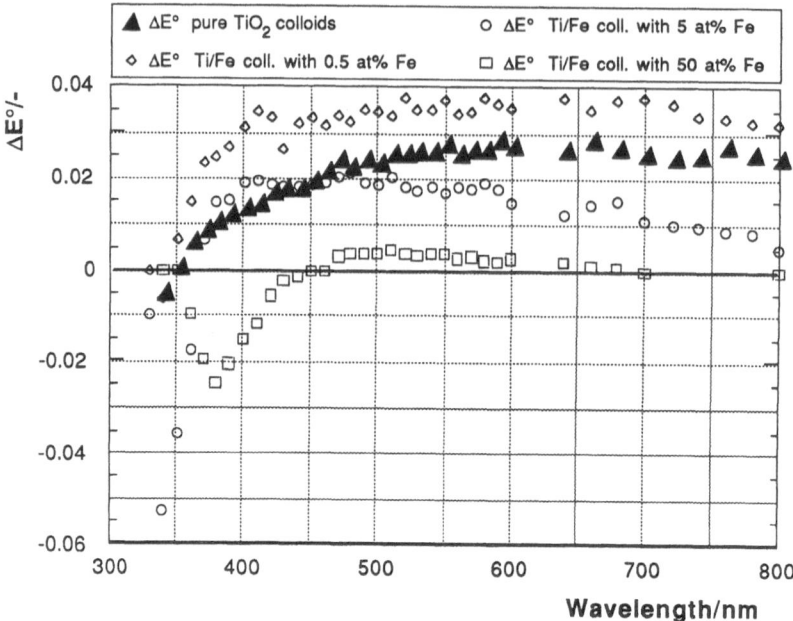

Figure 5: Transient absorption spectrum of Ti/Fe-mixed oxide colloids containing no, 0.5, 5 and 50 at% of iron at a pH of ~2.3. The concentration of the colloids was 0.16 g/l in 20% Methanol, $5*10^{-3}$ M $HClO_4$, the solution was deaerated with argon.

The latter's absorption band around 600 nm disappears with increasing iron content and simultaneously a bleaching of the Ti/Fe-colloid spectrum occurs in the wavelength range between 330 and 370 nm .

It is intriguing to note that Ti/Fe-colloids containing 50 at% of iron (□) exhibit a strong bleaching effect (negative ΔE values) in the whole wavelength range between 350 and 450 nm. The bleaching effect obtained at Ti/Fe-colloids containing 50 at% of iron is much greater than that observed at colloids with 5 at% iron content.

The transient spectrum obtained with Ti/Fe-colloids containing 50 at% of iron at pH~2.3 and taken approximately 2 ms after the flash (*cf.* fig. 5) still exhibits an increased intensity above 400 nm. This is taken as evidence that the photogenerated electrons are initially located at Ti^{3+}-centers. 200 ms after the flash (*cf.* fig. 7) the transient spectrum has significantly changed. Bleaching is now encountered for wavelengths up to 550 nm while no absorption change is observed above 550 nm compared to the original absorption spectrum before the UV-flash (base line). Here the transient spectrum due to Ti^{3+} is no longer exhibited. In agreement with the steady-state UV-illumination experiments (*cf.* fig. 4) we explain this bleaching effect by a trapping of the electrons at energetically more favored iron centers. Bleaching effects have been observed for all Ti/Fe-colloids and their extents are dependent on the portion of iron incorporated in the TiO_2-lattice.

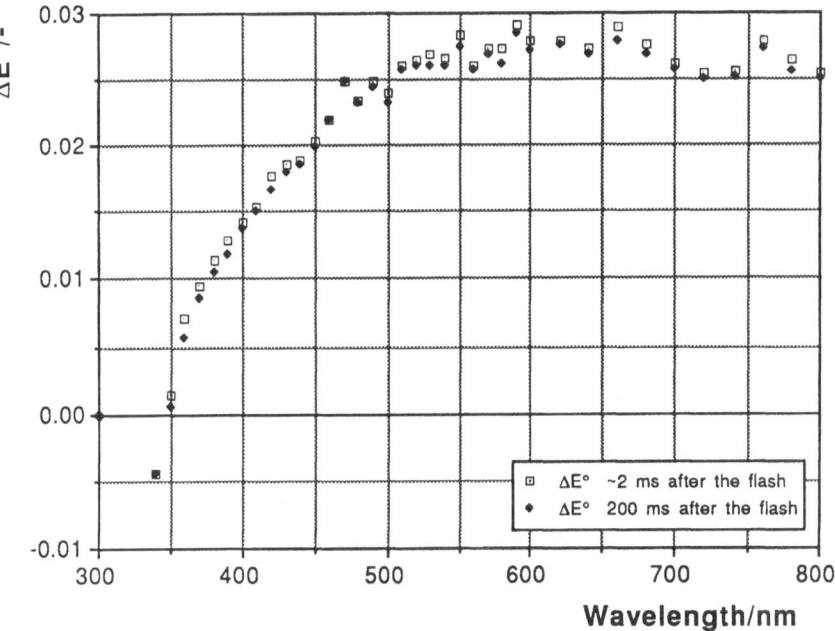

Figure 6:   Transient spectra of a solution of pure TiO_2-colloids at a pH of ~2.3. The concentration of the colloids was 0.16 g/l in 20% Methanol, $5*10^{-3}$ M $HClO_4$, the solution was deaerated with argon.

Experimental results obtained in alkaline solution at a pH of 11.3 are shown in figure 8. In this case the surfaces of the colloidal particles are negatively charged. Nevertheless the transient spectrum of pure TiO_2 colloids is almost identical to the spectrum recorded in acidic solution. Thus the surface charge seemingly does not influence the trapping of conduction band electrons at titanium-centers.

However a significant change in the shape of the transient absorption spectra can be observed in figure 8 for Ti/Fe-colloids containing 0.5 and 5 at% iron. Here a peak with a more pronounced maximum at 400 nm can be seen compared to acidic solutions. The most surprising difference between the transient absorption spectra of the mixed oxide colloids containing 50 at% iron is that at pH 11.3 there is no bleaching effect at all. The transient spectrum changes with time but a broad band indicating Ti^{3+} still remains in the range of 400...650 nm.

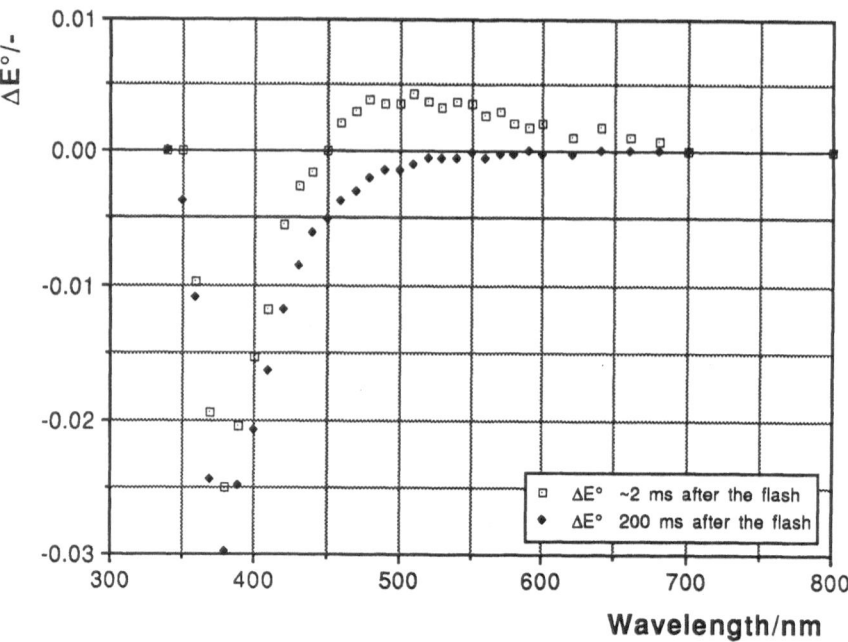

Figure 7: Transient spectra of a solution of Ti/Fe-mixed oxide colloids (50 at% Fe) at a pH of ~2.3. The concentration of the colloids was 0.16 g/l in 20% Methanol, $5*10^{-3}$ M $HClO_4$, the solution was deaerated with argon.

Because no bleaching effect could be seen in alkaline solution it can be concluded that here the photogenerated electrons in the Ti/Fe-colloids are not localized at iron centers forming Fe^{2+}. This is in agreement with the chemical behaviour of iron in homogeneous systems where Fe^{2+}-ions are by no means thermodynamically stable in alkaline media [93]. Taking into account that the particles consist of approximately 1000 metal atoms (estimated density ~ 3 g/cm^3) they therefore cannot exhibit properties of solid materials.

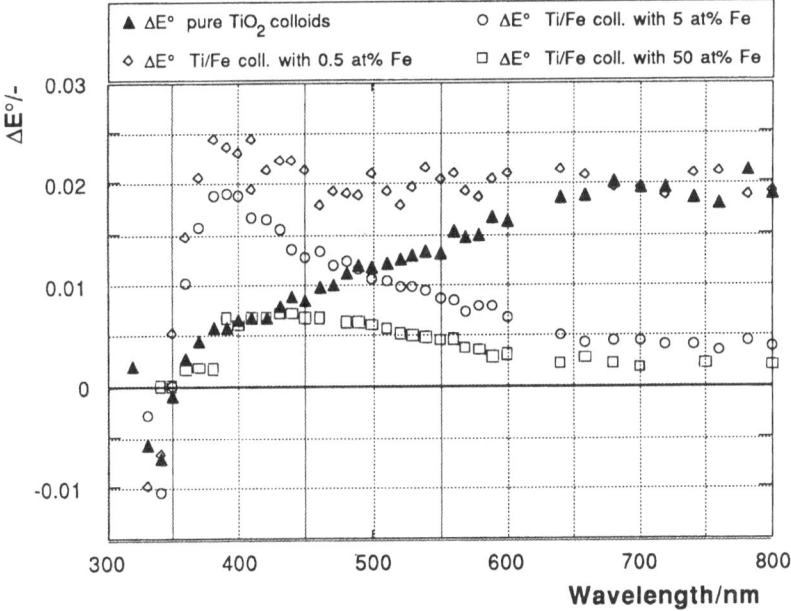

Figure 8: Transient absorption spectrum of Ti/Fe-mixed oxide colloids containing no, 0.5, 5 and 50 at% of iron at a pH of ~11.3. The concentration of the colloids was 0.16 g/l in 20% Methanol, $5*10^{-3}$ M $HClO_4$, the solution was deaerated with argon.

4.2 Detoxification properties

The photophysical properties of the 50 at% Fe-colloids are promising with respect to the goal of employing a wider range of wavelengths of the incident sunlight as compared to pure TiO_2. We used dichloroacetic acid as a model compound for all detoxification experiments. Dichloroacetic acid (Cl_2HC-COOH, DCA) is an organic molecule which is easily soluble in water even at high concentrations. Due to its low vapor pressure no detectable concentration changes will be encountered in aqueous solutions under constant bubbling with oxygene.

During a typical detoxification experiment we recorded the proton, chlorine and dichloroacetic acid concentration. The results are shown in figure 9. It can be seen that there is no dark reaction before the light is switched on (t=1200 s). Then the detoxification process starts with the evolution of Cl^- (□) and H^+ (-) and the simultaneous disappearance of dichloroacetic acid (●) according to (cf. eqn. (d)):

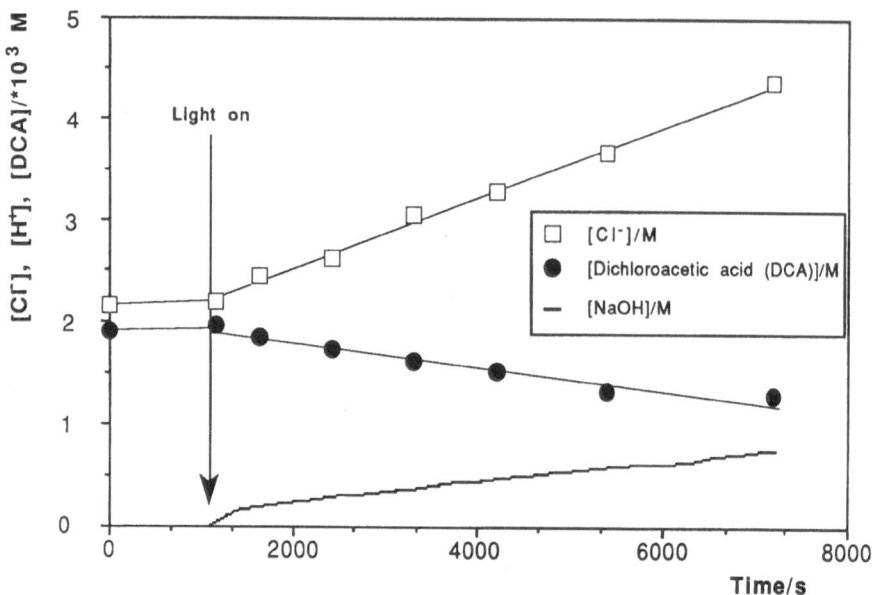

Figure 9: Degradation of dichloroacetic acid (2.5 mM) and evolution of chloride and protons in the presence of 0.5 g/l Ti/Fe-mixed oxide colloids (50 at% Fe) under UV-illumination (pH 2.6, in the presence O_2).

$$HCCl_2\text{-}C\text{-}OO^- + O_2 \xrightarrow{\text{2 hv /TiO}_2} 2\,CO_2 + H^+ + 2\,Cl^- \qquad (n)$$

Experiments with TiO_2-suspensions (P25) show a good agreement with the general stoichiometry whereas most of the detoxification experiments carried out with colloids do not exhibit such an exact correlation. As can be seen in figure 9 for a detoxification experiment with Ti/Fe-colloids (50 at% Fe) the slope of the dichloroacetic acid curve is nearly the same than the slope of the H^+-curve. This result reflects the stoichiometry given in eqn. (n). However the slope of the Cl^--curve is found to be more than three times greater than the slope of the H^+-curve. This result does not correlate with the stoichiometry and may be due to the very large and active surface of the colloidal particles. A significant amount of reaction products or educts may adsorb on this surface. Considering the conclusion from figure 4 that upon UV-irradiation in acidic solution the photogenerated electrons in colloidal particles with high iron content reduce Fe^{3+} to Fe^{2+}, we assume that the adsorption conditions might also be changed by the UV-irradiation itself, because these photoreduced surface atoms have probably other adsorption properties than the Fe^{3+} surface atoms.

The photocatalytic activity of the colloids is shown in figure 10 for particles with various iron contents. Here the quantum yields of the photocatalytic detoxification of dichloroacetic acid are plotted as bars for the different Ti/Fe-colloids at pH~2.6 and pH~11.3.

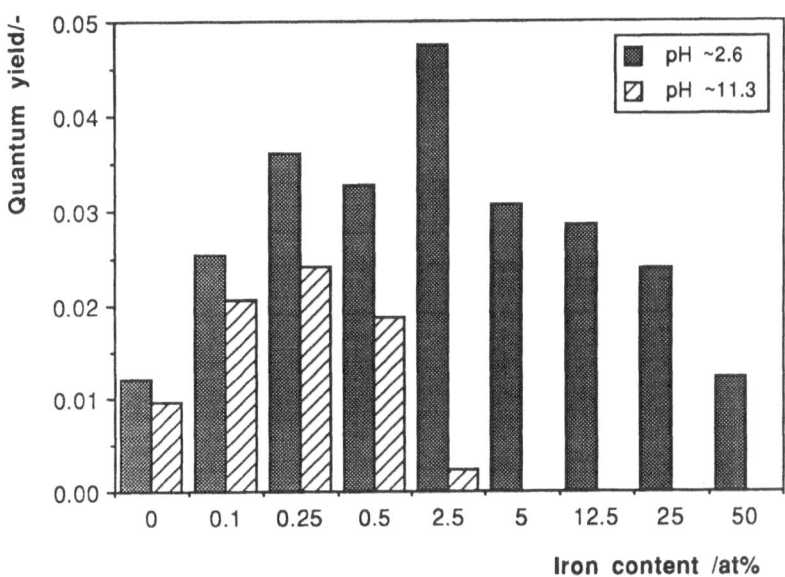

Figure 10: Photochemical quantum yields of the photocatalytic degradation of dichloroacetic acid with Ti/Fe mixed oxide colloids of different iron content.

At pH~2.6 it is shown that all quantum yields of the destruction of dichloroacetic acid obtained with the Ti/Fe-mixed oxide particles were higher than for pure TiO_2 colloids. Besides that the quantum yields increase strongly at low iron contents. They reach a maximum at 2.5 at% iron content. In this case the quantum yield is almost four times as efficient as for pure TiO_2 colloids. Therefore we assume that small iron contents inhibit the recombination of photogenerated charge carriers.

At pH~11.3 the quantum yields of the destruction of dichloroacetic acid are generally smaller than at pH~2.3. This effect might be due to the pH-dependence of the adsorption of dichloroacetic acid on the TiO_2 surface. In acidic solution the metal oxide surface is positively charged whereas dichloroacetic acid is deprotonated ($pK_s=1.29$) and therefore negatively charged. Hence DCA can interact with the charged surface and adsorb on it. A further explanation can be derived by a photoelectrochemical point of view. According to a »nernst'ian« behaviour of the surface potential of semiconducting metal oxides the energy levels of the band edges are shifting cathodically with increasing pH (59 mV per pH unit at 300 K). From a thermodynamical point of view the reduction potential of the conduction band electrons increases with increasing pH while the oxidation potential of the valence band holes decreases simultaneously. Assuming the oxidative attack is the rate determining step in the photocatalytic degradation of DCA, its quantum yield should decrease with increasing pH. This is in perfect agreement with the results presented.

It must be pointed out that at pH~11.3 the quantum yields of the detoxification of dichloroacetic acid decrease with increasing iron content of theTi/Fe-mixed oxide photocatalyst. Already at a content of 5 at% iron there is no detectable detoxification. A possible explanation is that the recombination of charge carriers is the predominant process in alkaline solution. We suppose that the trapping of the photogenerated electrons in surface states forming Ti^{3+} or Fe^{2+} suppress the recombination rate. In regard with flash photolytic measurements described in a previous chapter it must be considered that in alkaline solution the iron on the TiO_2 surface cannot be reduced to Fe^{2+} by the photogenerated electrons. Besides that it has to be considered that with an increasing iron content the amount of titanium-atoms on the surface is getting smaller compared to pure TiO_2. So the probability for the photogenerated electrons to be trapped in Ti-surface states (Ti^{3+}) is also reduced resulting in an enhanced recombination rate.

5. Conclusions and Outlook

It has demonstrated that the photocatalytic destruction of halogenated hydrocarbons offers a valuable method for the complete mineralization of these compounds from polluted aquifers. Titanium dioxide, the photocatalyst most frequently employed in these studies is a semiconductor with a very wide bandgap (3.2 eV). Therefore only 5% of the direct solar irradiation reaching the earth could in principle be utilized by this material. Consequently, we have synthesized mixed titanium/iron oxide particles with an iron content between 0.05 and 50 at%. Thereby the absorption threshold of pure TiO_2 could be shifted by more than 1 eV towards the visible part of the terrestrial solar spectrum. Detailed photophysical and photochemical studies have shown that upon illumination of aqueous suspensions of these particles in the presence of hole scavengers electrons are initially trapped on titanium sites where they are long-lived in the absence of powerful oxidants such as hydrogen peroxide. It could also be demonstrated that the presence of iron in the material enhances its photocatalytic activity in comparison to pure TiO_2 prepared the same way especially in acidic media. Detoxification studies using dichloroacetic acid as the probe molecule had been performed in acidic and alkaline colloidal solutions in order to compare the photocatalytic activity of the different Ti/Fe mixed oxides.

At pH~2.6 a maximum activity was observed with an iron content of 2.5 at% while at pH 11.3 the optimum iron content was found to be 0.25 at%.

Thinking in terms of an universal application of the photocatalytic detoxification employing Ti/Fe mixed oxides as photocatalysts an optimum iron content has to be found. Considering an application in acidic as well as in alkaline waste waters an optimum iron content for the photocatalyst seems to be 0.25 to 1 at% iron.

Future activities in this area will include the testing of the photocatalytic activity upon illumination with light of selected wavelengths (e.g. ~500 nm) which cannot be utilized by pure TiO_2. Also new synthetic pathways should be developed allowing the preparation of larger quantities of photocatalyst material for scale-up experiments which will, for example, be performed in parabolic trough solar reactors (i. e. at the SOTA test site in Almeria (Spain)). Moreover, we shall synthesize and test other mixed oxides, e. g. Al/TiO_x or Cr/TiO_x, with the aim to produce photocatalysts with long-term stability and an optimal match with the solar spectrum. Finally, our investigations will be directed towards a more detailed mechanistic understanding, i. e. to answer the question of the rate-limiting step (by performing photoelectrochemical experiments which allow a separation of anodic and cathodic processes) and to get further insight into the light intensity and the temperature dependence of the overall process.

Acknowledgment. We are grateful of J. Jirkovsky (Heyrovsky Institute/Prague) for the support by the colloid synthesis and their flash photolytic examinations. Besides we thank N. Saizev (Institute for Chemical Physics/Moscow) for the X-ray-fluorescense data and H. Weller (HMI) for the TEM measurements.

6. Literature

[1] Häfner, M.   Wasser ist kein Naturprodukt mehr.
                                  Natur 10 (1989), S. 20-24.

[2] Dieter, H. H.   Gebt dem Grenzwert eine Chance! Pestizide im
                                  Trinkwasser.
                                  Wechselwirkung 43 (1989), S. 4-9.

[3] Nerger, M.;   Biologischer Abbau von leichtflüchtigen
    Mergler-Völkl, R.   Chlorkohlenwasserstoffen in Grund- und
                                  Abwasser.
                                  Z. Wasser-Abwasser-Forsch. 21 (1988), S. 16-19.

[4] Patil, S. S.;   Biodegradation Studies of Aniline and
    Shinde, V. M.   Nitrobenzene in Aniline Plant Waste Water by Gas
                                  Chromatography.
                                  Environ. Sci. Technol. 22 (1988), S. 1160-1165.

[5] Woods, S. L.;   Characterization of Chlorphenol and
    Fergusen, J. F.;   Chlormethoxybenzene Biodegradation during
    Benjamin, M. M.   Anaerobic Treatment.
                                  Environ. Sci. Technol. 23 (1989), S. 62-68.

[6] Moore, A.T.;   Biodegradation of trans-1,2-Dichloroethylene by
    Vira, A.;   Methane-Utilizing Bacteria in a Aquifer Simular.
    Fogel, S.   Environ. Sci. Technol. 23 (1989), S. 403-406.

[7] Hristu, O.   Photobiological Production of Hydrogen by the
                                  Waste Water Treatment with Anaerobic Photo-
                                  synthetic Purple Bacteria.
                                  Toxicol. Environ. Chem. 20-21 (1989), S.495-
                                  500.

[8] Ollis, D. F.   Process Economics for Water Purification: A
                                  Comparative Assessment.
                                  In: Schiavello, M. (Ed.) Photocatalysis and Envi-
                                  ronment: Trends and Applications NATO-ASI
                                  Series C 237, S. 663-680.
                                  Dordrecht: Kluwer Academic Publishers 1988.

[9] Drechsler, W. D.   Destruction of PCDD/PCDF by Non-Thermal
                             Methods.
                             Chemosphere 15 (1986), S. 1529-1534.

[10] Ulrich, H.-J.;   Oxidation of Chlorphenols Adsorbed to Manganese
     Stone, A.T.   Oxide Surfaces.
                             Environ. Sci. Technol. 23 (1989), S. 421-428.

[11] Buser, H. R.;   Decomposition of Toxic and Environmentally
     Zehnder, H. J.   Hazardous 2,3,7,8-Tetra-chlorodibenzo-p-dioxin
                             by Gamma Irradiation.
                             Experientia 41 (1985), S.1082-1084.

[12] Tölgyessy, P.   The Degradation of Benzene and Chlorotoluron in
                             Aqueous Solutions by Gamma Radiation;
                             Biodegradability and Toxicity to Tubifex Tubifex
                             of Radiolysis Products.
                             J. Radioanal. Nucl. Chem. 128 (1988), S.321-329.

[13] Gehringer, P.;   Decomposition of Trichloroethylene and
     Proksch, E.;   Tetrachloroethylene in Drinking Water by a
     Szinovatz, W.;   Combined Radiation/Ozone Treatment.
     Eschweiler, H.   Wat. Res. 22 (1988),   S. 645-646.

[14] Getoff, N.   Advancements of Radiation Induced Degradation
                             of Pollutants in Drinking and Waste Water.
                             Appl. Radiat. Isot. 40 (1989), S. 585-594.

[15] Getoff, N.   Solid and Liquid Waste Treatment.
                             Radiat. Phys. Chem. 35 (1990), S. 432-439.

[16] Bandemer, T.;   Abbau chlororganischer Schadstoffe im Wasser
     Thiemann, W.   durch Ultraviolettbestrahlung und Zusatz von
                             Wasserstoffperoxid.
                             bbr 11 (1986), S. 413-417.

[17] Thiemann, W.;   Kombination von UV-Bestrahlung und Wasser-
     Bandemer, T.   stoffperoxid-Zusatz zur Beseitigung organischer
                             Substanzen aus Rohwasser.
                             DVGW-Schriftenreihe Wasser 107 (1988), S. 129-
                             145.

[18] Valerio, F.;
Lazzarotto, A.

Photochemical Degradation of Polycyclic
Aromatic Hydrocarbons (PAH) in Real and
Laboratory Conditions.
Intern. J. Environ. Anal. Chem. 23 (1985), S. 135-
151.

[19] Zepp, R. G.

Factors affecting the photochemical treatment of
hazardous Waste.
Environ. Sci. Technol. 22 (1988), S. 256.

[20] Wolfe, R. L.

Ultraviolet disinfection of potable water.
Environ. Sci. Technol. 24 (1990), S. 768-773.

[21] Hwang, H. M.;
Hodson, R. E.;
Lee, R. F.

Photochemical and Microbial Degradation of
2,4,5-Trichloroaniline in Freshwater Lake.
Appl. Environ. Microbiol. 50 (1985), S. 1177-
1180.

[22] Miller, R. M.;
Singer, G. M.;
Rosen, J. D.;
Bartha, R.

Sequential Degradation of Chlorophenols by
Photolytic and Microbial Treatment.
Environ. Sci. Technol. 22 (1988), S. 1215-1219.

[23] Carey, J. H.;
Lawrence, J.;
Tosine, H. M.

Photodechlorination of PCB's in the Presence of
Titanium Dioxide in Aqueous Suspensions.
Bull. Eviron. Contam. Toxicol. 16 (1976), S. 697-
701.

[24] Robin, E.

Sunlight found to help break down organic
pollutants.
In: San Francisco Examiner, May 20, (1989), D-3.

[25] Fox, B.

Audio tape makes a sound way of storing data.
In: New Sci., June 10, (1989), S. 36.

[26] Thornton, J.

Sunlight Destroys Hazardous Waste.
SERI S &T, in Review, 1989.

[27] Hecht, J.

Sunlight gives toxic waste a tanning.
New Sci., April 14, (1990), S. 28.

[28] Serpone, N.;   Photocatalysis : Fundamentals and Applications.
Pelizzetti, E. (Eds.)   New York: J., Wiley & Sons, 1989.

[29] Gerischer, H.   Solar Photoelectrolysis with Semiconductor Electrodes.
Topics Appl. Phys. 31 (1979), S. 115-169.

[30] Bahnemann, D. W.   Photocatalytic Formation of Sulfur-Centered Radicals by One-Electron Redox Processes on Semiconductor Surfaces.
In: Chatgilialoglu, C.; Asmus, K.-D. (Eds.): Sulfurcentered Reactive Intermediates in Chemistry and Biology.
NATO-ASI Series A: Life Sciences, S. 103-120.
London: Plenum Press, 1991.

[31] Bahnemann, D. W.   Mechanisms of Organic Transformations on Semiconductor Particles.
In: Pelizzetti, E., Schiavello, M. (Eds.): Photochemical Conversion and Storage of Solar Energy.
Proc. 8th Int. Conf. on Photochemical Conversion and Storage of Solar Energie, S. 251-276.
Dordrecht: Kluwer Academic Publishers, 1991.

[32] Memming, R.   Photochemical Utilization of Solar Energy.
Photochem. Photophys. 11 (1990), S. 143-189.

[33] Kormann, C.;   Photolysis of Chloroform and Other Organic
Bahnemann, D. W.;   Molecules in Aqueous TiO_2 Suspensions.
Hoffmann, M. R.   Environ. Sci. Technol. 25 (1991), S. 494-500.

[34] Ollis, D. F.   Solar-Assisted Photocatalysis for Water purification: Issues, Data, Questions.
In: Pelizzetti, E., Schiavello, M. (Eds.): Photochemical Conversion and Storage of Solar Energie. Proc. 8th Int. Conf. on Photochemical Conversion and Storage of Solar Energie, S. 593-622.
Dordrecht: Kluwer Academic Publishers, 1991.

[35] Pruden, A. L.;   Photoassisted Heterogeneous Catalysis: The
Ollis, D. F.   Degradation of Trichloroethylene in Water.
J. Catal. 82 (1983), S. 404-417.

[36] Hsiao, C.-J.;
Lee, C.-L.;
Ollis, D. F.

Heterogeneous Photocatalysis: Degradation of Dilute Solutions of Dichloromethane (CH_2Cl_2), Chloroform ($CHCl_3$), and Carbon Tetrachloride (CCl_4) with Illuminated TiO_2 Photocatalyst.
J. Catal. 82 (1983), S. 418-423.

[37] Pruden, A. L.;
Ollis, D. F.

Degradation of Chloroform by Photoassisted Heterogeneous Catalysis in Dilute Aqueous Suspensions of Titanium Dioxide.
Environ. Sci. Technol. 17 (1983), S. 628-634.

[38] Nguyen, T.;
Ollis, D. F.

Complete Heterogeneously Photocatalyzed Transformation of 1,1 - and 1,2 - Dibromoethane to CO_2 and HBr.
J. Phys. Chem. 88 (1984), S. 3386-3388.

[39] Ahmed, S.;
Ollis, D. F.

Solar Photoassisted Catalytic Decomposition of the Chlorinated Hydrocarbons Trichloroethylene and Trichloromethane.
Sol. Energy 32 (1984), S. 597-601.

[40] Ollis, D. F.;
Hsiao, C.-Y.;
Budiman, L.;
Lee, C.-L.

Heterogeneous Photoassisted Catalysis: Conversions of Perchloroethylene, Dichloro-ethane, Chloroacetic Acids and Chloro-benzenes.
J. Catal. 88 (1984), S. 89-96.

[41] Ollis, D. F.

Contaminant Degradation in Water: Heterogeneous Photocatalysis Degrades Halo-genated Hydrocarbon Contaminants
Environ. Sci. Technol. 19 (1985), S.480-484.

[42] Ollis, D. F.

Reply
J. Catal. 97 (1986), S. 569.

[43] Kormann, C.

Synthesis and Characterization of Quantum Size Metal Oxide Colloidal Particles. Photocatalytic Peroxide Formation on ZnO and TiO_2.
PhD-Thesis, California Institute of Technology Pasadena, California 1989.

[44] Kawaguchi, H.

Photocatalytic Decomposition of Phenol in the Presence of Titanium Dioxide.
Environ. Technol. Lett. 5 (1984), S. 471-474.

[45] Okamoto, K.;
Yamamoto, Y.;
Tanaka, H.;
Tanaka, M.;
Itaja, A.

Heterogeneous Photocatalytic Decomposition of Phenol over TiO_2 Powder.
Bull. Chem. Soc. Jpn. 58 (1985), S. 2015-2022.

[46] Okamoto, K.;
Yamamoto, Y.;
Tanaka, H.;
Tanaka, M.;
Itaja, A.

Kinetics of Heterogeneous Photocatalytic Decomposition of Phenol over Anatase TiO_2 Powder.
Bull. Chem. Soc. Jpn. 58 (1985), S. 2023-2028.

[47] Augugliaro, V.;
Palmisano, L.;
Sclafani, A.;
Minero, C.;
Pelizetti, E.

Photocatalytic Degradation of Phenol in Aqueous Titanium Dioxide Dispersions.
Toxicol. Environ.Chem. 16 (1988), S. 89-109.

[48] Matthews, R. W.

An Adsorption Water Purifier with in Situ Photocatalytic Regeneration.
J. Catal. 113 (1988), S. 549-555.

[49] Barbeni, M.;
Pramauro, E.;
Pelizzetti, E.;
Borgarello, E.;
Grätzel, M.;
Serpone, N.

Photodegradation of 4-Chlorophenol Catalyzed by Titanium Dioxide Particles.
Nouv. J. Chim. 8 (1984), S. 547-550.

[50] Pelizzetti, E.;
Borgarello, E.;
Serpone, N.

Photocatalytic reactions of organic compounds; Hydrogen generation from organics and degradation of wastes.
In: Schiavello, M. (Ed.): Photoelectrochemistry, Photocatalysis and Photoreactors; NATO-ASI Series C: Mathematical and Physical Science 146, S. 305-319.
Dordrecht: D. Reidel Publishing Company, 1985.

[51] Borgarello, E.;
Serpone, N.;
Barbeni, M.;
Minero, C.;
Pelizzetti, E.;
Pramauro, E.

Putting Photocatalysis to Work.
Chim. Ind. 68 (1986), S. 53-58.

[52] Barbeni, M.;
Pramauro, E.;
Pelizzetti, E.

Photochemical Degradation of Chlorinated
Dioxins, Biphenyls, Phenols, and Benzene on
Semiconductor Dispersion.
Chemosphere 15 (1986), S.1913-1916.

[53] Al-Ekabi, H.;
Serpone, N.

Kinetic Studies in Heterogeneous Photocatalysis.
1. Photocatalytic Deradation of Chlorinated
Phenols in Aerated Aqueous Solutions over
TiO_2 Supported on a Glass Matrix.
J. Phys. Chem. 92 (1988), S. 5726-5731.

[54] Al-Ekabi, H.;
Serpone, N.

Kinetic Studies in Heterogeneous Photocatalysis.
1. Photocatalytic Degradation of Chlorinated
Phenols in Aerated Aqueous Solutions over
TiO_2 Supported on a Glass Matrix.
Langmuir 5 (1989), S. 250-255.

[55] Matthews, R. W.

Photo-oxidation of Organic Materials in Aqueous
Suspensions of Titanium Dioxide.
Wat. Res. 20 (1986), S. 569-578.

[56] Matthews, R. W.

Photocatalytic Oxidation of Chlorobenzene in
Aqueous Suspensions of Titanium Dioxide.
J. Catal. 97 (1986), S. 565-568.

[57] Matthews, R. W.

Photooxidation of Organic Impurities in Water
Using Thin Films of Titanium Dioxide.
J. Phys. Chem. 91 (1987), S. 3328-3333.

[58] Matthews, R. W.

Kinetics of Photocatalytic Oxidation of Organic
Solutes over Titanium Dioxide.
J. Catal. 111 (1988), S. 264-272.

[59] Mills, G.; Hoffmann, M. R.

Breakdown of the Photoelectrochemical Model: Reactions of Pentachlorophenol on Illuminated TiO$_2$ Particles.

In: Extended Abstracts. Spring Meeting Electro chem. Soc. 89 (1989), S. 529.

[60] Draper, R. B.; Fox, M. A.

Pulse Radiolysis of 2,4,5-Trichlorophenol: Formation, Kinetics, and Properties of Hydroxy-trichlorocyclohexadienyl, Trichlorophenoxyl, Di-hydroxytrichlorocyclohexadienyl Radicals.

J. Phys. Chem. 93 (1989), S. 1938.-1944.

[61] Matthews, R. W.

Photocatalytic Oxidation and Adsorption of Methylene Blue on Thin Films of Near-ultraviolet-illuminated TiO$_2$.

J. Chem. Soc., Faraday Trans. I 85 (1989), S. 1291-1302.

[62] Tennakone, K.; Punchihewa, S.; Wickremanayaka, S.; Tantrigoda, R. U.

Titanium Dioxide Catalysed Photo-Oxidation of Methyl Violet.

J. Photochem. Photobiol. A: Chemistry 46 (1989), S. 247-252.

[63] Hustert, K.; Moza, P. N.

Photokatalytischer Abbau von Phthalaten an Titandioxid in wässriger Phase.

Chemosphere 17 (1988), S. 1751-1754.

[64] Borello, R.; Minero, C.; Pramauro,E.; Pelizzetti, E.; Serpone, N.; Hikada, H.

Photocatalytic Degradation of DDT Mediated in Aqueous Semiconductor Slurries by Simulated Sunlight.

Environ. Toxicol. Chem. 8 (1989), S. 997-1002.

[65] Hidaka, H.; Yamada, S.; Suenaga, S.; Kubota, H.; Serpone, N.; Grätzel, M.; Pelizzetti, E

Photodegradation of Surfactants: V. Photocatalytic Degradation of Surfactants in the Presence of Semiconductor Particles by Solar Exposure.

J. Photochem. Photobiol. A: Chemistry 47 (1989), S. 103-112.

[66] Pelizzetti, E.;   Photocatalytic Degradation of Nonylphenol
     Minero, C.;   Ethoxylated Surfactants.
     Maurino, V.;   Environ. Sci. Technol. 23 (1989), S. 1380-1385.
     Sclafani, A.;
     Hidaka, H.;
     Serpone, N.

[67] Barbeni, M.;   Photodegradation of Pentachlorophenol Catalyzed
     Pramauro, E.;   by Semiconductor Particles.
     Pelizzetti, E.;   Chemosphere 14 (1985), S. 195-208.
     Borgarello, E.;
     Serpone, N.

[68] Pelizzetti, E.;   Sunlight Photodegradation of Haloaromatic
     Barbeni, M.;   Pollutants Catalyzed by Semiconductor Particulate
     Pramauro, E.;   Materials.
     Serpone, N.;   Chim. Ind. 67 (1985), S. 623-625.
     Borgarello, E.;
     Jamieson, M.;
     Hidaka, H.

[69] Filby, W. G.;   Heterogeneous Catalytic Degradation of Chloro-
     Mintas, M.;   fluoromethanes on Zinc Oxide Surfaces.
     Güsten, H.   Ber. Bunsenges. Phys. Chem. 85 (1981), S. 189-
                          192.

[70] Sehili, T.;   Transformation des Composes Aromatiques
     Bonhomme, G.;   Chlores Photocatalysee Par De L'Oxide De Zinc:
     Lemaire, J.   I-Comportement Du Chloro-3 Phenol.
                          Chemosphere 17 (1988), S. 2207-2218.

[71] Sehili, T.;   Photocatalysed Transformation of Chloroaromatic
     Boule, P.;   Derivatives on Zink Oxide II: Dichlorobenzenes.
     Lemaire, J.   J. Photochem. Photobiol., A: Chemistry 50 (1989),
                          S. 103-116.

[72] Sehili, T.;   Photocatalysed Transformation of Chloroaromatic
     Boule, P.;   Derivatives on Zink Oxide III: Chlorophenols.
     Lemaire, J.   J. Photochem. Photobiol. A: Chemistry 50 (1989),
                          S. 117-125.

[73] Bahnemann, D. W.;   Preparation and Characterization of Quantum Size
     Kormann, C.;   Zinc Oxide: A Detailed Spectroscopic Study.
     Hoffmann, M. R.   J. Phys. Chem. 91 (1987), S. 3789-3798.

[74] Sclafani, A.;
Palmisano, L.;
Schiavello, M.

Influence of the Preparation Methods of TiO_2 on the Photocatalytic Degradation of Phenol in Aqueous Dispersion.
J. Phys. Chem. 94 (1990), 829-832.

[75] Matthews, R. W.

Adsorption Photocatalytic Oxidation: a new Method of Water Purification
Chem. Ind. January 4 (1988), S. 28-30.

[76] Kormann, C.;
Bahnemann, D. W.;
Hoffmann, M. R.

Preparation and Characterization of Quantum Size Titanium Dioxide (TiO2).
J. Phys. Chem. 92 (1988), S. 5196-5201.

[77] Ibusuki, T.;
Takeuchi, K.

Toluene Oxidation on U.V.-Irradiated Titanium Dioxide with and without O_2, NO_2 or H_2O at Ambient Temperature.
Atmos. Environ. 20 (1986), S. 1711-1715.

[78] Kormann, C.;
Bahnemann, D. W.;
Hoffmann, M. R.

Peroxide Production on Illuminated Suspensions of TiO_2, ZnO, and Desert Sands.
Environ. Sci. Technol. 22 (1988), S. 798-806.

[79] Tanaka, K.;
Hisanaga, T.;
Harada, K.

Photocatalytic Degradation of Organohalide Compounds in Semiconductor Suspensions with Added Hydrogen Peroxide.
New J. Chem. 13 (1989), S. 5-7.

[80] Turchi, C. S.;
Ollis, D. F.

Comments: Photocatalytic Reactor Design: An Example of Mass-Transfer Limitations with an Immobilized Catalyst.
J. Phys. Chem. 92 (1988), S. 6852-6853.

[81] Matthews, R. W.

Response to the Comment »Photocatalytic Reactor Design: An Example of Mass-Transfer Limitations with an Immobilized Catalyst«.
J. Phys. Chem. 92 (1988), S. 6853-6854.

[82] Matthews, R. W.

A Comparison between Ultraviolet Illuminated TiO_2 and ^{60}Co Gamma Rays for the Destruction of Organic Impurities in Water.
Appl. Radiat. Isot. 37 (1986), S. 1247-1248.

[83] Stumm, W.;
Morgan, J. J.

Aquatic Chemistry, 2nd edition, S. 599-640.
NewYork, Chichester, Brisbane, Toronto,
Singapure: John Wiley & Sons, 1981.

[84] Fojtik, A.;
Jirkovsky, J.

Flash photolysis of Q-particles of Cd_3P_2 in
aqueous solution.
Chem. Phys. Lett. 137 (1987), S. 226-228.

[85] Bockelmann, D.

Photokatalytischer Abbau Halogenierter Kohlen-
wasserstoffe an Halbleiteroberflächen.
Diplomarbeit, TU Clausthal-Zellerfeld 1989.

[86] Heller, H. G.;
Langan, J. R.

A New Reusable Chemical Actinometer.
EPA Newslett. (1981), S. 71-73.

[87] Weiß, J.

Einführung in die Ionenchromatographie:
Grundlagen, Instrumentation und Anwendung
Teil 1.
Chemie in Labor und Betrieb 34 (1983),S. 293-
297.

[88] Moser, J.;
Grätzel, M.;
Gallay, R.;

Inhibition of Electron-hole Recombination in
Sustitutionally doped Colloidal Semiconductor
Crystallites.
Helv. Chim. Acta 70 (1987), S. 1596-1604.

[89] Faust, B. C.;
Hoffmann, M. R.;
Bahnemann, D. W.

Photocatalytic Oxidation of Sulfur Dioxide in
Aqueous Suspensions of α-Fe_2O_3.
J. Phys. Chem. 93 (1989], S. 6371-6381.

[90] Koch, U.;
Fojtik, A.;
Weller, H.;
Henglein, A.

Photochemistry of Semiconductor Colloids:
Preparation of Extremely Small ZnO Particles,
Fluorescence Phenomena and Size Quantization
Effects.
Chem. Phys. Lett. 122 (1985), S. 507-510.

[91] Henglein, A.;
Kumar, A.;
Janata, E.;
Weller, H.

Photochemistry and Radiation Chemistry of
Semiconductor Colloids: Reaction of the Hydrated
Electron with CdS and Non-Linear Optical
Effects.
Chem. Phys. Lett. 132 (1986), S. 133-136.

[92] Spanhel, L.;
Weller, H.;
Fojtik, A.;
Henglein, A.

Photochemistry of Semiconductor Colloids: Strong luminescing CdS and CdS-Ag$_2$S Particles.
Ber. Bunsenges. Phys. Chem. <u>91</u> (1987), S. 88-94.

[93] Datta, N. C.

Chemistry of Iron (III) Oxides and Oxyhydroxides.
J. Sci. Ind. Res. <u>40</u> (1981), S. 571-583.

Model Compounds for the Oxidation
of Water in Photosynthesis

H. W. Roesky,

Universität Göttingen, Institut für Anorganische Chemie

Abstract

Two novel mononuclear manganese(II)complexes,
$[(CF_3)_3C_6H_2O]Mn \cdot 3THF$ and $Mn(CF_3COO)_2 \cdot 4pyridin$ and furthermore
two trinuclear manganese(II)complexes $Mn_3(CF_3COO)_6(benz)_6$
(benz=benzonitrile) and $Mn_3(NHAr)_4(N(SiMe_3)_2)_2$, where Ar=2,6-i-
$Pr_2C_6H_3$, have been prepared and their crystal structures have been
determined by X-ray structure analysis.

Introduction

The role of polynuclear complexes containing oxygen and nitrogen
donor ligands in biological systems has received increased
attention in the recent literature. [1-4]
Water oxidation during photosynthesis of green plants represents a
source of electrons for the light-driven reaction of Photosystem 2
(PS2) and leads also to the evolution of molecular oxygen.

$$2 \ H_2O \longrightarrow O_2 + 4 \ H^+ + 4 \ e^-$$

It is generally believed that tetranuclear manganese complexes in
various oxidation states (II,III,IV) are involved in the oxygen-
evolving center (OEC) in Photosystem 2.[5-11]
In recent years a number of two-and multinuclear manganese
complexes containing O,N-atoms have been prepared [5-7,12-14].
We were also interested in the synthesis of multinuclear complexes
as model compounds for Photosystem 2.
In this report we describe the synthesis and structures of two
mononuclear manganese(II)complexes $[(CF_3)_3C_6H_2O]Mn \cdot 3THF$ (1) and
$Mn(CF_3COO)_2 \cdot 4pyridine$ (3) and two trinuclear manganese(II)-
complexes $Mn_3(CF_3COO)_6(benz)_6$ (benz=benzonitrile) (4) and
$Mn_3(NHAr)_4(N(SiMe_3)_2)_2$, (5) (Ar=2,6-i-$Pr_2C_6H_3$).

Results and discussion

The reaction of $(CF_3)_3C_6H_2ONa$ with $MnCl_2 \cdot 2THF$ results in the formation of the paramagnetic manganese complex $[(CF_3)_3C_6H_2O]_2Mn \cdot 3THF$ (Fig.1).

$$MnCl_2 \cdot 2THF \ + \ 2(CF_3)_3C_6H_2ONa \ \longrightarrow \ [(CF_3)_3C_6H_2O]_2Mn \cdot 3THF$$
$$(1)$$

The single crystal X-ray structure analysis shows a five co-ordinated manganese atom bound to two $(CF_3)_3C_6H_2O$ groups and three THF molecules. The ligands are arranged to the Mn in a tetragonal pyramidal geometry.

In further attempts to model the compounds in Photosystem 2, we have used manganese trifluoroacetate as a starting material. This is obtained by reacting powdered manganese metal with an excess of trifluoroacetic acid.

$$Mn \ + \ 2 \ CF_3COOH \ \longrightarrow \ H_2 \ + \ Mn(CF_3COO)_2 \quad (2)$$

When the product (2) is treated with pyridine (py), the mononuclear complex $Mn(CF_3COO)_2 \cdot 4 \ py$ (3) is obtained (Fig.2). The ligands at the Mn atom in this complex have an octahedral geometry, with each of the trifluoroacetate groups bound through one oxygen atom to the Mn. When the reaction is carried out in benzonitrile instead of pyridine, the trinuclear $Mn_3(CF_3COO)_6 \cdot benz_6$ (benz=benzonitrile) (4) is formed. This complex has been structurally characterized by single crystal X-ray techniques (Fig.3). Unlike $Mn(CF_3COO)_2 \cdot 4 \ py$, the manganese atoms in $Mn_3(CF_3COO)_6 \cdot benz_6$ are bridged by trifluoroacetate ligands, without changing the coordination side of the manganese atoms.

Another multinuclear Mn complex is formed from the reaction of $Mn[N(SiMe_3)_2]_2$ with H_2NAr (Ar=2,6-i-$Pr_2C_6H_3$). The product $Mn_3(NHAr)_4(N(SiMe_3)_2)_2$ (5), yields ruby-red crystals.

$$3 \ Mn[N(SiMe_3)_2]_2 \ + \ 4 \ H_2NAr \ \longrightarrow \ Mn_3(NHAr)_4(N(SiMe_3)_2)_2 \quad (5)$$

Single crystal X-ray analysis shows a backbone of three linear Mn
atoms, bridged by four N atoms (NHAr) and capped at each end by
two nitrogens (N(SiMe$_3$)$_2$) (Fig.4). the atoms around the central Mn
form a distorted tetrahedron, and the two outer Mn atoms are in a
trigonal planar environment.
The complexes mentioned here are sensitive to moisture.

The details of the structure determination, crystal data,
intensity measurements, refinements, atomic coordinates, bond
lenghts and angles have been published by
H.W.Roesky, M.Scholz, M.Noltemeyer
Chemische Berichte 123 (1990) 2303-2309.
H.W.Roesky, K.Hübner, M.Noltemeyer, R.Bohra
Chemische Berichte 124 (1991) 515-517.

Experimental section

All reactions were performed with Schlenk techniques (under N_2).
Solvents were freshly destilled from drying agents and degassed
three times immediatly before use.
Manganese powder, manganese dichloride, trifluoroacetic acid,
pyridine, benzonitrile, toluene and 2,6-di-i-propylaniline were
used as received. $Mn[N(SiMe_3)_2]_2$ was prepared from $MnCl_2$ and
$LiN(SiMe_3)_2$ by modification of the procedure described by Bürger
and Wannagat [15]. Sodium-2,4,6-tris(trifluoromethyl)phenolate was
prepared by reaction of 2,4,6-tris(trifluoromethyl)phenole, which
was made from tris(trifluoromethyl)benzene and trimethylsilyl-
peroxide in presence of hydrochloric acid, and sodium hydride.
IR: Perkin-Elmer 180 and 325.
Elemental microanalysis were performed by the analytical
laboratory of the Institute of Inorganic Chemistry, University of
Göttingen and by the microanalytical laboratory Beller, Göttingen.
MS data were obtained on a Finnigan MAT 8230 spectrometer.

Bis[2,4,6-tris(trifluoromethyl)phenoxy]manganese(II), $Mn[(CF_3)_3C_6H_2O]_2 \cdot 3THF$ (1)

To a suspension (1.0g,3.7mmol) of $MnCl_2 \cdot 2THF$ in 40ml THF was added
dropwise a solution (3.1g,7.4mmol) of $Na[(CF_3)_3C_6H_2O] \cdot 1.5THF$ in
40ml THF (THF=tetrahydrofuran). The reaction mixture was stirred
for 15h, than the solvent was removed and finally 100ml of n-
hexane were added. The turbid solution was filtered over a thin
layer of Celite and concentrated to yield 50ml. After some time at
room temperature colorless crystals have been formed.

Yield: 2.0g (62%), mp.133 °C
IR(Nujol): ν = 1634cm^{-1}, 1520, 1270, 1191, 1125, 1031, 917.
MS(FI): m/z= 297 [$(CF_3)_3C_6H_2O$], 72(THF)
Anal.Calcd. for $C_{18}H_4F_{18}MnO_2 \cdot 3THF$ (MW=864.9)
　　　　C 41.6 , H 3.2
Found　C 41.6 , H 3.2.

Manganese bis(trifluoroacetate), Mn(CF₃COO)₂ (2)

This compound was prepared from manganese powder (10.0g,0.18mol) and trifluoroacetic acid (28ml,0.36mol) in 30ml trifluoroacetic acid. The reaction started after a few minutes by evolving hydrogen and a white precipitate was formed. After stirring the reaction mixture at room temperature for 3-4 h the white product was filtered off and dried in vacuum.

Yield: 50.7g (100%), mp. 63 °C

IR(Nujol): ν =1686cm^{-1}, 1204, 1147, 1033, 799, 724.

Anal.Calcd. for $C_4F_6MnO_4$ (MW=281.0)

      C 17.1 , F 40.6

Found C 17.2 , F 40.4.

Pyridine adduct of Manganese bis(trifluoroacetate),
Mn(CF₃COO)₂ · 4py (3)

Compound (2) was recrystallised from pyridine (py) solution at room temperature to give colorless crystals of (3).

mp.190 °C

IR(Nujol): ν =1688cm^{-1}, 1600, 1446, 1377, 1188, 1119, 1099, 1008, 832, 795, 762, 705, 625, 416.

MS(EI) m/z=518 (M-C₅H₅N)

Anal.Calcd. for $C_{24}H_{20}F_6MnN_4O_4$ (MW=597.4)

      C 48.2 , H 3.4 , F 19.1 , N 9.4

Found C 47.9 , H 3.3 , F 18.9 , N 9.2.

Benzonitrile adduct of Manganese bis(trifluoroacetate),
Mn(CF₃COO)₆ · (benz)₆ (4)

Colorless crystals of (4) were obtained from a solution of (2) in benzonitrile (benz) at room temperature.

mp.153 °C

IR(Nujol): ν =2251cm^{-1}, 2245, 1700, 1203, 1144, 933, 836, 797, 717, 684, 556

Anal.Calcd. for $C_{54}H_{30}F_{18}Mn_3N_6O_{12}$ (MW=1461.7)

      C 44.2 , H 2.1 , F 23.4 , Mn 11.3 , N 5.7

Found C 43.9 , H 2.1 , F 23.0 , Mn 11.4 , N 5.7.

Mn$_3$(NHAr)$_4$(N(SiMe$_3$)$_2$)$_2$ (5)

Compound (5) was readily prepared by directly combining
Mn[N(SiMe$_3$)$_2$]$_2$ (1.61g,4.29mmol) and H$_2$NAr (Ar=2,6-i-Pr$_2$C$_6$H$_3$)
(0.76g,4.29mmol) and warming the mixture to 90 °C for 10 minutes.
After cooling to room temperature the volatile product HN(SiMe$_3$)$_2$
was removed in vacuum. The remaining solid was recrystallised from
toluene to afford large ruby-red crystals.

Yield: 1.3g, mp.169-172 °C

Anal. Calcd. for C$_{67}$H$_{116}$Mn$_3$N$_6$Si$_4$ (MW=1281.8)

        C 62.43 , H 9.11 , N 6.55

Found   C 62.4 , H 9.1 , N 6.6 .

References

1. R.L. Rardin, A. Bino, P. Poganiuch, W.B. Tolman, S. Liu, and S.J. Lippard, Angew.Chem., 1990, 102, 842; Angew.Chem. Int. Ed. Engl. 1990, 29.

2. S. Brooker and V. McKee, J. Chem. Soc., Chem. Commun. 1989, 619

3. K. Wieghardt, K. Pohl, and U. Bossek, Z. Naturforsch. 1988, 43b, 1184.

4. J.B. Vincent, C. Christmas, J.C. Hoffman, C. Christou, H.-R. Chang, and D.N. Hendrickson, J.C.S., Chem. Commun. 1987, 236.

5. K. Wieghardt, Angew.Chem. 1989, 101, 1179; Angew. Chem. Int. Ed. Engl. 1989, 28, 1153.

6. C. Christou, Acc. Chem. Res. 1989, 22, 328.

7. G. Renger, Angew. Chem. 1987, 99, 660; Angew.Chem. Int. Ed. Engl. 1987, 26, 643.

8. G.I. Babcock in New Comprehensive Biochemistry; Photosynthesis (J.Amesz.Ed) S.125, Amsterdam, 1987.

9. V.L. Pecoraro, Photochem.Photobiol. 1988, 48, 249.

10. M. Kuzunoki in The Oxygen Evolving System of Photosynthesis (Y. Inoue, G.R. Crafts, R. Govindjee, N. Murata, G. Renger, K. Satoh) p.165, Academic Press Tokyo, 1983.

11. G. Renger in Photosynthesis Oxygen Evolution (H. Metzner,Ed.) p.229, Academic Press London, 1978

12. R.D. Cannon and R.P. White, Prog. Inorg. Chem. 1988, 36, 195.

13. R. Cammach, A. Chapman, W.P. Lu, A. Karagonni, and D.P. Kelly, FEBS Lett. 1989, 253, 239.

14. G.N. George, R.C. Prince, and S.P. Cramer, Science 1989, 243, 789.

15. H. Bürger and U. Wannagat, Mh.Chem. 1964, 95, 1099.

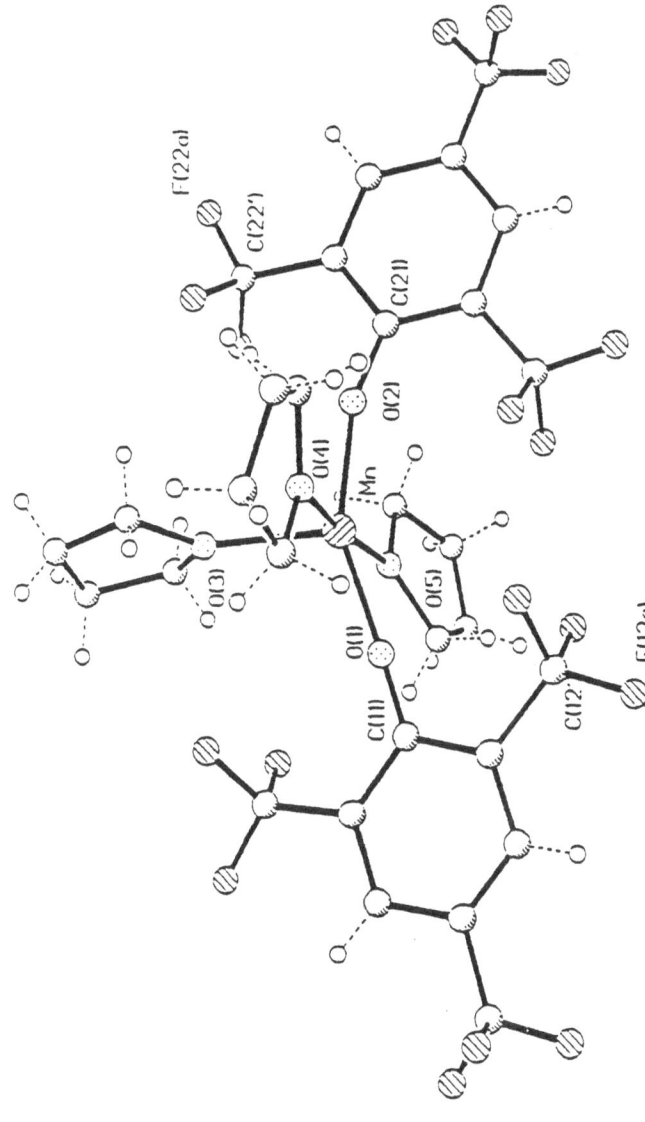

Fig. 1: The molecular structure of $((CF_3)_3C_6H_2O)_2Mn \cdot 3THF$ in the crystal

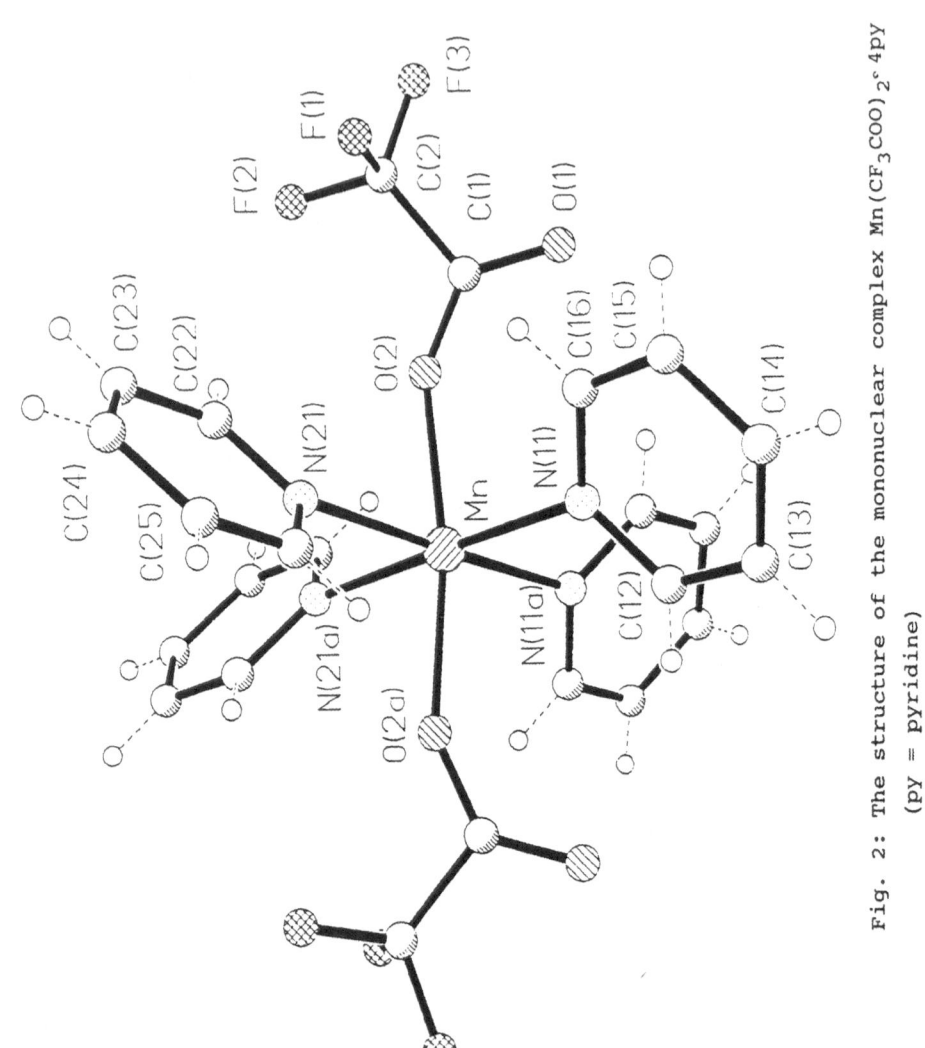

Fig. 2: The structure of the mononuclear complex Mn(CF$_3$COO)$_2$·4py
(py = pyridine)

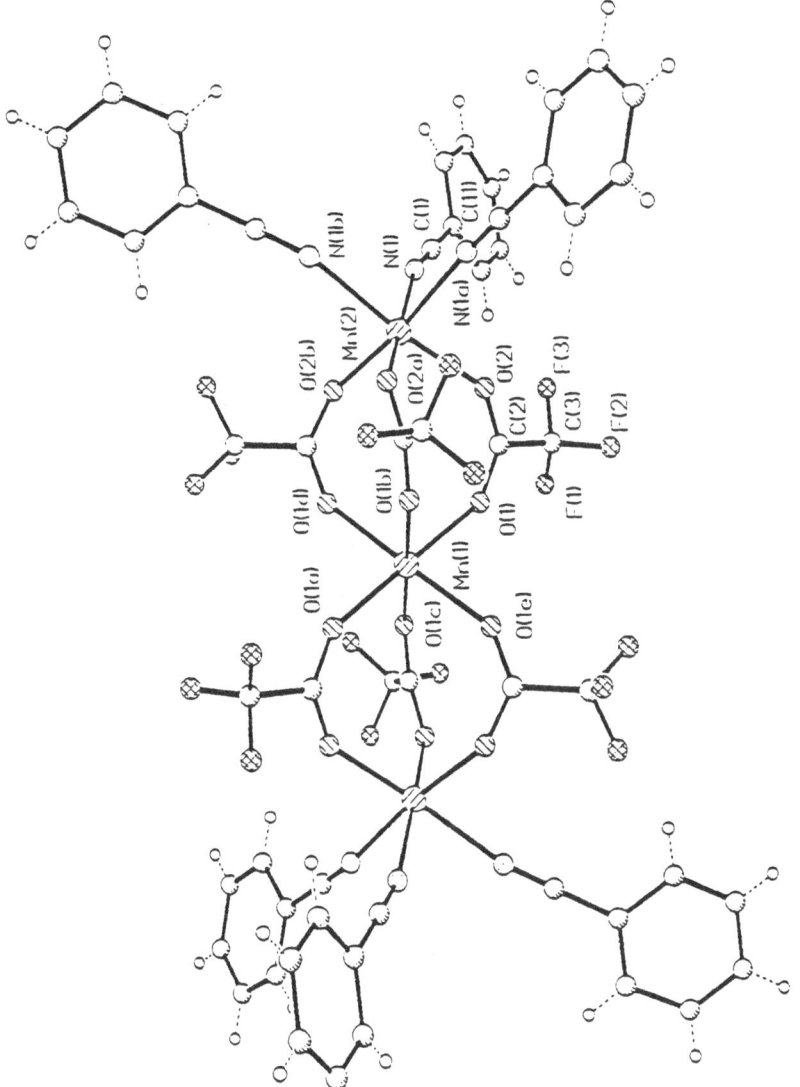

Fig. 3: The molcular structure of the trinuclear complex
$Mn_3(CF_3COO)_6 \cdot benz_6$ (benz = benzonitrile)

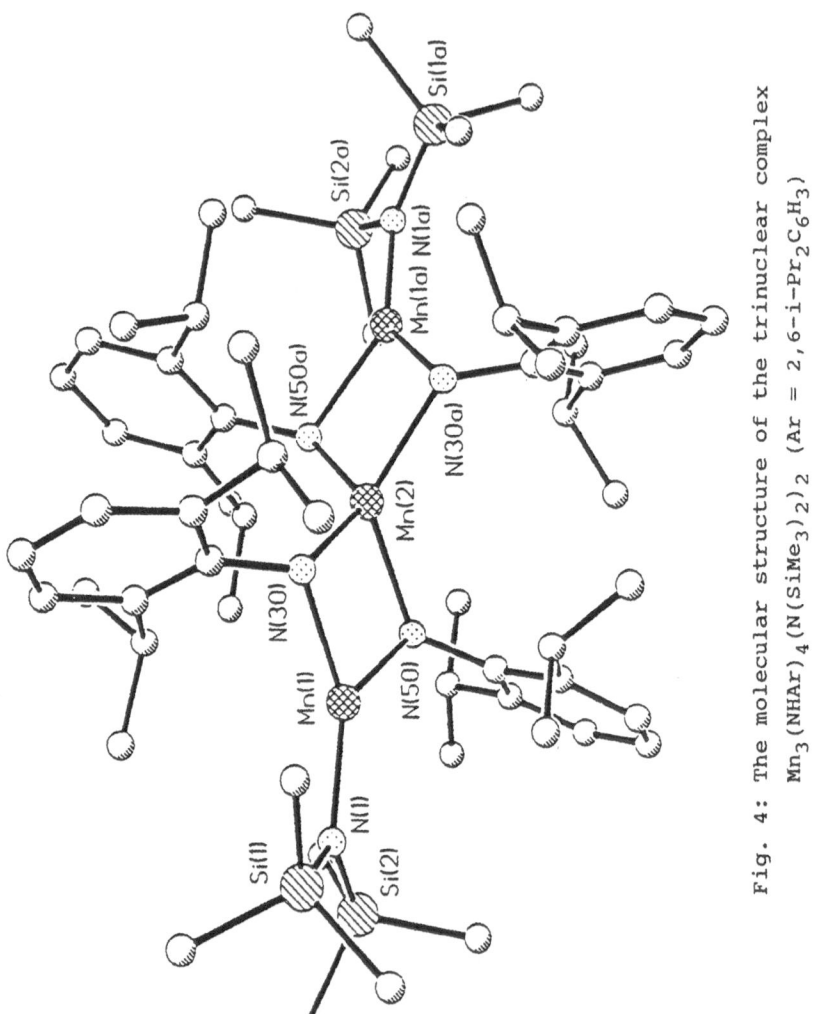

Fig. 4: The molecular structure of the trinuclear complex
$Mn_3(NHAr)_4(N(SiMe_3)_2)_2$ $(Ar = 2,6-i-Pr_2C_6H_3)$

Chemical Use of Photonic Solar Radiation -
Photochemical Syntheses (I)

A. Hülsdünker, A. Ritter, M. Demuth,

Max-Planck-Institut für Strahlenchemie,
Mülheim a.d. Ruhr

1. Introduction

There exists already a remarkable number of examples documenting the successful use of photoreactions in complex natural product syntheses [1-3]. Phototransformations serve in this context as key steps for the preparation of versatile intermediates en route to biologically active materials; such reaction schemes, involving photochemical methodologies, were shown in many cases to be more efficient and shorter than the parent preparations using classical chemical steps only.

Quite generally it has been documented that phototransformations proceed with high levels of chemo- and stereoselection upon proper choice of the wavelength of excitation. This particularly important but frequently neglected parameter can be satisfactorily tuned in preparative runs by using filters. Furthermore, an appreciably strong advantage of photochemistry lies in its cleanness; waste problems related to the work-up procedures of photoreactions are rarely encountered because negligible amounts or even no chemical auxiliaries and additives are usually necessary.

However, one major drawback has always been the high costs of photons for industrial purposes, i.e. the cost of producing fine chemicals when employing artificial irradiation sources. We have therefore been seeking a solution to the problem by running the above mentioned reactions in sun light and have devised a pilot reactor which concentrates solar radiation 100 fold.

2. Pilot Reactor

For our project we have used a prototype of a linearly focussing concentrator which in its basic form has been developed by *Bomin–Solar* (Lörrach, Germany); we have made major modifications which were necessary for our purpose, i.e. to run photochemical reactions in a temperature range from ambient to elevated (see figure 1). The main feature of this reactor is the unique geometry of the reflector sphere (**A** in figure 1) being a hybrid between parabolic and cylindrical. The reflector material is an organofluor polymer foil (*Miraflext*, Hoechst AG; size: 100 x 100 cm) on the surface of which aluminum is deposited. In contrast to comparable mirrors made of glass, this material reflects the entire solar radiation even at its shortest wavelength region of 295–300 nm. The flexible polymer foil, in its function as a stretched membrane, allows an optimal geometrical adaptation of the surface and hence adjustment of the focus upon gradual evacuation of an aluminum chamber onto which the polymer is attached. At the same time this given variability of the focus allows an appreciably large control of the temperature at the reaction site, which is a quartz tube of dimensions 0.8 x 100 cm being positioned in the focus of about the same size (**B** in figure 1: rear side and in figure 2: front view). Maximum adjustment of this linear focus results in a concentration of the solar radiation by a factor of approximately 100. This unit, consisting of reflector and reaction tube, is fixed on an axle and adjustable to the sun by way of a synchronous motor (**C**) with a speed of 15°/h.

The reaction medium is continuously pumped into the reaction tube from a 2 L–reservoir (**D**) by a compressed air–driven pump (Gather Ind.) at a rate of 90–100 L/h. Additionally, a cooling tower (**E**) has been installed for reactions which need to be run at lower temperatures, i.e. in the range of 5–20° C. In order to save cooling water, this tower operates on the basis of a stream of compressed air in which the heat is removed by evaporation cooling from the closed cycle of cooling water. This set–up, however, needs further testing. The temperature of the solution of reactants is measured at the entry and exit of the reaction tube and is continuously recorded. A further addition to this pilot reactor is a solarimeter/recorder unit noting the global solar radiation continuously.

FIGURE 1: Solar Reactor installed on the roof of the Max–Planck
Institute in Mülheim

FIGURE 2: Solar Reactor installed on the roof of the Max–Planck
Institute in Mülheim: front view onto the reaction
tube (quartz glass) and linear focus

3. Preliminary Experiments with Photoreactions in Aqueous Solution: Adaptation of Conditions to Runs in the Solar Reactor

For preliminary exploration of the reaction conditions, phototrans-formations have been chosen which proceed at high chemical and quan-tunm efficiencies. They should preferably also be useful as key steps in the synthesis of natural products, especially of biologically active target materials [1]. Very prominent reactions satisfying these requirements are certain cases of di-π-methane rearrangements [3-5] and, almost without exception, oxadi-π-methane photoisomerizations [1-3]. For safety reasons we aim at running such phototransformations in aqueous rather than in organic media since water has an appreciably higher heat capacity than any organic solvent. Furthermore, environmental aspects, i.e. reduction of waste, are again in favor of water rather than organic solvents. Having employed in our earlier work exclusively organic media, we need now to check whether complex photoreactions, mostly involving radical in-termediates, can also be carried out under these new conditions. Both the high polarity of water and oxygen dissolved in it may be limiting factors. Especially the latter point is of eminent interest since oxygen is a notorious quencher of electronically excited states and hence inhibitor of photochemical transformations.

The water soluble 5,10-dihydro-5,10-biscarboxyethenoanthracene (1b, 2% w/w dissolved in water; figure 3) together with p-hydroxyaceto-phenone as sensitizer was irradiated with 300 nm-light in a Rayonet reactor for 7 h. Solely 18% of the photoproduct 2b could be isolated after

FIGURE 3: Di-π-methane photoisomerizations of 1a and 1b, respectively

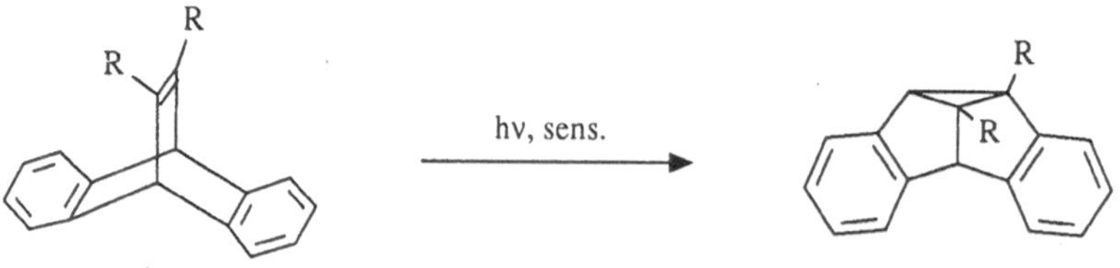

1a	hν, sens.	2a
1b	R = -COOMe, -COOH	2b

HPLC purification of the product mixture which contained numerous side products. However, if the analogous methyl ester **1a** is irradiated in an aqueous micellar acetone (= sensitizer) solution, the photoisomer **2a** is cleanly formed and isolated in 80% yield ($\lambda_{irr.}$ 300 nm, 5.5 h). The micellar solution serves in this case to solubilize the substrate **1a** (addition of sodium laurylsulfate until the critical micelle concentration is reached). The same yield of **2a** is achieved upon irradiation of **1a** in aqueous solution containing acetophenone (20% w/w) as the sensitizer ($\lambda_{irr.}$ 350 nm). Notably, under these conditions the reaction is already completed after 2 h.

It should be noted that all these preliminary experiments have been carried out without prior degassing of the reaction solutions and hence the high yields (80%) achieved are mechanisticly surprising. At least some quenching by the dissolved oxygen was anticipated by us. Further studies of di-π-methane and oxadi-π-methane isomerizations under similar conditions are in progress.

4. References

[1] Demuth, M.   New Developments in the Field of
     Mikhail, G.   Photochemical Syntheses.
                         Synthesis (1989), 145.

[2] Demuth, M.   Oxadi-π-methane Photoisomerizations.
                         In: Comprehensive Organic Synthesis, Vol. 5,
                         B.M. Trost (Ed.).
                         Oxford: Pergamon, 1991.

[3] Demuth, M.   Synthetic Aspects of the Oxadi-π-methane
                         Rearrangement.
                         In: Organic Photochemistry, Vol. 11,
                         A. Padwa (Ed.).
                         New York: Marcel Dekker, 1991.

[4] Adam, W.   Di-π-methane Photoisomerizations.
     De Lucchi, O.   In: Comprehensive Organic Synthesis, Vol. 5,

B.M. Trost (Ed.).
Oxford: Pergamon, 1991.

[5] Zimmerman, H.E.   The Di-π-methane Rearrangement.
                       In: Organic Photochemistry, Vol. 11,
                       A. Padwa (Ed.).
                       New York: Marcel Dekker, 1991.

Asea Brown Boveri AG
Postfach 10 13 32
6900 Heidelberg 1
Telefon: 06221/778-0
Telefax: 06221/77 86 31

Ruhr-Universität Bochum
Institut für Thermo- und Fluiddynamik
Postfach 10 21 48
4630 Bochum
Telefon: 0234/700-3038
Telefax: 0234/700-2001

Sektion Physik der Ludwig-Maximilians Universität
Schellingstr. 4/IV
8000 München 40
Telefon: 089/2180-0
Telefax: 089/21 80 33 91

N.U.Tech GmbH
Postfach 22 28
2350 Neumünster
Telefon: 04321/306-20
Telefax: 04321/30631

CESIWID Elektrowärme GmbH
Neumühle 4
8520 Erlangen
Telefon: 09131/4 10 25
Telex: 620816

Forschungszentrum Jülich GmbH
Postfach 19 13
5170 Jülich
Telefon: 02461/61-0
Telefax: 02461/61 53 27

Rheinisch-Westfälische Technische Hochschule
Lehrstuhl für Technische Thermodynamik
Schinkelstr. 8
5100 Aachen
Telefon: 0241/801
Telex: 08 32 704

Krupp Industrietechnik GmbH
Systemtechnik
Münchner Straße 100
4300 Essen 1
Telefon: 0201/188-0
Telefax: 0201/188-2250

Deutsche Montan Technologie (DMT)
Institut für Kokserzeugung und Kohlenchemie
Franz-Fischer-Weg 61
4300 Essen 13
Telefon: 0201/172-0
Telefax: 0201/172-1462

Hahn-Meitner-Institut Berlin GmbH
Glienicker Str. 100
1000 Berlin 39
Telefon: 030/8009-1
Telex: 185 763

Institut für Solarenergieforschung GmbH
Sokelantstr. 5
3000 Hannover 1
Telefon: 0511/3 58 50-0
Telefax: 0511/3 58 50-10

Institut für anorganische Chemie
der Universität Göttingen
Tammannstr. 4
3400 Göttingen
Telefon: 0551/39-391

Max-Planck-Institut für Strahlenchemie
Stiftstr. 34 - 36
4330 Mühlheim an der Ruhr
Telefon: 0208/304-3662
Telefax: 0208/3 04 39 51

Solar Thermal Central Receiver Systems

Volume 1 and 2: **M. Becker** (Ed.)

Proceedings of the Third International Workshop,
June 23–27, 1986, Konstanz, Federal Republic of Germany

1986. (Not available separately) Softcover DM 148,–
ISBN 3-540-17052-9

Solar Thermal Central Receiver Systems

Volume 3: **M. Carasso, M. Becker** (Eds.)

**Performance Evaluation Standards
for Solar Central Receivers**

1990. Softcover DM 58,– ISBN 3-540-53270-6

M. Becker, M. Böhmer (Eds.)

GAST – The Gas-Cooled Solar Tower Technology Program

Proceedings of the Final Presentation

1988. Softcover DM 148,–
ISBN 3-540-50121-5

**C.-J. Winter, R. Sizmann,
L. L. Vant-Hull** (Eds.)

Solar Power Plants

**Fundamentals – Technology –
Systems – Economics**

1989. Hardcover DM 148,–
ISBN 3-540-18897-5

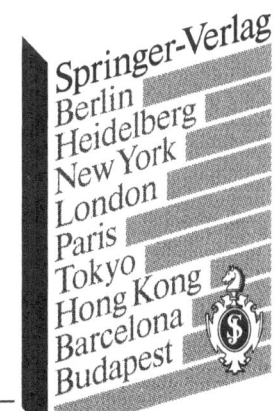

Springer-Verlag
Berlin
Heidelberg
New York
London
Paris
Tokyo
Hong Kong
Barcelona
Budapest

Solar Thermal Energy Utilization

German Studies in Technology and Applications

Volume 1: **M. Becker** (Ed.)

General Investigations on Energy Availability

1987. Softcover DM 85,- ISBN 3-540-18028-1

Volume 2: **M. Becker** (Ed.)

Technologies of Heat Exchangers and Storage

1987. Softcover DM 85,- ISBN 3-540-18031-1

Volume 3: **M. Becker** (Ed.)

Solar Thermal Energy for Chemical Processes

1987. Softcover DM 170,- ISBN 3-540-18032-X
1987. Volumes 1–3 (as a set). Softcover DM 295,- ISBN 3-540-18033-8

M. Becker, K.-H. Funken (Eds.)

Solar Thermal Energy Utilization IV

German Studies on Technology and Applications

Final Reports 1988
1990. Softcover DM 98,- ISBN 3-540-53268-4

**M. Becker, K.-H. Funken,
G. Schneider** (Eds.)

Solar Thermal Energy Utilization V

*German Studies on Technology
and Applications*

Final Reports 1989
1990. Softcover DM 118,- ISBN 3-540-53269-2

Springer-Verlag
Berlin
Heidelberg
New York
London
Paris
Tokyo
Hong Kong
Barcelona
Budapest